Carlos Martín-Fernández • Stuart A. Macgregor
Editors

Computational Methods for the Analysis of Non-Covalent Interactions

With contributions by

S. Alvarez · D. Barrena-Espés · G. Bistoni · K. Carter-Fenk ·
M. Colucci · J. Echeverría · M. Gray · J. M. Herbert ·
F. Jiménez-Grávalos · I. F. Leach · K.-Y. Liu · T. Lu ·
M. Mandado · Á. Martín Pendás · Y. Mo · J. Munárriz ·
G. Regni · P. Su · C. Wang · Z. Wu

Springer

Editors
Carlos Martín-Fernández
EaStCHEM School of Chemistry, North Haugh
University of St. Andrews
St. Andrews, UK

Stuart A. Macgregor
EaStCHEM School of Chemistry, North Haugh
University of St. Andrews
St. Andrews, UK

ISSN 0081-5993 ISSN 1616-8550 (electronic)
Structure and Bonding
ISBN 978-3-032-01542-6 ISBN 978-3-032-01543-3 (eBook)
https://doi.org/10.1007/978-3-032-01543-3

This Springer imprint is published by the registered company Springer Nature Switzerland AG
The registered company address is: Gewerbestrasse 11, 6330 Cham, Switzerland

Preface

As we move into the second quarter of the 21st century, computational chemistry is now firmly established in the toolbox of the modern chemist. This key discipline provides insight into the structure and bonding of molecules and the intra- and intermolecular interactions that also underpin their reactivity. Within this, non-covalent interactions (NCIs) are increasingly recognized as playing a central role: the accumulative effect of (individually weak) atom···atom NCIs can have a profound impact on the energetics of the large systems that are now amenable to routine calculation on ever-more powerful computing platforms. Traditionally a challenge to theory, advanced methodologies coupled with smart algorithms and software can often now provide "chemical accuracy" even for these difficult systems.

In parallel with these advances, a range of tools to analyze NCIs (and other weak interactions) has also been developed. These are the subject of this volume which aims to highlight how these tools allow us not only to describe the structure and energetics of non-covalent interactions, but also to dissect and analyze the nature of these interactions between the atoms and molecules involved. To that end, the present volume covers a broad range of methodologies that can be applied to diverse chemistries.

The contributions to this volume are organized under two broad themes. Firstly, Chapters "Interpreting Non-covalent Interactions with the Block-Localized Wavefunction (BLW) Method", "Local Energy Decomposition of Coupled Cluster Energies: Principles and Applications", "Electron Density-Based Energy Decomposition Analysis: From QM to QM/MM Calculations", "GKS-EDA Method for Intermolecular Interactions in Complex Systems", and "Extended Symmetry-Adapted Perturbation Theory (XSAPT): A Cubic-Scaling Platform for Computing Accurate Intermolecular Interaction Energies and *Ab Initio* Energy Decomposition Analysis" feature various types of energy decomposition analyses (EDAs), which can partition the interaction energy into different contributions and thus provide further insight into the forces behind the NCIs under study. In Chapter "Interpreting Non-covalent Interactions with the Block-Localized Wavefunction (BLW) Method", Mo and Wang review the block-localized wavefunction (BLW) methodology and demonstrate its application for very diverse chemical systems. In Chapter "Local Energy Decomposition of Coupled Cluster Energies: Principles and Applications", Bistoni and colleagues discuss some recent developments on the localized energy decomposition (LED) method, which is based on high-level local coupled-cluster calculations. In Chapter "Electron Density-Based Energy Decomposition Analysis: From QM to QM/MM Calculations", Mandado presents the deformation density EDA (DD-EDA) and its application from small to large systems, incorporating QM/MM calculations. In Chapter "GKS-EDA Method for Intermolecular Interactions in Complex Systems", Su and Wu examine recent extensions of the generalized Kohn-Sham EDA (GKS-EDA), established as a family of methods, and extending this to more complex systems. Lastly, in Chapter "Extended Symmetry-Adapted Perturbation Theory (XSAPT): A Cubic-Scaling Platform for Computing Accurate Intermolecular Interaction Energies and *Ab Initio* Energy Decomposition Analysis", Herbert and co-workers provide an overview of extended symmetry-adapted perturbation theory (XSAPT) methodologies, providing a pedagogical discussion of some key factors to be taken into account when running these calculations and their interpretation.

The second general theme, in Chapters "A Survey of Contemporary Applications of Quantum Chemical Topology", "The Interpenetration Index and Its Applications in Chemistry", and "Graphically Revealing Weak Interactions in Dynamic Environments Using amIGM Method", showcases methodologies based on geometrical considerations and topological analyses accompanied by real-space partitions of the molecular electron density. In Chapter "A Survey of Contemporary Applications of Quantum Chemical Topology", Martín-Pendás and colleagues review some of the key concepts used in real-space partition techniques and their applications to different bonding regimes, from covalent bonds to NCIs. In Chapter "The Interpenetration Index and Its Applications in Chemistry", Echeverría and Álvarez discuss their recently developed interpenetration index, a broadly applicable geometrical tool based on their van der Waals crust concept. Finally, Lu describes the average modified independent gradient model (amIGM) methodology and its application to dynamic environments in Chapter "Graphically Revealing Weak Interactions in Dynamic Environments Using amIGM Method".

We hope that this volume will be useful both to the expert computational chemist wanting to discover new methodologies and tools to analyze NCIs as well as to the non-expert, who might encounter these methods through the literature or via collaboration—or who may even be inspired to use these tools themselves.

We would also like to take this opportunity to thank all the authors who have committed their time and expertise to contribute such comprehensive and contemporary accounts of these important methodologies.

St. Andrews, UK Carlos Martín-Fernández
St. Andrews, UK Stuart A. Macgregor
June 2025

Contents

Struct Bond (2026) 190: 1–46
https://doi.org/10.1007/430_2025_97

Published online: 11 April 2025

Interpreting Non-covalent Interactions with the Block-Localized Wavefunction (BLW) Method

Changwei Wang and Yirong Mo

Contents

C. Wang (✉)
Key Laboratory for Macromolecular Science of Shaanxi Province, School of Chemistry and Chemical Engineering, Shaanxi Normal University, Xi'an, China
e-mail: snnu.changweiwang@snnu.edu.cn

Y. Mo (✉)
Department of Nanoscience, Joint School of Nanoscience and Nanoengineering, University of North Carolina at Greensboro, Greensboro, NC, USA
e-mail: y_mo3@uncg.edu

Abstract The block-localized wavefunction (BLW) method combines the advantages of both molecular orbital (MO) and valence bond (VB) theories and can be regarded as the simplest ab initio VB method. It can derive the wavefunction for an electron-localized (diabatic) state and thus can critically evaluate the impacts of electron transfer on molecular geometry and energetic and spectral properties. The energy decomposition approach based on the BLW method, or BLW-ED, can decompose non-covalent interactions in terms of a series of physically meaningful energy components, including polarization and charge transfer energy terms. In this chapter, we showcase the applications of the BLW method and BLW-ED approach to a broad range of non-covalent interactions, notably including hydrogen and halogen bonding, in an attempt to provide a unified picture for the forces governing various non-covalent interactions.

Keywords Anion receptor · Block-localized wavefunction (BLW) · Charge transfer · Directionality · External electric field · Frequency shift · Inter-anion attraction · Non-covalent interaction

1 Overview

Non-covalent interactions play a fundamental role in the architecture and property tuning of molecular systems at all scales [1–7], owing to their ubiquitous existence and extensive applications. As the most extensively studied case, hydrogen bonding is an illuminating reference for understanding non-covalent interactions [8–10]. Generally, hydrogen bonds are interpreted through the electrostatic paradigm, with additional stability gained from covalency. Importantly, the scope of hydrogen bonding has significantly expanded with the identification of newly recognized types [11–14], several of which are counterintuitive [15–19], prompting the need to update our understandings of this long-standing concept. Alternatively, halogen bonding refers to the interaction between the nucleophilic site of a Lewis base (the halogen bond acceptor) and the halogen atom in a molecule. Recently, halogen bonding has garnered significant attention and been applied in various fields, from chemistry to the rational design of novel materials and drugs [20–27]. Superficially, halogen bonding appears counterintuitive, as both of the halogen atom and Lewis base are electron rich, which could lead to a repulsion between them. However, the stability of halogen bonding can be intuitively elucidated through the σ-hole concept [28, 29], which highlights a specific region with positive electrostatic potential around the negatively charged halogen atom. This σ-hole occurs at covalently bonded halogens, chalcogens, and other atoms, including elements from both the p- and d-blocks of the periodic table, leading to the recognition of various non-covalent interactions [4]. Furthermore, similar concepts such as the π-hole and δ-hole have been proposed [30, 31], providing theoretical clues for expanding the landscape of non-covalent interactions.

The diverse patterns of non-covalent interactions necessitate comprehensive studies. Moreover, comparable strengths and numerous analogous properties are evident in the most prominent and representative types, namely, the hydrogen and halogen bonds [20, 28, 32–34], calling for in-depth studies to uncover the underlying factors behind these similarities. In this regard, theoretical studies are essential, as they can reveal the contributions of various physical forces, providing a comprehensive understanding of interactions. For example, the presence of hybridization was demonstrated in a state-of-the-art experimental study of hydrogen bonds [35, 36], but clarifying the precise extent of the contribution of covalency to the interaction remains experimentally challenging. Conversely, the exact roles of physical factors in chemical interactions can be directly assessed through energy decomposition methods or intuitively visualized using reduced density gradient approaches [37, 38]. Thus, innovative theoretical investigations into non-covalent interactions remain desirable for gaining novel insights, even a century after the discovery of hydrogen bonds.

In this chapter, a novel theoretical approach, namely the block-localized wavefunction (BLW) method [39–42] developed by us, is introduced and employed, aiming to provide a chemical explanation for non-covalent interactions. The energy decomposition (BLW-ED) scheme based on the BLW method was also utilized to specify the governing factors for the nature and properties of non-covalent interactions. The potential analysis based on BLW and BLW-ED was also developed, offering us new insights into non-covalent interactions in terms of force and force constant. We focused on hydrogen bonds and σ-hole interactions and uniformly represented them as D–Y···A, where Y denotes hydrogen, halogen, and other atoms that exhibit a σ-hole, while A indicates a Lewis base, for the sake of convenience.

2 Methodologies

2.1 Brief Introduction of the Block-Localized Wavefunction (BLW) Method

Quantum chemical methods possessing chemically intuitive characteristics are crucial for understanding the nature of chemical interactions. In molecular orbital (MO) theory, all orbitals are expanded across the whole system and constrained to be orthogonal. While the computational efficiency is high, the fundamental concept that any chemical bond results from orbital overlap is lost. To address this issue, various post-SCF analytical approaches have been developed to bridge MO results with chemical insights. Alternatively, the well-developed ab initio valence bond (VB) theory [43–46] defines a many-electron wavefunction as a combination of chemically meaningful VB functions, each corresponding to an electron-localized Lewis structure. It should also be noted that many key concepts in chemistry, such as orbital overlap, atomic hybridization, and resonance, are inherently incorporated into

VB theory. In contrast to MOs, VB orbitals are strictly localized and nonorthogonal, leading to nonzero orbital overlaps. The latter provides a chemically meaningful description of bond formation. Moreover, resonance energy can be conveniently evaluated as the energy difference between the most stable Lewis (VB) structure and the full VB wavefunction, which is a superposition of all VB functions. Notably, the localized and nonorthogonal characteristics of VB orbitals provide appealing insights into chemical interactions while increasing computational costs as well.

The block-localized wavefunction (BLW) method combines the advantages of MO and VB theories, by using nonorthogonal localized orbitals which is the characteristic of VB theory and a single Slater determinant which is the feature of MO theory and usually corresponds to the most stable Lewis structure. Specifically, within the BLW method, all electrons and basis functions are partitioned into several subgroups (blocks), with each orbital (BL-MO) expanded using the basis functions belonging to only one subgroup. Orbitals in the same block are constrained to be orthogonal, but orbitals from different blocks are nonorthogonal. The BLW method is essentially the simplest variant of VB theory, with the efficiency of MO theory. All BL-MOs or the BLW itself are self-consistently optimized by minimizing the energy of the electron-localized diabatic state. The BLW method has been extended to the density functional theoretical (DFT) level with Grimme's dispersion correction incorporated [47]. Furthermore, both geometrical optimization and spectral property calculations of the diabatic state (namely BLW state) are available. The role of charge transfer interactions (or resonance), a fundamental concept in chemistry, can be quantitatively examined by comparing the electron-delocalized (with a regular DFT computation) and electron-localized (with a BLW computation at the same theoretical level) states in terms of energy, electron density distribution, and vibrational modes. This comparison can be further assisted by combining the BLW method with quantum theory of atoms in molecules (QTAIM) theory [48, 49], reduced density gradient (RDG) analysis [50], and local vibrational mode analysis [51–58].

2.2 *Energy Decomposition (BLW-ED) Scheme Based on the BLW Method*

Energy decomposition analyses (EDAs) can reveal the contributions of chemically and physically meaningful factors to chemical interactions, offering clear explanation from an energetic perspective. Numerous EDA schemes, such as the Kitaura and Morokuma (KM) method [59, 60], symmetry-adapted perturbation theory (SAPT) [61–64], EDA-NOCV [65, 66], and GKS-EDA [67, 68], have been developed and extensively applied. Energy decomposition based on the BLW method (BLW-ED in short) has also been developed, offering a chemically intuitive tool for exploring chemical interactions. The two-block model, specifically for a Lewis acid A and a Lewis base B, provides the simplest description of non-covalent interactions and is adopted here to introduce the BLW-ED approach. The binding energy is defined as the overall energy variation upon the formation of a complex AB, from isolated and fully relaxed monomers A and B. The binding energy consists of the deformation energy and the interaction energy (Eq. 1). The deformation energy is the energy cost to deform the optimized monomers to their geometries in the complex and can be calculated following Eq. (2), where $\Psi_{A0/B0}$ and $\Psi_{A/B}$ stand for the fully optimized and deformed monomers, respectively. The interaction energy refers to the energy difference between the complex (Ψ^{MO}) and energetic sum of deformed and isolated monomers as expressed in Eq. (3). It should be noted that the definitions of binding energy, deformation energy, and interaction energy are the same in all the above EDA schemes, and the difference lies in the further decomposition of interaction energy.

$$\Delta E_b = \Delta E_{def} + \Delta E_{int} \tag{1}$$

$$\Delta E_{def} = E(\Psi_A) + E(\Psi_B) - E(\Psi_{A0}) - E(\Psi_{B0}) \tag{2}$$

$$\Delta E_{int} = E\left(\Psi^{MO}\right) - E(\Psi_A) - E(\Psi_B) + \text{BSSE} \tag{3}$$

In the BLW-ED approach, the interaction energy is divided into the frozen, polarization, and charge transfer energy components (Eq. 4). The frozen energy represents the energy change associated with bringing deformed and infinitely separated monomers to their positions in the complex, without altering their electron densities. The expression for the frozen energy is Eq. (5), where Ψ^{BLW0} refers to the initial electron localized state, constructed with optimal orbitals of isolated monomers at their geometries in the complex (Eq. 6). The complex can be stabilized through adjusting the electron densities within each monomer, driven by the influence of the electric field and Pauli exchange repulsion from the other monomer. This energy-lowering effect corresponds to the self-consistent optimization of the electron-localized wavefunction and is referred to as polarization, whose stability is measured through the energy difference between the optimized electron-localized state (Ψ^{BLW}) and the initial diabatic state (Eq. 7). Subsequently, additional stability can be achieved by allowing all orbital expansion spaces to extend from a specified subgroup to the entire set of basis functions of the complex. This energy change is defined as the charge transfer energy (Eq. 8), in which the basis set superposition error (BSSE) is included. Notably, the electron-delocalized state is denoted with Ψ^{MO}, representing the regular Hartree-Fock (HF) or DFT wavefunction:

$$\Delta E_{int} = \Delta E_F + \Delta E_{pol} + \Delta E_{CT} \tag{4}$$

$$\Delta E_F = E\left(\Psi^{BLW0}\right) - E(\Psi_A) - E(\Psi_B) \tag{5}$$

$$\Psi^{BLW0} = \hat{A}(\Psi_A \Psi_B) \tag{6}$$

$$\Delta E_{pol} = E\left(\Psi^{BLW}\right) - E\left(\Psi^{BLW0}\right) \tag{7}$$

$$\Delta E_{CT} = E\left(\Psi^{MO}\right) - E\left(\Psi^{BLW}\right) + BSSE \tag{8}$$

Although DFT already encompasses most short-range electron correlation, other advanced methods including additional long-range electron correlation are essential for studying non-covalent interactions. Post-SCF methods are ideal options, providing the electron correlation component (ΔE_{ec} in Eq. 9), which is defined as the difference in correlation energy between the complex and the sum of the deformed monomers. It is important to emphasize that the BSSE derived from post-SCF calculations is divided into HF and correlation energy components and incorporated into the charge transfer and electron correlation terms, respectively. Methods such as spin component-scaled MP2 (SCS-MP2) [69–72] and domain-based local pair natural orbital CCSD (DLPNO-CCSD) [73–75] are usually employed in the BLW-ED scheme to balance accuracy and computational efficiency. Additionally, the frozen component can be further divided into the electrostatic interaction and Pauli exchange repulsion at the HF level. Estimates of electrostatic interaction and Pauli exchange repulsion are notably similar across most EDA approaches, where rigid wavefunctions or electron densities of free and distorted monomers are used [61–64, 67, 68, 76]. In summary, the BLW-ED scheme based on post-SCF calculations is expressed as Eq. (9):

$$\Delta E_{int} = \Delta E_{ele} + \Delta E_{Pauli} + \Delta E_{pol} + \Delta E_{CT} + \Delta E_{ec} \tag{9}$$

Density functionals offer alternative approaches for incorporating electron correlation (ΔE_{corr}) with notable computational efficiency. In this context, Grimme's D3 dispersion correction plays an indispensable role. Accordingly, the difference in Grimme's D3 dispersion correction between the complex and the sum of deformed monomers is defined as the dispersion correction component (ΔE_{disp} in Eq. 10), which is a density-independent energy component. Furthermore, the electrostatic interaction, Pauli exchange repulsion, and partial electron correlation are included in the frozen component at DFT level and can be obtained using the XEDA method [77] developed by Su et al. due to the low computational cost. Notably, the electron correlation and dispersion correction terms are usually combined as the total electron correlation component (ΔE_{ec} in Eq. 11), yielding the same expression for the BLW-ED scheme as the post-SCF version (Eq. 9). In addition, it is also intuitive to explain the overall intermolecular interaction in straightforward terms of electronic (charge transfer) and steric interactions, in line with conventional stereoelectronic theory as expressed in Eq. (12) [78–82]:

$$\Delta E_{int} = \Delta E_F + \Delta E_{pol} + \Delta E_{CT} + \Delta E_{disp} \tag{10}$$

$$\Delta E_{ec} = \Delta E_{corr} + \Delta E_{disp} \tag{11}$$

$$\Delta E_{int} = \left(\Delta E_F + \Delta E_{pol} + \Delta E_{disp}\right) + \Delta E_{CT} = \Delta E_s + \Delta E_{CT} \tag{12}$$

2.3 Potential Analysis via BLW and BLW-ED

Potential energy surface (PES) analysis can provide intuitive interpretations for chemical interactions in terms of energy, force, and force constant, which are directly correlated with the intrinsic bond strength, vibrational spectrum, and mechanisms of physical and chemical processes. In detail, the force exerted by a specific chemical interaction determines bond elongation or contraction. The bond force constant is an intrinsic indicator of bond strength and correlates directly with the stretching vibrational frequency. The force on the bond and the bond force constant can be obtained from the first (gradient) and second derivatives (curvature) of energy with respect to bond length, respectively. Moreover, exploring the potential energy surface is also the conventional way for evaluating the barrier heights, which are widely used to explain the mechanisms of physical processes, chemical reactions, and metastability. Importantly, all these physical properties derived from potential energy profiles can be decomposed by employing the BLW-ED approach to reveal their governing factors:

$$F_{\mathrm{b}} = \frac{\mathrm{d}\Delta E_{\mathrm{b}}}{\mathrm{d}R} = \frac{\mathrm{d}\Delta E_{\mathrm{def}}}{\mathrm{d}R} + \frac{\mathrm{d}\Delta E_{\mathrm{F}}}{\mathrm{d}R} + \frac{\mathrm{d}\Delta E_{\mathrm{pol}}}{\mathrm{d}R} + \frac{\mathrm{d}\Delta E_{\mathrm{CT}}}{\mathrm{d}R} + \frac{\mathrm{d}\Delta E_{\mathrm{disp}}}{\mathrm{d}R} = F_{\mathrm{def}} + F_{\mathrm{F}} + F_{\mathrm{pol}} + F_{\mathrm{CT}} + F_{\mathrm{disp}} \quad (13)$$

$$k_{\mathrm{b}} = \frac{\mathrm{d}^2\Delta E_{\mathrm{b}}}{\mathrm{d}R^2} = \frac{\mathrm{d}^2\Delta E_{\mathrm{def}}}{\mathrm{d}R^2} + \frac{\mathrm{d}^2\Delta E_{\mathrm{F}}}{\mathrm{d}R^2} + \frac{\mathrm{d}^2\Delta E_{\mathrm{pol}}}{\mathrm{d}R^2} + \frac{\mathrm{d}^2\Delta E_{\mathrm{CT}}}{\mathrm{d}R^2} + \frac{\mathrm{d}^2\Delta E_{\mathrm{disp}}}{\mathrm{d}R^2} = k_{\mathrm{def}} + k_{\mathrm{F}} + k_{\mathrm{pol}} + k_{\mathrm{CT}} + k_{\mathrm{disp}} \quad (14)$$

In the BLW-ED-based potential analysis, the energy profile along a specified bond is examined through constrained geometry optimizations. Correlations between each energy component and bond length were determined using BLW-ED analyses. Gradients and curvatures of each energy profile are subsequently evaluated via numerical differentiation with polynomial approximations. As a result, both the bond force and force constant are derived from the gradient and curvature of the binding energy profile and are decomposed according to the BLW-ED scheme, as expressed in Eqs. (13) and (14), where F and k represent the force and force constant, respectively, with subscripts denoting contributions of specific physical factors. The potential analysis based on the BLW-ED approach has been utilized in a series of studies of non-covalent interactions as will be exemplified in the following. Notably, geometrical optimization of the electron-localized state is available, enabling a diabatic potential analysis for a critical examination of charge transfer interactions. Geometrical variations caused by charge transfer interactions are explicitly included in this potential analysis based on the BLW method, which has also been broadly employed in the study of non-covalent interactions [41, 82–86]. BLW computations in this chapter were performed with the in-house version of GAMESS (US) [87], where the BLW code was ported to. It is also available in the XMVB software [88, 89].

3 Hydrogen and Halogen Bonding: Unified Theory for Red- and Blue-Shift Phenomena in the IR Spectrum

3.1 Electrostatics vs. Covalency: A Long-Standing Debate

It is a general consensus that hydrogen bonding D–H···A is driven by the electrostatic attraction between the positively charged hydrogen and the electron-rich center on the Lewis base [60, 90]. However, the significance of the charge transfer from the lone pair orbital n(A) on the hydrogen bond acceptor A to the antibonding orbital σ*(D–H) of D–H has also been emphasized [91–97]. Obviously, the σ-hole concept favors an electrostatic explanation for the formation of hydrogen bonding. Interestingly, we replicated the electrostatic attraction in 60 tetrel bonding complexes by substituting the tetrel donor with point charges, demonstrating that the anisotropy of electron density is not a prerequisite for all σ-hole interactions [86]. On the other hand, the charge transfer theory was initially proposed by Mulliken in 1950, attributing the formation of what are now known as halogen bonds to the charge transfer from the lone pair orbital of A to the antibonding orbital of D–X [98]. This n(A) → σ*(D–X) model was validated by our study of 55 representative cases of halogen bonding, utilizing both the ab initio VB theory and BLW method [99]. In summary, it is reasonable to conclude that both the electrostatic and charge transfer interactions are important to the stability of hydrogen bond and σ-hole interactions.

The controversy over the electrostatic and charge transfer interactions is also a focal point in the explanation of the IR spectra of hydrogen and halogen bonds. For example, a red shifting upon the formation of a hydrogen bond can be predicted based on both the electrostatic and charge transfer models. First, the nucleophilic A attracts the positively charged hydrogen atom, leading to the elongation and weakening of the D–H bond [100]. Second, the n(A) → σ*(D–H) charge transfer injects electrons into the antibonding orbital, resulting in the reduction in bond strength and the red-shifting phenomenon in the D–H stretching vibration [101, 102]. Similarly, weakening of the D–X bond and the accompanying red shifting can also be anticipated according to the σ-hole explanation and n(A) → σ*(D–X) charge transfer model. Still, there is an ongoing debate regarding whether charge transfer or electrostatic interaction primarily governs the red shifting. The situation becomes more complex with the observation of "improper" blue-shifting hydrogen bonds, where the D–H bond length is shortened and its stretching vibrational frequency is shifted to a higher value [103–110]. Notably, the stretching vibration of the D–H bond strongly couples with other vibrations, resulting in relatively few observations of pure D–H stretching modes experimentally. Clearly, a unified theory for the red- and blue-shifting phenomena in the IR spectra of both hydrogen and halogen bonds is of fundamental importance, given the significance of these prototypical non-covalent interactions. A series of general theories have been developed for this ongoing topic [111–122], yet a chemically and physically intuitive explanation is still lacking.

3.2 *Force Analysis Based on the BLW-ED Approach*

Explaining the variation in the D–Y bond length serves as a straightforward approach to elucidate the IR spectra, although any monotonic relationship between the bond length and strength does not universally exist [123, 124]. Importantly, the shortening or elongation of the D–Y bond can be directly explained through the direction of force exerted by the Lewis base to the D–Y bond. The force can be obtained and explained from the previously introduced PES analysis and divided by employing the BLW-ED approach. Here, the energy profile along the D–Y distance was generated through restricted geometric optimizations to correlate the bond length with overall stability. Subsequently, the binding energy between D–Y and A was calculated to interpret the variation in the bond length through the intermolecular interactions. Scheme 1 shows the correlation between the binding energy and $\Delta R_{D\text{–}Y}$, which denotes the deviation of the D–Y bond length from its value in the optimized monomer. When $\Delta R_{D\text{–}Y}$ equals zero, a negative force indicates an elongation of the D–Y bond upon the formation of a non-covalent interaction, whereas a positive force characterizes a bond contraction. Finally, the first derivative of binding energy with respect to the D–Y distance was calculated numerically as the force and further decomposed using BLW-ED to identify the physical factors determining the force direction. In this section, all calculations were carried out at the M06-2X/cc-pVTZ theoretical level augmented by Grimme's D3 dispersion correction. The BLW-ED scheme expressed in Eq. (10) was selected. Moreover, geometrical optimization and vibrational analysis of the electron-localized state were also performed to rigorously examine the role of charge transfer interactions in the IR spectrum.

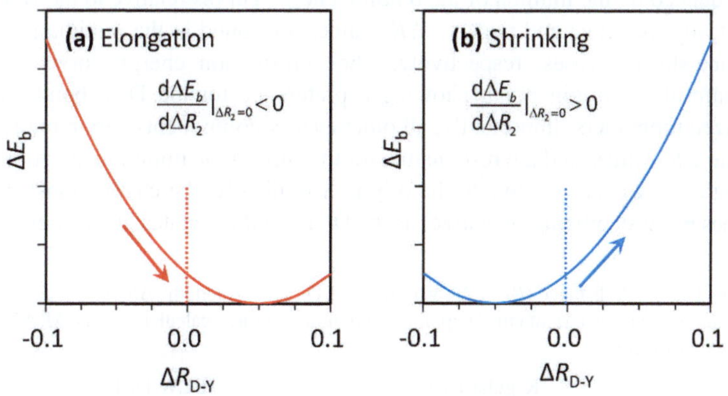

Scheme 1 Correlation between the binding energy of the non-covalent interaction in complex D–Y···A and the variation of its D–Y bond length ($\Delta R_2 = 0.0$ Å corresponds to the parent monomer DY). (Reused with permission from Ref. [85]. Copyright (2017) American Chemical Society)

3.3 A Unified Theory for the Red- and Blue-Shifting Phenomena

The Y···A distances (R_1), variations in the D–Y bond length (ΔR_2), and stretching vibrational frequencies due to the formation of hydrogen (Y = H) or halogen (Y = Cl) bonds were computed using both the regular DFT and BLW for four representative complexes, with the results summarized in Table 1. Complexes **1** and **3** exhibit blue-shifting hydrogen and halogen bonds, respectively, as indicated by the contracted D–Y bond (negative ΔR_2 values) and blue-shifted frequencies (positive $\Delta \nu$ values). Complexes **2** and **4** are red-shifting cases linked by hydrogen and halogen bonds, respectively. All these systems follow the conventional trend that the red shifts align with bond elongation, while blue shifts correspond to bond contraction. Thus, the red- and blue-shifting phenomena can be understood by interpreting the variation in bond lengths upon the formation of hydrogen and halogen bonds. Two conclusions can be obtained by comparing the results derived from regular DFT and BLW calculations. First, the value of ΔR_2 increases from the BLW geometry to regular DFT geometry, suggesting that the D–Y bond is elongated by the charge transfer interactions, consistent with the n(A) → σ*(D–Y) charge transfer explanation. Second, reductions in vibrational frequencies were observed from the BLW geometry to regular DFT geometry, aligning with the elongation of the D–Y bond caused by charge transfer interactions. In summary, the charge transfer interactions tend to elongate the D–Y bond and lower the stretching vibrational frequency, contributing positively to the red shifts.

The potential energy surface analysis based on BLW was performed to clarify the role of each physical factor in the frequency shifts. Profiles of all energy components along ΔR_2 are represented in Fig. 1. The contraction and elongation of D–Y bonds are evidenced by the minima of the binding energy curves relative to the zero point. Specifically, negative and positive ΔR_2 values are found at the minima in the red- and blue-shifting cases, respectively. The deformation energy curve reaches a minimum at the origin point, showing a preference for the D–Y bond length in optimized monomers. Importantly, all other energy components vary monotonically with the ΔR_2 values and are responsible for the shift in the minimum position on the binding energy curve relative to the origin. Specifically, the charge transfer energy becomes more stabilizing (negative) as the D–Y bond elongates in all cases, aligning

Table 1 The Y···A distance (R_1 in Å), variations of D–Y bond length (ΔR_2 in Å), and stretching vibrational frequencies ($\Delta \nu$ in cm^{-1}) upon formation of complex, calculated at the M062X-D3/cc-pVTZ level of theory

		Regular DFT			BLW-DFT		
No.	Complex	R_1	ΔR_2	$\Delta \nu$	R_1	ΔR_2	$\Delta \nu$
1	$F_3CH \cdots OH_2$	2.195	−0.002	34.2	2.274	−0.002	46.2
2	$FH \cdots OH_2$	1.693	0.017	−409.9	1.871	0.006	−73.4
3	$F_3CCl \cdots OH_2$	2.910	−0.006	11.9	3.002	−0.008	10.9
4	$FCl \cdots OH_2$	2.535	0.015	−67.5	2.744	0.006	−37.1

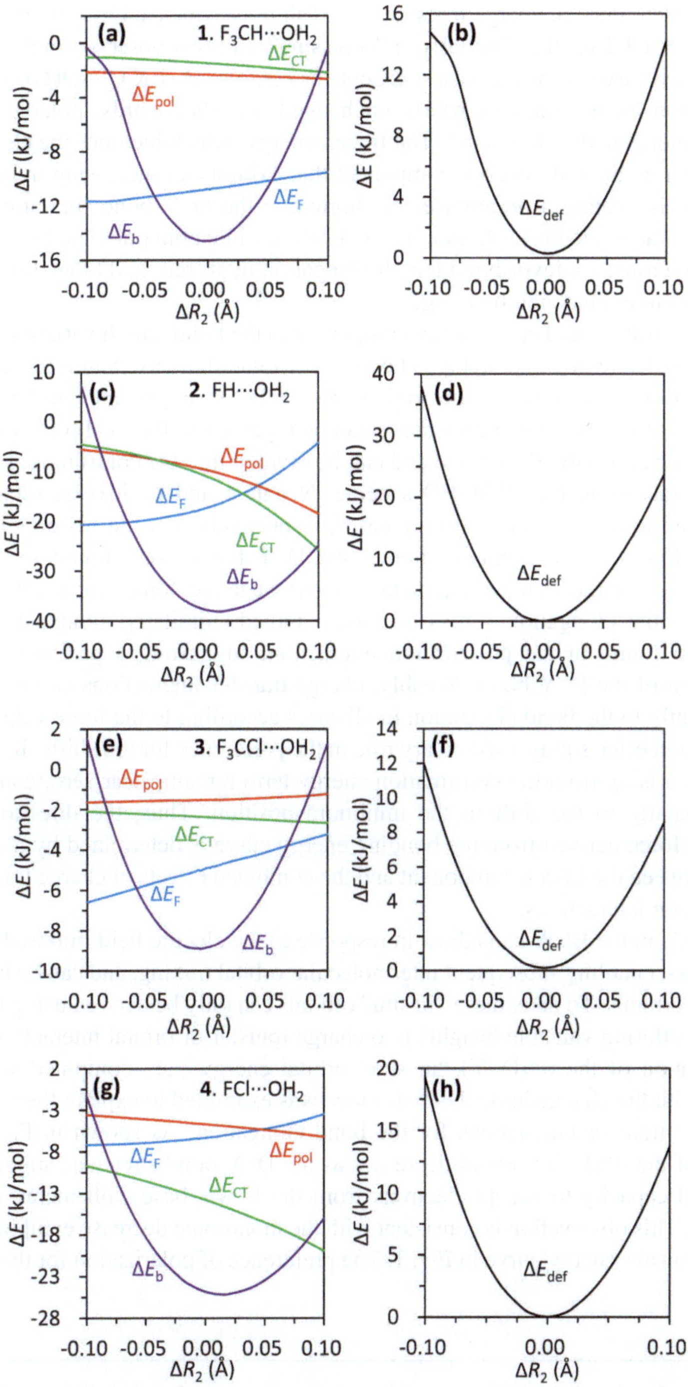

Fig. 1 Energy profiles along the variation of the Y···A distance (ΔR_2) based on the DFT optimal geometries at the M06-2X-D3/cc-pVTZ level. Right curves represent the variation of the destabilizing deformation energy term, while left curves are the profiles of frozen energy,

with the preference of charge transfer for red shift as evidenced by comparing BLW and regular DFT results. Similarly, a decreasing trend was observed in the polarization energy curve across all cases, except for complex **3** ($F_3CCl \cdots OH_2$), where the polarization energy remains nearly unchanged. In other words, polarization also tends to elongate the D–Y bond. The frozen energy, which includes the electrostatic interaction, is the only energy component that exhibits an increasing trend as ΔR_2 increases, indicating a preference for shortening the D–Y bond. In summary, the frozen interaction tends to shorten the D–Y bond, while both polarization and charge transfer interactions favor bond length elongation in all red- and blue-shifting cases of hydrogen and halogen bonding.

The contribution of each energy component to the bond length variation upon the formation of a hydrogen or halogen bond can be quantitatively evaluated by its value of the first-order derivative at the origin ($\Delta R_2 = 0.00$ Å), as presented in Table 2. The derivative of the binding energy curve can be regarded as the total force exerted by the Lewis base on the D–Y bond and can be divided into the contributions from the physical factors in the BLW-ED scheme. Negative and positive resultant forces correspond to red- and blue-shifting cases, respectively. For the frozen interaction, positive forces, which tend to shorten the D–Y bond, were found in all cases. Conversely, charge transfer interactions exert negative forces on the D–Y bond and favor the elongation across all cases studied. Similarly, nonpositive forces were also found in the polarization interaction, suggesting a preference for the elongation of the D–Y bond. Notably, charge transfer interactions contribute most significantly to the bond elongation in all cases according to the force values, while polarization effects play a secondary role in the preference for red shifts. In addition, the force arising from the deformation energy term remains near zero, contributing insignificantly to the shift in the minimum position. Thus, the direction of the resultant force derived from the bonding energy curve is determined by the competition between the frozen component and the combined effects of charge transfer and polarization interactions.

Orbitals in the BLW can adjust in response to the electric field imposed by other monomers, enabling subsequent intermolecular orbital mixing, and can be labeled as "in situ" orbitals. To date, these "in situ" orbitals can only be derived using the BLW method, offering valuable insights into charge transfer or orbital interactions. Here, the variation of the σ*(D–Y) "in situ" orbital energy (i.e., computed within the dimer) with the change in the D–Y distance was examined to explain the preference of charge transfer interactions for the bond elongation. As shown in Fig. 2a, the energy of the σ*(D–Y) orbital decreases as the D–Y bond stretches, suggesting an enhanced capacity to accept electrons from the Lewis base with increasing D–Y distance. This observation is consistent with the monotonic decrease exhibited by the charge transfer energy curve in Fig. 1. The preference of polarization for the red shift

Fig. 1 (continued) polarization energy, charge transfer energy, and total binding energy. (Reused with permission from Ref. [85]. Copyright (2017) American Chemical Society)

Table 2 Derivatives of all BLW energy components with respect to ΔR_2 (ΔF in kJ/mol/Å) at the origin point ($\Delta R_2 = 0.00$ Å)

No.	Complex	ΔF_{def}	ΔF_F	ΔF_{disp}	ΔF_{pol}	ΔF_{CT}	ΔF_b
1	$F_3CH\cdots OH_2$	0.0	11.2	0.0	−2.5	−5.8	2.9
2	$FH\cdots OH_2$	0.0	78.1	0.0	−62.4	−102.7	−87
3	$F_3CCl\cdots OH_2$	0.0	16.7	0.0	0.0	−2.9	13.8
4	$FCl\cdots OH_2$	0.0	17.7	0.0	−12.9	−50.5	−45.7

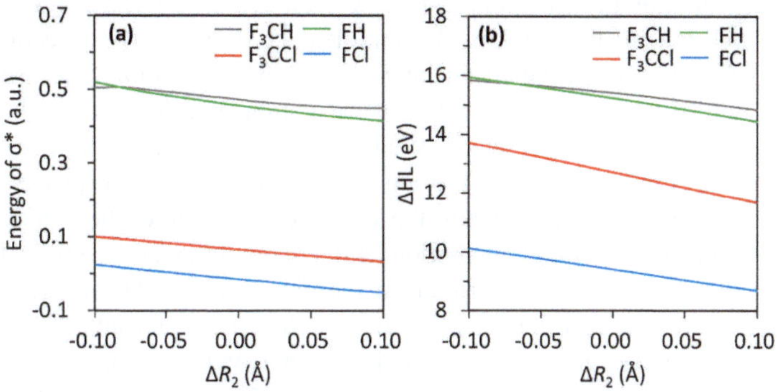

Fig. 2 (**a**) Correlation between the σ*(D–Y) "in situ" orbital energy (in a.u.) of the deformed and polarized hydrogen and halogen donors. (**b**) Correlation between the HOMO-LUMO energy gap (in eV) and ΔR_2 in distorted hydrogen and halogen donors

was explained by examining the correlation between HOMO-LUMO energy gap of the Lewis acid and the D–Y bond length. As represented in Fig. 2b, the HOMO-LUMO energy gap decreases with the elongation of the D–Y bond in each case studied. Theoretically, the polarization interaction corresponds to the orbital mixing within each monomer caused by the field effect of the other. These intramolecular orbital interactions happen between the occupied and unoccupied molecular orbitals within each monomer. Hence, the decrease in the HOMO-LUMO energy gap shown in Fig. 2b indicates a strengthened polarization effect as the D–Y bond stretches.

4 Directionality of Non-covalent Interactions

4.1 Directionality: Electrostatic and Charge Transfer Interpretations

Directionality refers to the precise orientation of a bonding interaction between atoms, leading to characteristic bond angles. It is a key feature of covalent bonds, where hybrid orbitals align favorably to maximize their orbital overlap. In contrast, ionic bonds are nondirectional, as they are governed by Coulombic interactions, which depend solely on the distance between ions. Hence, usually directionality is closely associated with the magnitude of covalency in a chemical bond. But the role of electrostatic interactions cannot be ignored as they can alter the local dipoles through polarization and affect the directionality. Furthermore, electrostatic interactions deviate from classical multipole expansion at short range (due to electrostatic penetration) and require analysis based on quantum mechanics. Notably, covalency contributes to both hydrogen bonds and σ-hole interactions, and so may be reflected in variations in directionality.

Halogen bonds are representative directional interactions, typically favoring a bond angle of around 180° for ∠D–Y–A (denoted as θ) [125, 126], whereas hydrogen bonds tend to be relatively nonlinear [127, 128]. The directionality of halogen bonds can be attributed to the σ-hole, which maximizes the electrostatic attraction with the Lewis base in a linear orientation. Importantly, a σ-hole is not only the preferred region for electrostatic interaction, but also the center of the antibonding σ* orbital of the D–Y bond (σ*(D–Y)) that can accept the lone pair electrons of A most favorably when the D–Y–A angle approaches 180°. In summary, both the electrostatic interaction and charge transfer energetically favor linearity and synergistically contribute to the directionality. However, it remains challenging to distinguish the electrostatic and charge transfer interactions experimentally, making theoretical analyses indispensable. Murray et al. were among the first to propose an electrostatic explanation for the directionality of halogen bonds [127, 129]. Nevertheless, the linearity of halogen bonds was analyzed using the SAPT method by Adhockery and Scheiner, who found that exchange repulsion is primarily responsible for the linearity, while the electrostatic interaction is relatively insensitive to the bond angle [130]. In other words, the destabilizing exchange repulsion is minimized in the linear D–Y–A arrangement, resulting in the observed directionality. Huber et al. employed the natural energy decomposition analysis (NEDA) to study the directionality of halogen bonds, emphasizing the importance of charge transfer and Pauli exchange repulsion [125]. As the focal point of the controversy surrounding the origin of directionality, charge transfer is absent in the SAPT framework [61–64] and often overestimated in NBO theory [131–133]. Thus, a comprehensive BLW study of the directionality in hydrogen bonds and classical σ-hole interactions is still desired to clarify the extent to which orbital interaction contributes to the preference for linearity.

4.2 Decomposition of Bond Angle Force Constant Using BLW-ED

Potential analysis based on the BLW method offers an insightful approach to explore the origin of directionality in non-covalent interactions. Specifically, the favorable bond angle can be identified by the position of the minimum point on the binding energy profile along θ. The strength of directionality can be quantitatively assessed by the curvature at this minimum position, represented by the bond angle force constant (k_b in Eq. 15), which describes the sensitivity of energy to the bond angle. Importantly, the binding energy can be decomposed at each point on the energy profile by employing the BLW-ED approach, detailing the contribution of each physical factor to the binding energy, and the bond angle force constant. In the study of directionality, the binding energy is divided into steric and charge transfer components, addressing the ongoing controversy surrounding the charge transfer interaction, which can be rigorously measured using the BLW method. Accordingly, the bond angle force constant is decomposed into the steric (k_s) and charge transfer (k_{CT}) components (Eq. 15), which are obtained by numerically calculating the second derivative of the corresponding energy profile along θ. Notably, the energy profile of the electron-localized state can be obtained by performing restricted geometrical optimizations with the BLW method. Hence, the role of the charge transfer interaction in the strength of directionality can be directly identified by comparing the bond angle force constant derived from MO calculations, and the diabatic BLW state (k_b'), in which the charge transfer is turned off:

$$k_b = \frac{2\Delta E_b}{d\theta^2} = \frac{2\Delta E_s}{d\theta^2} + \frac{2\Delta E_{CT}}{d\theta^2} = k_s + k_{CT} \tag{15}$$

4.3 *Origin of Directionality in Hydrogen and Halogen Bonds*

As shown in Scheme 2, in this section, we examine the directionality of hydrogen (**1**), halogen (**2**), chalcogen (**3**), and pnictogen (**4**) bonding systems, with ammonia selected as the universal Lewis base. The M06-2X functional, proven to be reliable for non-covalent interactions [71, 134], was employed in conjunction with the cc-pVTZ basis set. Grimme's D3 dispersion correction was also utilized in the calculations of both electron-delocalized and diabatic states. Energy profiles were explored by fixing θ at a series of values ranging from 120° to 240°, while fully optimizing other geometrical parameters in the hydrogen (**1**) and halogen (**2**) bonding complexes. For the remaining cases, the nitrogen atom was constrained to remain coplanar with SFH and the symmetry plane of PFH_2 in complexes **3** and **4**, respectively. Results obtained from the diabatic potential energy surfaces, where geometries of electron-localized states were optimized with the BLW method, are denoted with an apostrophe in the upper right corner.

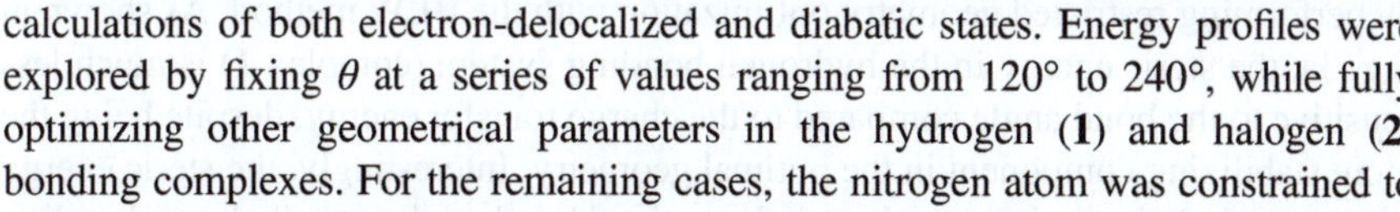

1. $FH\cdots NH_3$ **2.** $FCl\cdots NH_3$ **3.** $HFS\cdots NH_3$ **4.** $H_2FP\cdots NH_3$

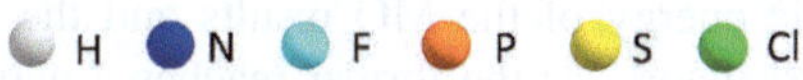

Scheme 2 Model complexes studied in this section

Table 3 Bond length (R in Å), bond angles (θ in degree), and binding energy (ΔE_b in kJ/mol) of non-covalent interactions in the fully relaxed structures of the electron-delocalized and -localized states (denoted with prime)

No.	Complex	Bond type	R	R'	θ	θ'	ΔE_b	$\Delta E'_b$
1	$FH\cdots NH_3$	Hydrogen	1.688	1.949	180.0	180.0	−52.2	−39.5
2	$FCl\cdots NH_3$	Halogen	2.342	2.840	180.0	179.9	−42.9	−21.1
3	$HFS\cdots NH_3$	Chalcogen	2.546	2.902	170.3	165.7	−34.6	−22.9
4	$H_2FP\cdots NH_3$	Pnictogen	2.687	2.955	166.5	163.1	−28.8	−21.3

Key geometrical parameters and binding energies of non-covalent interactions in the fully optimized structures of the electron-delocalized and -localized states are summarized in Table 3. The length of hydrogen bond and σ-hole interactions are elongated (0.2–0.5 Å) after quenching the charge transfer interaction. Moreover, the absolute values of bonding energy decrease by 24–51% when charge transfer interactions are deactivated. From the diabatic to the adiabatic state, the most notable variations in bond length and binding energy were observed in the halogen bonding system. Hence, we conclude that charge transfer interactions are crucial to the strength of the non-covalent interactions in the studied complexes, particularly in halogen bonding. Similar bond angles were observed in the optimized structures of the electron-delocalized and electron-localized states, indicating an insignificant contribution (0.0°–4.6°) of charge transfer interaction to the preferred bond angles. This is reasonable, as the linear arrangement of D–Y···A is favorable for both electrostatic and charge transfer interactions, as previously confirmed.

The binding energy profile was explored along the bond angle and analyzed using the BLW-ED approach to provide an intuitive explanation for the strength of directionality. Additionally, the energy profile of the diabatic state was obtained by performing restricted geometry optimization with the BLW method. As shown in Fig. 3a, the steric energy in the hydrogen bonding system (complex **1**) is much less sensitive to the bond angle compared to the charge transfer energy, despite being the more stabilizing component in the optimal geometry. Interestingly, the steric energy curves reach their maximum values at the optimal bond angle in the halogen bonding case (complex **2**), suggesting that this disfavors the linear arrangement and consequently has a negative contribution to the strength of directionality. Conversely, the curve of charge transfer energy reaches its minimum around a bond angle of 180°, dominating the directionality of halogen and chalcogen bonds (Fig. 3c, e). The charge transfer interaction becomes significantly more stabilizing as the bond angle approaches linearity in the chalcogen and pnictogen bonding cases (complexes **3** and **4**), while the curves of steric energy favor a bond angle of roughly 160°. The electron-localized states of each case studied also exhibit a preference for the approximately linear arrangement D–Y···A. However, the well depth is lower than the MO results, exhibiting decreased strengths of directionality. Seemingly, the tendencies in the steric energy of the MO results and the binding energy of the diabatic state are contradictory, as the former reaches a maximum while the latter reaches a minimum at an angle of approximately 180° in the halogen and chalcogen

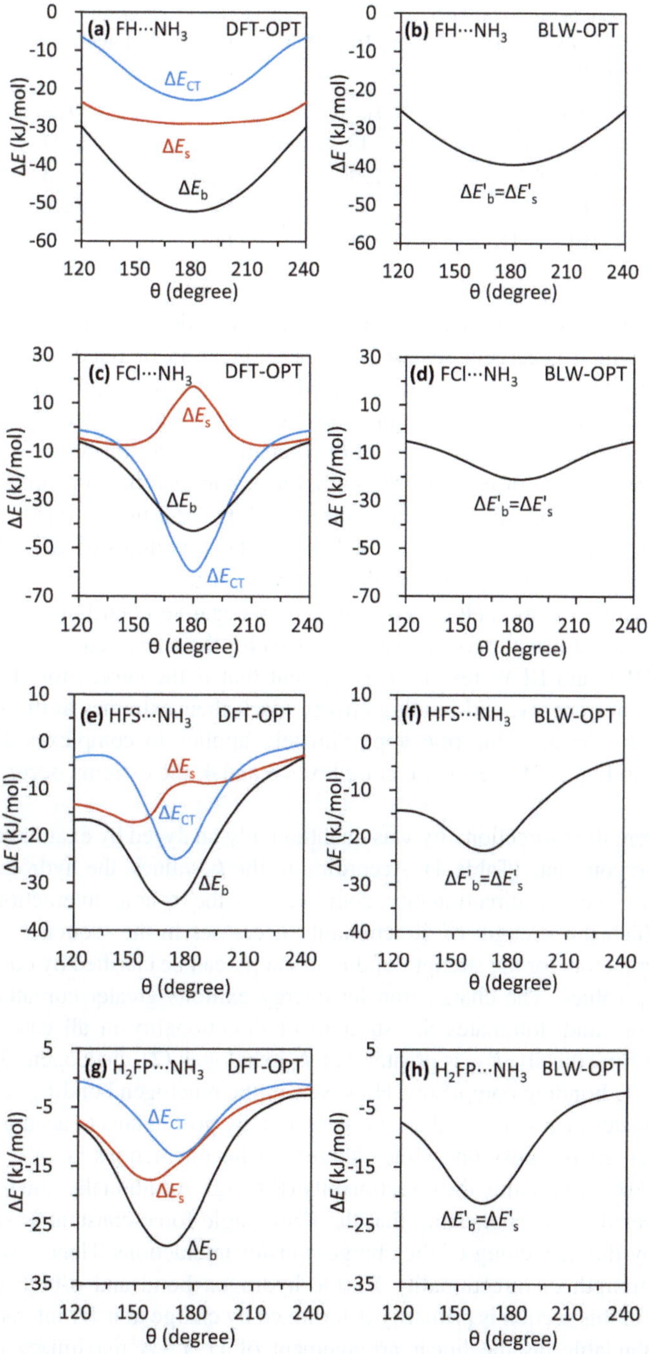

Fig. 3 Energy profiles of the adiabatic (left) and diabatic (right) states with the variation of bond angle of non-covalent interactions. (Adapted with permission from Ref. [84]. Copyright (2015) John Wiley and Sons)

Table 4 Bond angle force constants (k_b in kJ/mol/rad^2) of the electron-delocalized state and their decomposition results (k_s and k_{CT}) using BLW-ED, along with the force constants (k'_b) of the diabatic state for all complexes

No.	Complex	k_b	k_s	k_{CT}	k'_b
1	$FH\cdots NH_3$	48	7	47	28
2	$FCl\cdots NH_3$	228	−536	765	76
3	$HFS\cdots NH_3$	193	54	212	94
4	$H_2FP\cdots NH_3$	141	100	92	105

bonding complexes. This can be attributed to the destabilizing Pauli exchange repulsion, which is largely released by the elongation of bond length when the charge transfer interactions are deactivated in the electron-localized state. Consequently, the electrostatic interaction can overrule the diminished Pauli exchange repulsion, leading to the preference of the diabatic state for linearity.

The contributions of each energy component can also be compared using their force constants derived from the corresponding best-fitting quadratic curves. Accordingly, k_b, k_s, k_{CT}, and k'_b are defined as the curvatures of ΔE_b, ΔE_s, ΔE_{CT}, and $\Delta E'_b$, respectively. A positive value of force constant means a preference for the linearity, while a value below zero indicates a negative contribution to the directionality. Table 4 lists the force constant values for the energy curves obtained from both the DFT and BLW results. It is evident that if the curves for steric energy, charge transfer energy, and binding energy reach their extremes at the same point, then $k_s + k_{CT} = k_b$. This rule approximately applies to complexes **1** and **2**, as illustrated in Fig. 3. However, for complexes **3** and **4**, the extrema occur at different points.

The strength of directionality was quantitatively analyzed by examining the bond angle force constants (Table 4). According to the k_b values, the hydrogen bonding (**1**) exhibits weaker directionality compared to the σ-hole interactions studied, among which the strength of directionality decreases in the sequence **2** > **3** > **4**. Governing factors for the strength of directionality can be clarified by comparing the k_s and k_{CT} values. The charge transfer energy exhibits greater curvature than the steric energy and dominates the strength of directionality in all cases, with one exception. Specifically, k_{CT} is greater than k_s in halogen (**2**), chalcogen (**3**), and even hydrogen (**1**) bonding complexes. However, in the pnictogen bonding case (**4**), k_s is slightly greater than k_{CT}. Notably, the steric energy profile fails to accurately predict the minimal energy position, while charge transfer interactions favor a linear bond angle of 180°, dominating the directionality (Fig. 3g). Additionally, the k'_b values are much lower than k_b, suggesting that the bond angle force constant is significantly lowered by the quenching of the charge transfer interactions. Hence, we conclude that the strength of directionality for the hydrogen bond and σ-hole interactions discussed in this section is primarily determined by charge transfer interactions. This is understandable, as the linear arrangement of D–Y···A maximizes the overlap between the lone pair orbital of A and the σ-antibonding orbital of D–Y, resulting in the most favorable n(A) → σ*(D–Y) charge transfer, which dominates the directionality.

5 Hydrogen and Halogen Bonds in External Electric Fields

5.1 The BLW-ED Approach with External Electric Fields

Recent experiments demonstrate that the responses of hydrogen bonds to external electric fields (EFs) hold significant potential for applications such as piezoelectric effects [135], control of assembly behavior [136, 137], and phase transformations in supramolecular networks [78, 138–142]. The influence of electric fields on a range of non-covalent interactions has been recognized in both theoretical and experimental contexts [49, 136, 137, 142–157]. Alkorta et al. employed the quantum theory of atoms in molecules (QTAIM) [158] to analyze hydrogen bonds formed with hydrogen fluoride, finding that EFs can modulate bond length, strength, and covalency [49]. Moreover, a theoretical study by Magrioties and colleagues demonstrated that electric fields can tune both the strength and the number of hydrogen bonds within water clusters [144]. The response of bond strength to an electric field was also examined in non-covalent interactions involving π orbitals, such as π/π stacking [147, 152], ion/π interactions [149, 153, 154], as well as hydrogen and halogen bonds [147, 152, 156]. Notably, non-covalent interactions under an electric field have been suggested as suitable linkers for the design of "smart materials" [135–137, 157]. These studies underscore the need for comprehensive theoretical insights.

However, it is necessary to specify the definition of field direction before introducing any bonding analysis in electric fields. In physics, it is conventional to define the direction of force experienced by a positive charge as the positive direction of electric field (Scheme 3a). In contrast, the direction energetically favored by negative charges is usually taken as the positive one by chemists who are concerned with electrons. Here, the chemist's convention is adopted.

The responses of hydrogen and halogen bonds to external electric fields are of fundamental importance, because they are prototypical non-covalent interactions. The stability of non-covalent systems in electric fields can be understood by considering two physical mechanisms. First, electric fields can interact with the intrinsic multipoles of deformed monomers and stabilize the system. Second, orbital mixing

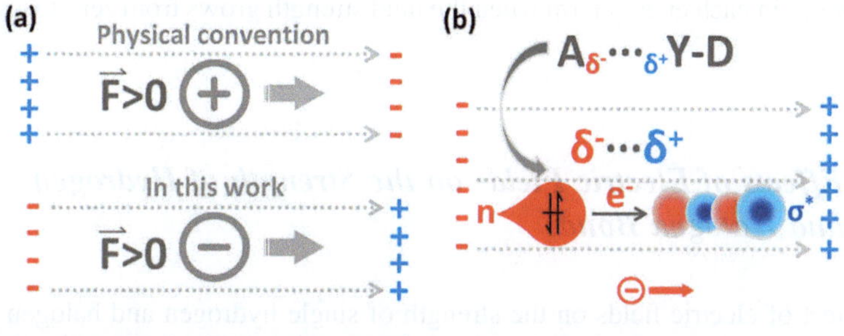

Scheme 3 (**a**) Conventions for the external electric field direction in this work and in physics. (**b**) Strengthened electrostatic and charge transfer interactions in the favorable field direction. (Reused with permission from Ref. [159]. Copyright (2021) John Wiley and Sons)

interactions within and between monomers can both be altered by the electric field imposed, affecting the interaction between monomers. In this section, we focus on the impact of the external electric field on the intrinsic strength of non-covalent interactions, making the interaction energy the focal point of our analysis. As shown in Scheme 3b, the electric field directs the electron density within A toward the electron-deficient Y while pushing the electron density on Y toward D. Thus, the electric field-induced polarization enhances both the electrophilicity of Y and the nucleophilicity of A, suggesting a strengthened electrostatic interaction. Between monomers, the electric field tends to shift electrons from A toward Y–D, aligning with the direction in which the key n(A) → σ*(Y–D) orbital interaction occurs. This suggests a coherent response of electrostatic and charge transfer interactions, even though their relative magnitudes remain to be clarified. Therefore, the BLW-ED scheme expressed in Eq. (16) was adopted to separate the interaction energy into the charge transfer energy and the steric energy, in which the electrostatic interactions are included:

$$\begin{aligned} \Delta E_{\mathrm{int}}^{\mathrm{I}} &= \Delta E_{\mathrm{int}}^{\vec{\mathrm{F}}} - \Delta E_{\mathrm{int}}^{0} \\ &= \left(\Delta E_{\mathrm{F}}^{\vec{\mathrm{F}}} - \Delta E_{\mathrm{F}}^{0}\right) + \left(\Delta E_{\mathrm{pol}}^{\vec{\mathrm{F}}} - \Delta E_{\mathrm{pol}}^{0}\right) + \left(\Delta E_{\mathrm{CT}}^{\vec{\mathrm{F}}} - \Delta E_{\mathrm{CT}}^{0}\right) + \left(\Delta E_{\mathrm{disp}}^{\vec{\mathrm{F}}} - \Delta E_{\mathrm{disp}}^{0}\right) \\ &= \Delta E_{\mathrm{F}}^{\mathrm{I}} + \Delta E_{\mathrm{pol}}^{\mathrm{I}} + \Delta E_{\mathrm{CT}}^{\mathrm{I}} + \Delta E_{\mathrm{disp}}^{\mathrm{I}} = \Delta E_{\mathrm{s}}^{\mathrm{I}} + \Delta E_{\mathrm{CT}}^{\mathrm{I}} \end{aligned} \tag{16}$$

Geometry optimizations were carried out both with and without external electric fields, followed by BLW-ED analyses. The energies of the intermediate states in the BLW-ED scheme were calculated in the presence of an electric field, with the energy components derived denoted with a superscript "$\vec{\mathrm{F}}$". It is important to note that we chose the monomers (at their geometries in the complex) polarized by electric fields as the initial states, because this study focused on variations in intrinsic bond strengths, while the background stability arising from interactions between the multipoles of the monomers and the electric fields was not considered. Importantly, we assessed and analyzed the sensitivity of bond strength to electric fields using the induced energy components (denoted as superscript "I" in Eq. 16), which represent the changes in each energy term when the field strength grows from zero to a specific value.

5.2 *Effects of Electric Fields on the Strength of Hydrogen and Halogen Bonds*

The effect of electric fields on the strength of single hydrogen and halogen bonds, where chlorine, bromine, or iodine atoms act as halogen bond donors, was investigated using model complexes **1**–**12**, as shown in Scheme 4, at the M062X-D3/cc-

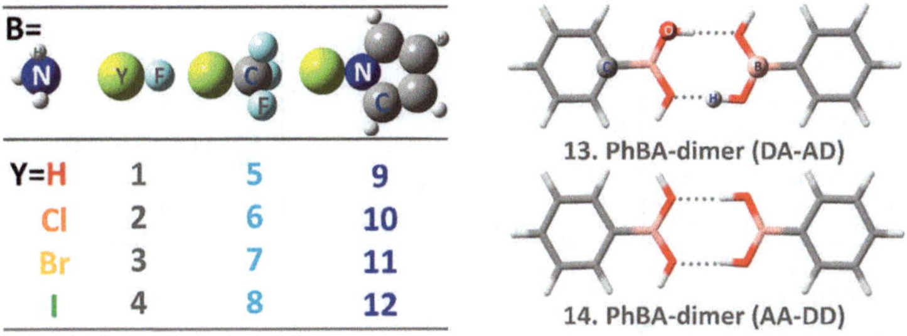

Scheme 4 Complexes analyzed in this section. (Reused with permission from Ref. [159]. Copyright (2021) John Wiley and Sons)

pVTZ level of theory [159]. Furthermore, a strategy for modulating the sensitivity of non-covalent interactions to electric fields was explored with the phenylboronic acid (PhBA) dimer, which exhibits conformation-dependent double-hydrogen bonds. Specifically, hydrogen bond donors (D) may align with hydrogen bond acceptors (A) on opposite sides (the DA-AD conformer in Scheme 4) or on the same sides (AA-DD conformer). The parallel alignment of adjacent hydrogen bonds in the AA-DD configuration allows a uniform response to the electric field, suggesting enhanced sensitivity of bonding strength to the electric field. In contrast, the anti-parallel orientation of hydrogen bonds in the DA-AD conformer could lead to opposite responses and thereby reduced sensitivity. The electric fields were imposed in the direction pointing from Y to D in all cases.

The strengthening of hydrogen and halogen bonds in electric fields is evidenced by the negative values of induced interaction energies, which become more negative as the field strength increases (Fig. 4a–m). This increase in bond strengths aligns with the shortening of H- and X-bonds in electric fields (Fig. 4o). Both the charge transfer and steric interactions become more stabilizing as the field strength increases except for complexes **2** and **3**, in which the Pauli exchange repulsion dominates the steric component and becomes more repulsive as the bond length shrinks under external electric fields. The electric field shifts electrons from A to Y–D, thereby leading to the strengthened n(A) → σ*(Y–D) orbital interaction and the monotonically decreasing curves of charge transfer interactions in Fig. 4. Enhanced steric interaction can be understood by inspecting its stabilizing components. The electrophilicity of Y and the nucleophilicity of A are both enhanced under an electric field as illuminated in Scheme 3b. This polarization of monomers induced by the electric field also tends to enlarge the dipole moment of the monomers, suggesting enhanced intermolecular polarization. Additionally, the electric field shortens the hydrogen and halogen bonds, leading to enlarged orbital overlaps between monomers and thereby enhanced short-range dispersion interactions. It is necessary to emphasize that the short-range dispersion is largely included in the frozen interaction, while the contributions of Grimme's D3 correction for long-range correlations are residual changes and insignificant.

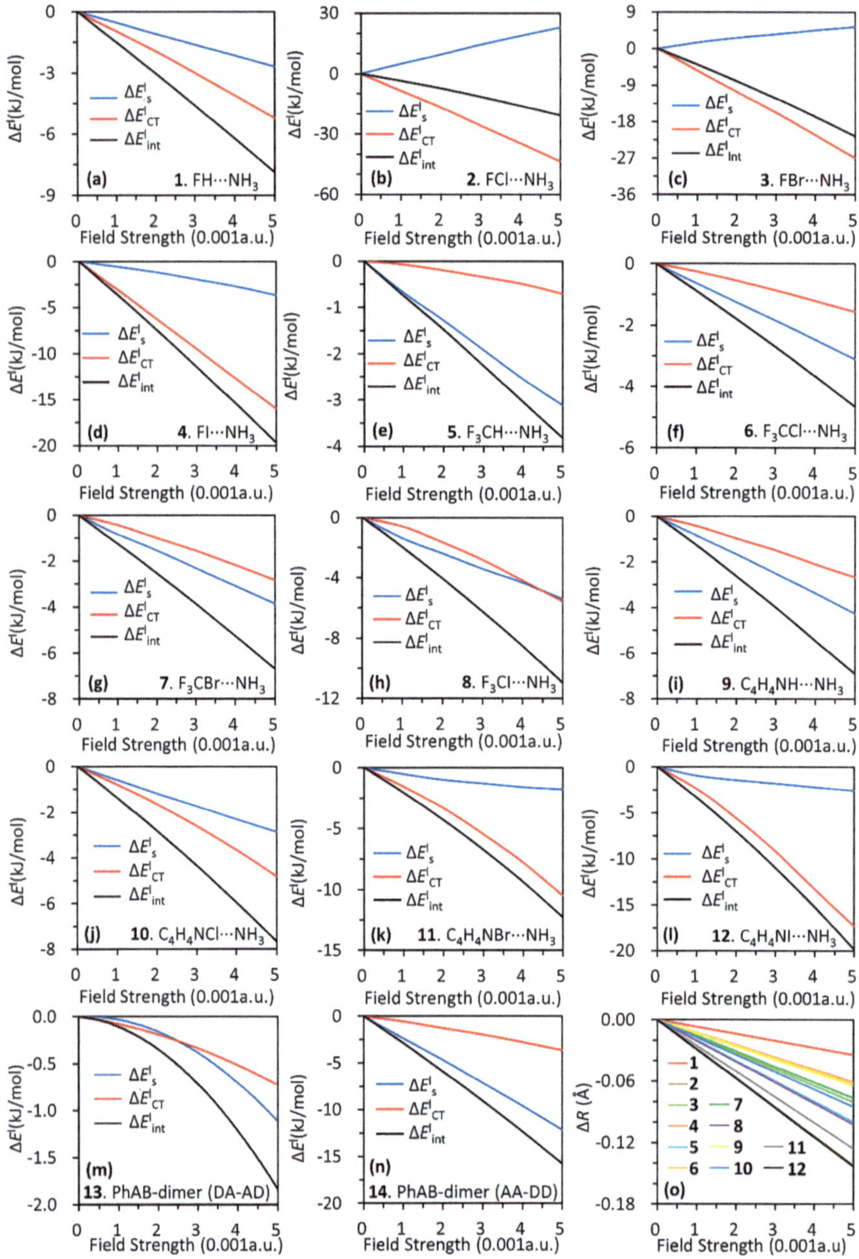

Fig. 4 (**a**–**m**) Correlations between the induced energy components (in kJ/mol) and EF strengths (in 10^{-3} a.u.) for complexes **1**–**14**. (**o**) Variations in lengths (Å) of H/X bonds as field strength grows from zero to 0.005 a.u. for complexes **1**–**12**. (Reused with permission from Ref. [159]. Copyright (2021) John Wiley and Sons)

Table 5 Induced energy components (in kJ/mol) with the electric field strength of 0.005 a.u.

No.	Complex	$\Delta E_{\mathrm{F}}^{\mathrm{I}}$	$\Delta E_{\mathrm{pol}}^{\mathrm{I}}$	$\Delta E_{\mathrm{disp}}^{\mathrm{I}}$	$\Delta E_{\mathrm{s}}^{\mathrm{I}}$	$\Delta E_{\mathrm{CT}}^{\mathrm{I}}$	$\Delta E_{\mathrm{int}}^{\mathrm{I}}$
1	FH···NH_3	0.0	−2.7	0.0	−2.7	−5.2	−7.9
2	FCl···NH_3	32.8	−10.0	0.0	22.8	−43.7	−20.9
3	FBr···NH_3	15.7	−10.4	0.0	5.3	−26.9	−21.6
4	FI···NH_3	4.6	−8.3	0.0	−3.7	−16.0	−19.7
5	F_3CH···NH_3	−2.4	−0.7	0.0	−3.1	−0.7	−3.8
6	F_3CCl···NH_3	−2.4	−0.7	0.0	−3.1	−1.6	−4.7
7	F_3CBr···NH_3	−2.5	−1.3	0.0	−3.8	−2.8	−6.6
8	F_3CI···NH_3	−2.0	−3.4	0.0	−5.4	−5.6	−11.0
9	C_4H_4NH···NH_3	−2.4	−1.9	0.0	−4.3	−2.7	−7.0
10	C_4H_4NCl···NH_3	−1.2	−1.5	0.0	−2.7	−4.9	−7.6
11	C_4H_4NBr···NH_3	2.8	−4.1	0.0	−1.3	−10.9	−12.2
12	C_4H_4NI···NH_3	7.1	−9.7	0.0	−2.6	−17.2	−19.8
13	PhAB-dimer (DA-AD)	−0.4	−0.7	0.0	−1.1	−0.7	−1.8
14	PhAB-dimer (AA-DD	−7.3	−4.8	0.0	−12.1	−3.6	−15.7

A quantitative assessment of sensitivity was achieved by analyzing the induced energy component values obtained under an electric field of 0.005 a.u. As summarized in Table 5, the induced interaction energy becomes higher when the hydrogen atom is replaced by a halogen atom (**1** vs. **2–4**, **5** vs. **6–8**, and **9** vs. **10–12**), suggesting a more sensitive response of halogen bonds to the electric field compared with the corresponding hydrogen bonds. Importantly, this greater sensitivity of halogen bonds originates from the induced charge transfer interactions. For instance, from complex **1** to **2**, the induced charge transfer energy increases most significantly (from −5.2 to −43.7 kJ/mol) among all energy components, accounting for the variation in induced interaction energy (−13.0 kJ/mol). The strengthening of hydrogen bonds in electric fields is primarily governed by steric interactions across all cases studied except for complex **1**, which is a strong hydrogen bonding case. Complexes **6–7** exhibit weak halogen bonds and respond to the electric field largely through the induced steric interactions. Conversely, the sensitivity of other halogen bonding cases is predominantly ruled by charge transfer interactions. Thus, it is reasonable to conclude that steric interactions govern the response of hydrogen bonds to electric fields, whereas halogen bonds are strengthened in electric fields due to increased covalency.

Guidelines for modulating the sensitivity of hydrogen and halogen bonds can be derived from the analysis of multiple bonding cases (**13–14**). As listed in Table 5, complexes **13** and **14** serve as illustrative examples: the strength of the double-hydrogen bond remains largely unaffected by the external electric field (0.005 a.u.) in the DA-AD conformer of the phenylboronic acid dimers (complex **13**), whereas it varies significantly in the AA-DD conformer (complex **14**). Specifically, the induced interaction energies are −1.8 and − 15.7 kJ/mol in DA-AD and AA-DD conformers, respectively. The insensitive response of hydrogen bonds in DA-AD conformer is attributed to the antiparallel arrangement of adjacent hydrogen bonds, which respond

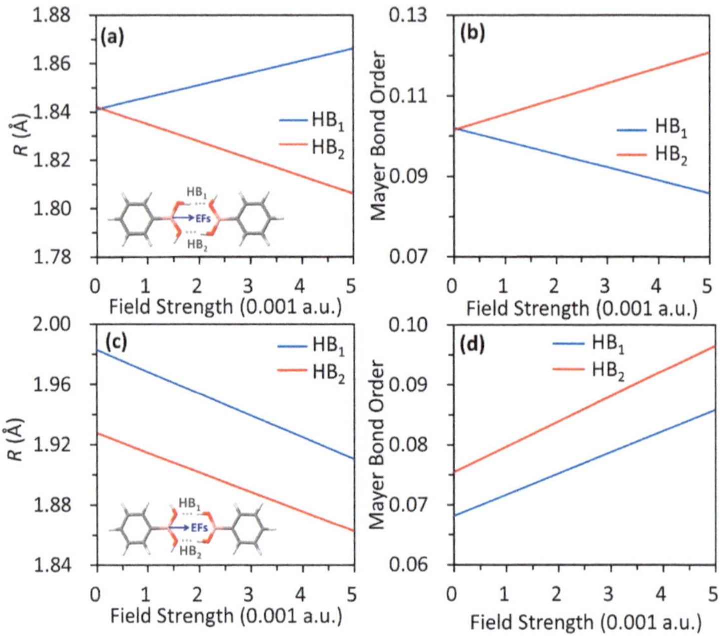

Fig. 5 Correlations between key bond lengths, Mayer bond orders of complexes **13** and **14**, and field strengths. (Reused with permission from Ref. [159]. Copyright (2021) John Wiley and Sons)

to EFs in opposing directions. This offset is evidenced by the variations in bond lengths and bond orders depicted in Fig. 5. Specifically, the upper hydrogen bond (HB_1) elongates (Fig. 5a) and experiences a decrease in the bond order (Fig. 5b), while the lower hydrogen bond (HB_2) shortens (Fig. 5a) with an increase in the bond order (Fig. 5b). Consequently, HB_2 strengthens while HB_1 weakens, leading to an overall insensitivity of bonding strength to EFs. In addition, the DA-AD geometry goes from C_{2h} symmetry to C_s symmetry in the presence of an electric field. In contrast, both hydrogen bonds in the AA-DD conformer strengthen coherently in EFs, as evidenced by decreased bond lengths (Fig. 5c) and increased bond orders (Fig. 5d), leading to a sensitive response of bond strength to the electric fields.

6 Selectivity of Accumulated Hydrogen Bonds for Anions Inside Cryptands

6.1 Analyzing Selectivity Using BLW-ED

Non-covalent bonds with varying strengths can form between a specified molecule and different entities, resulting in the selective characteristics of interactions from a thermodynamic perspective. In this context, the concept of anion receptors [160, 161] serves as an illuminating example, with diverse applications spanning chemistry, biology, and industrial separations [162–175]. Various non-covalent interactions function as elementary "linkers" in the anion-binding processes, contributing to the design of selective anion receptors. Typically, strong hydrogen bond donors, such as hydroxyl (–OH) and amine (–NH) groups, are employed to capture anionic species. A notable example is the polyamide cryptand synthesized by Kang et al., which selectively traps anions using endo-oriented bridgehead amines [176]. Interestingly, unconventional –CH hydrogen bonds have emerged as a significant driving force for anion attraction [177–183], gaining prominence following the noteworthy work of Liu et al. in 2019 [184]. Moreover, the importance of receptor rigidity in anion affinity was also highlighted, in addition to the fundamental linker and the size of binding pocket [184]. This is reasonable, as host-guest interactions occur within the confined cavities of receptors. Thus, rigidity affects the deformation of the acceptor and consequently influences the magnitude of the confinement effect, ultimately contributing to the preference for specific anions [185–187]. Therefore, factors such as the anion preference for individual hydrogen bonds, the confinement effect, and the rigidity of acceptor all affect the selectivity of interactions, which calls for systematic theoretical studies.

The selectivity can be accessed by comparing the strengths of interactions between a receptor and different anions. Hence, the relative values of interaction energies serve as a measure of selectivity, for which the governing factor can be identified by employing the BLW-ED approach. Specifically, relative values of each energy component can be calculated with the case exhibiting the most attractive interaction selected as the reference zero point. Thus, values above zero mean positive contributions to the selectivity. Here, the interaction energy is decomposed into the electrostatic, Pauli exchange repulsion, polarization, charge transfer, and electron correlation components as expressed in Eq. (9) to provide a comprehensive understanding of the selectivity. Notably, representative conformers of anion receptors should be included in the study of selectivity for a comprehensive understanding and can be derived from MD simulations. In this section, the initial structures of typical conformers were derived from MD simulations using the GFN2-xTB method [188, 189] incorporated in ORCA [190], followed by geometric optimizations at the M06-2X/def2-SVP(-PP) theoretical level, combined with Grimme's D3 dispersion correction. The subsequent BLW-ED analyses were performed at the SCS-MP2/def2-TVZP(-PP) theoretical level.

6.2 Governing Factors in the Selectivity of C–H Hydrogen Bonds Inside Cryptands

In this section, the cryptand cage synthesized by Liu et al. is selected as the prototype of the anion receptor featuring inward unconventional CH hydrogen donors [184]. We simplified the molecular cage by substituting the *N*,*N*-

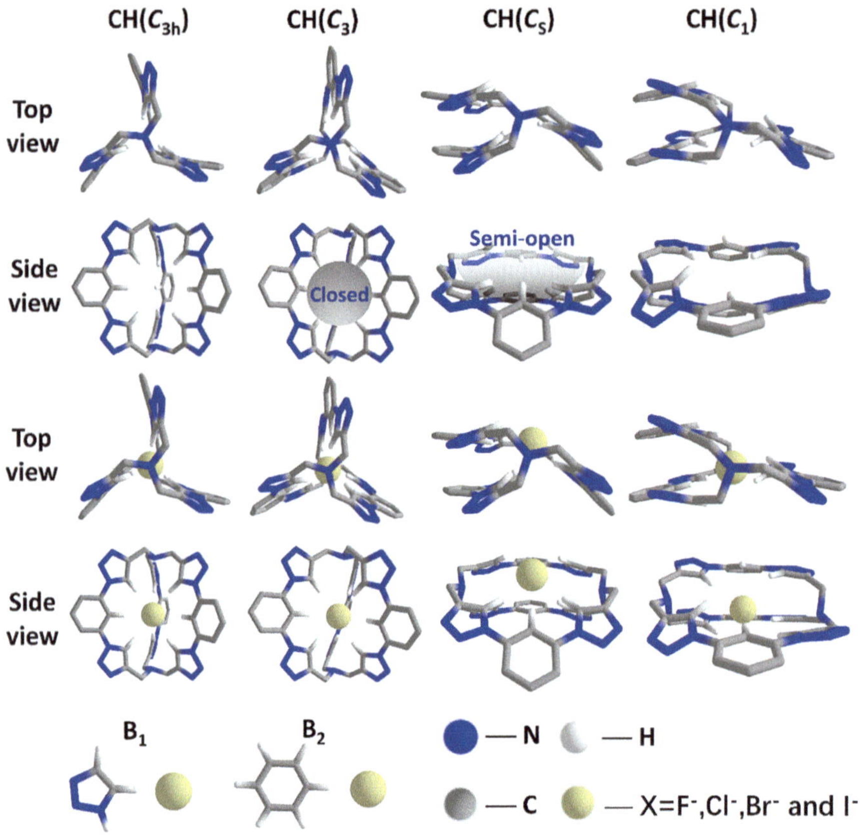

Scheme 5 Anion receptors and complexes studied in this section. (Reused with permission from Ref. [191]. Copyright (2023) John Wiley and Sons)

dicyclohexylformamide groups in the experimental structure with hydrogen atoms for computational efficiency [191]. Scheme 5 displays the geometries of selected conformers with C_{3h}, C_3, C_S, and C_1 symmetries, along with the corresponding complexes formed with F^-, Cl^-, Br^-, and I^- anions. We also investigated the selectivity of the key building blocks (1,2,3-triazole and benzene, denoted as B_1 and B_2 in Scheme 5), which provide inward CH hydrogen donors for the molecular cage. Conformers with a C_3 axis trap each halide anion near the centroid of the cage, forming a closed structure. In contrast, the C_S and C_1 conformers have semi-open structures and exhibit shifted bonding sites for different halide anions. Based on the interaction energies, all conformers exhibit the same sequence of preference for halogen anions, following the order $F^- > Cl^- > Br^- > I^-$, which is consistent with the selectivity observed experimentally.

The nature of the accumulated CH hydrogen bonds was examined prior to exploring the selectivity. Table 6 summarizes the BLW-ED results of complexes formed with the most favorable F^- anion and all conformers. The significant

Table 6 Energy components (in kJ/mol) based on the BLW-ED analyses

Complexes	ΔE_{ele}	ΔE_{Pauli}	ΔE_{pol}	ΔE_{CT}	ΔE_{ec}	ΔE_{int}
$CH(C_{3h})\cdot F^-$	−298.5	46.1	−128.9	6.7	5.0	−369.7
$CH(C_3)\cdot F^-$	−341.3	75.3	−147.8	7.5	4.7	−401.6
$CH(C_S)\cdot F^-$	−334.9	101.8	−147.9	5.2	3.4	−372.5
$CH(C_1)\cdot F^-$	−343.7	104.7	−149.4	4.9	4.7	−378.9
$B_1\cdot F^-$	−334.7	618.7	−327.2	−157.7	6.6	−194.3
$B_2\cdot F^-$	−415.6	598.1	−359.0	−179.4	23.1	−332.8

strength of the hydrogen bonds is evidenced by the interaction energy, which reaches −401.6 kJ/mol in the complex formed with the C_3 conformer. This observation is interesting because an individual CH is usually considered as a weak hydrogen donor, while the accumulated CH hydrogen bonds can remarkably attract the fluoride anion. The interaction energies are governed by the electrostatic interaction and enhanced by the polarization. The electron correlation is the third important contributor to the interaction energy, and the charge transfer interaction serves as the least stabilizing factor. The unreasonably positive values for the charge transfer and electron correlation terms are caused by the overestimation of the BSSE correction using the counterpoise method. The dominating role of electrostatic interaction is consistent with the electrostatic model for conventional hydrogen bonds. Similarly, electrostatic interaction also rules the elementary CH hydrogen bonds between the F^- and the building blocks providing the inward CH hydrogen donors. Notably, the charge transfer interaction plays a more important role in the elementary hydrogen bonds, implying stronger covalency than in the confined multiple hydrogen bonds. This is reasonable, as the encapsulated anion must balance among the inner bonding sites, resulting in stretched hydrogen-bonding distances compared to individual elementary cases and consequently reduced covalency.

The physical factor determining the selectivity can be identified by examining the relative values of each energy component, with the results in the complexes formed with fluoride anion, which is preferred over all the other halogen anions, as the reference zero points (Fig. 6). Notably, positive values of relative interaction energies are all dominated by the usually overlooked Pauli exchange repulsion, with positive contributions also gained from the polarization interaction. In contrast, electron correlation and charge transfer interactions both favor the bigger halogen anions, contributing negatively to the selectivity. The orbital overlap in the host-guest interaction increases as the size of the anion grows, leading to enhanced charge transfer, electron correlation, and Pauli exchange repulsion, among which the former two are stabilizing physical factors and prefer larger anions. Importantly, the Pauli exchange repulsion is destabilizing and varies most significantly as the size of anion grows, exhibiting the preference for F^- and dominating the selectivity. Additionally, F^- has the most concentrated electron density compared with the other halogen anions and imposes the strongest electric field to the cage, resulting in the strongest polarization of the receptor. The above explanations are based on the condition that the size of the pocket inside the receptor is insignificantly varied when different

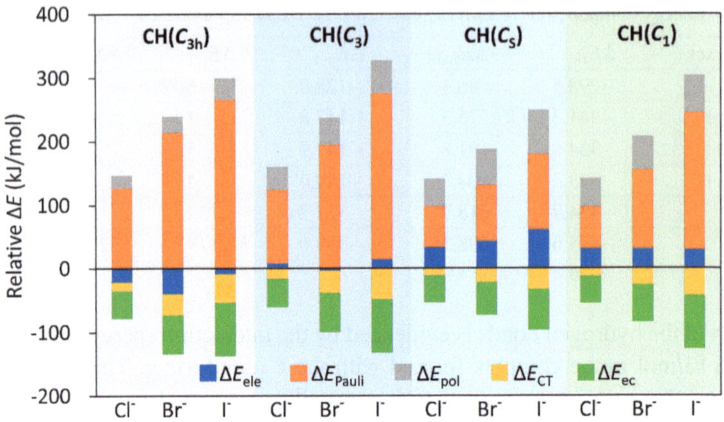

Fig. 6 Relative values (in kJ/mol) of all energy components, with the complexes constructed using F^- and the specified conformer of receptors CH as the reference zero points. (Reused with permission from Ref. [191]. Copyright (2023) John Wiley and Sons)

Fig. 7 Relative values of the energy components (kJ/mol), with the complexes formed with F^- and each specified building block of cages as the reference. (Reused with permission from Ref. [191]. Copyright (2023) John Wiley and Sons)

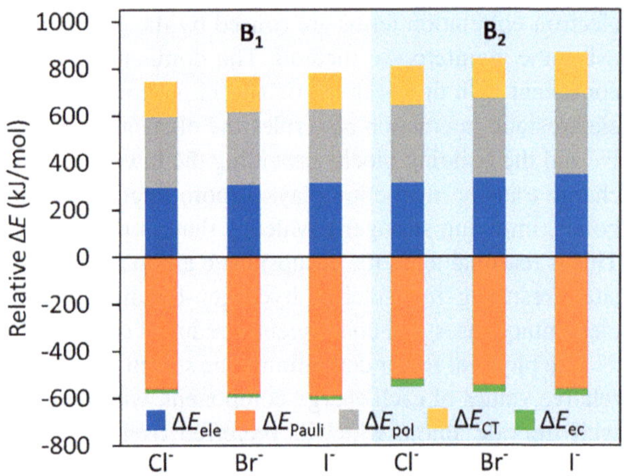

halogen anions enter. It was proved by inspecting the deformation of the molecular cage and further interpreted through the remarkable rigidity of this CH cryptand, which responds insensitively to external mechanical forces.

The selectivity of the key building blocks of the molecular cage was examined, revealing the same anion preference ($F^- > Cl^- > Br^- > I^-$) as observed for the CH cryptand (Fig. 7). Interestingly, this selectivity is primarily ruled by electrostatic interactions, with a negative contribution gained from Pauli exchange repulsion. This observation contrasts sharply with the accumulated CH hydrogen bonds inside the pocket of the molecular cage, where the anion preference is governed by Pauli exchange repulsion, with minimal contribution gained from electrostatic interactions. This is reasonable, as single hydrogen bonds can adjust their bond lengths more freely due to the absence of confinement, allowing for the strengthening of stabilizing factors like electrostatic interactions while reducing repulsive effects.

7 Interactions Between Ions of Like Charges

7.1 Metastability in Inter-anion Interactions

Unconventional patterns of non-covalent interactions offer novel linkers for constructing desired structures while also pushing the boundaries of our understanding of chemical bonds. A notable example is the inter-anion hydrogen bond, which has already been utilized as a new elementary linker in experimental studies [19, 192–196]. Importantly, this interaction challenges both the electrostatic model and the force field description, as it involves significant Coulombic repulsion. Moreover, inter-anion halogen bonds have also been experimentally observed and applied [81, 197–199], following pioneering theoretical predictions [81, 197, 200]. To date, the concepts of inter-anion/cation hydrogen bonds, halogen bonds, and other bonding modes have been recognized [16, 19, 192, 195, 196, 198, 201–224], broadening the scope of counterintuitive non-covalent interactions. Thus, a unified theory explaining the origin of these inter-anion/cation non-covalent interactions is expected.

The inter-anion hydrogen bond, which has been extensively studied, can serve as an illuminating example for the theoretical investigation of other counterintuitive interactions. The inter-anion hydrogen bonds can be analyzed in terms of energy, force constant, and electron density. Mata et al. first introduced the energy indicator by exploring the energy profile along the inter-anion distance. Despite the overall repulsive interaction, a local minimum is observed, followed by a transition state, which prevents the dissociation of the anion adduct [225]. Thus, the inter-anion bonding exhibits metastability, which can be measured by the energy difference between the minimum and transition state. The energy difference is often referred to as the well depth. Alternatively, the strength of inter-anion hydrogen bonds can also be qualified by the local bond force constant [226], which can be regarded as an indicator describing the intrinsic strength of chemical bonds [51–58]. The QTAIM theory [158] provides criteria for both the conventional and inter-anion non-covalent interactions from the perspective of electron density distribution [227].

Like the debate over the nature of conventional hydrogen bonds, whether electrostatic or charge transfer interactions govern the non-covalent interactions between like-charged ions remains a center of controversy. The charge transfer model was highlighted by Weinhold and Klein [16], who contested the idea that electrostatic interactions account for the inter-anion hydrogen bond, given the Coulombic repulsion between the charged species. Instead, they advocated for the charge transfer model, based on the results from their NBO method. Nevertheless, Frenking et al. asserted that it is a fallacious oversimplification to consider the inter-anion electrostatic interaction as merely Coulombic repulsion between two point charges and that a correct description of the electrostatic interaction is required [228, 229]. Surprisingly, despite this, these same interactions were still referred to as "antielectrostatic" phenomena [198, 199]. This terminology of "antistatic" is outdated and does not capture the true nature of the interactions between like-charged ions, which we will discuss below.

The potential surface analysis based on the BLW-ED approach rigorously describes the variation of each physical factor along the inter-anion distance and can thus identify the key factors governing the well depth, a widely adopted indicator of metastability. Additionally, the energy profile of the diabatic state can be obtained by performing a restricted geometrical optimization with the BLW method, offering direct evidence for the role of charge transfer interactions in metastability. To date, the BLW method has been successfully applied to the interpretation of the nature of inter-anion hydrogen, halogen, chalcogen, coinage bonds, and π–π stacking between ions of like charges [80–82, 86, 99, 223, 230, 231]. In this section, the origin of inter-anion hydrogen and halogen bonds will be discussed.

7.2 *The Physical Origin of the Metastability*

The bicarbonate dimer, extensively studied for its characteristic inter-anion hydrogen bonds, was selected as the prototypical complex in this section [226]. Additionally, the experimentally observed inter-anion halogen bond between the chloride anion and iodinated bis(dicyanomethylene) cyclopropanide anion is analyzed here. Neutral counterparts featuring conventional hydrogen or halogen bonds were constructed and studied alongside each inter-anion case to elucidate the impact of additional electrons on these non-covalent interactions. Specifically, the carbonic acid dimer serves as the counterpart for the inter-anion hydrogen bonding complex, while the halogen bonding counterpart was prepared by substituting one negatively charged carbon atom with boron (Scheme 6). The M06-2X-D3/def2-TZVP(-PP) theoretical level was adopted for all geometry optimizations and vibrational frequency analyses for the computational efficiency. The subsequent BLW-ED calculations were accomplished at the SCS-MP2/def2-TZVP(-PP) theoretical level. Energy profiles of the bicarbonate dimer and the corresponding counterparts were explored by solely fixing the distance between central carbon atoms at various values. Dissociation profiles for the halogen bonding cases were obtained through constrained geometrical optimizations, where the length of halogen bond was fixed.

Profiles of energy components along with the variation of inter-anion distances are represented in Fig. 8 for all counterintuitive cases and their conventional counterparts. For each inter-anion complex, the interaction energy curve (black) revealed a local minimum with positive interaction energy, exhibiting the characteristic of metastability. Similar to the conventional references, polarization, charge transfer, and dispersion terms remain almost zero at long ranges but become increasingly attractive at closer distances, stabilizing the minimal structure while minimally contributing to the stability of the transition state. As the sole

Scheme 6 Complexes **1–4** in this section. (Adapted with permission from Ref. [226]. Copyright (2021) American Chemical Society)

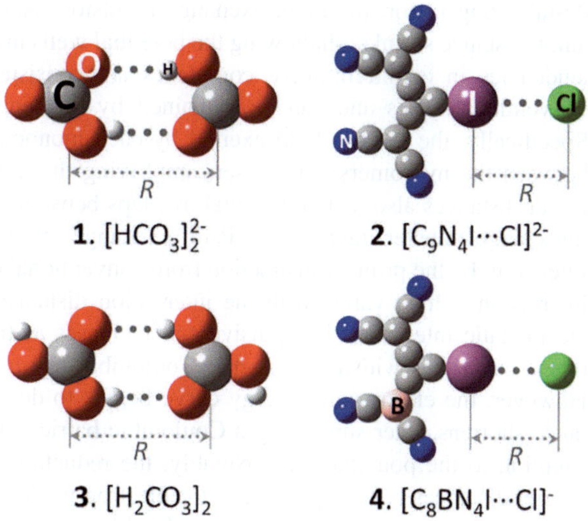

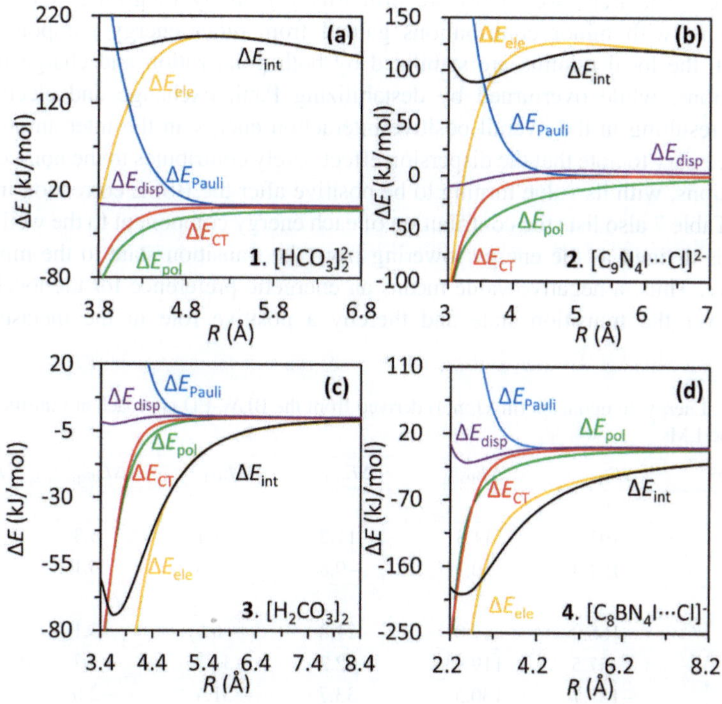

Fig. 8 Variations of energy components along with the inter-anion/molecular distances. (Adapted with permission from Ref. [226]. Copyright (2021) American Chemical Society)

destabilizing factor, the Pauli exchange repulsion rises exponentially as the inter-anion distance shrinks, shallowing the potential wells in the inter-anion cases. These tendencies in counterintuitive complexes are consistent with those observed in conventional cases and can be explained by the length of non-covalent bonds. Specifically, the electric field exerted by each monomer increases as the distance between the monomers gets closer, amplifying the polarization effect. Moreover, closer distances also increase orbital overlaps between monomers, thereby enhancing charge transfer interactions, Pauli exchange repulsion, and dispersion forces. Interestingly, the primary distinction from conventional cases lies in the electrostatic interaction, which varies with the inter-anion distance. At larger separations, the electrostatic interaction is repulsive and increases as the anions approach to each other, consistent with the expected Coulombic repulsion between point charges. However, the electrostatic energy curve begins to decrease as the inter-anion distance shortens, after surpassing a Coulombic barrier, suggesting a stabilizing contribution to the potential well. Notably, the reduction in electrostatic repulsion at shorter distances is unsurprising, as the electron-nucleus attractions between monomers become competitive with the Coulombic repulsion between the monomers.

Table 7 summarizes the BLW-ED results of the minima and the transition state structures. The transition states are primarily driven by long-range electrostatic repulsions, with minor contributions gained from other energy components. In contrast, the local minima are stabilized by both polarization and charge transfer interactions, while overturned by destabilizing Pauli exchange and electrostatic forces, resulting in the overall positive interaction energy in the inter-anion cases. It is necessary to note that the dispersion effect barely contributes to the non-covalent interactions, with its value turning to be positive after the BSSE correction in some cases. Table 7 also lists the contribution of each energy component to the well depth, which is defined as the energy lowering from the transition state to the minimum structure. Thus, a negative value means an energetic preference for the local minimum over the transition state and thereby a positive role in the metastability.

Table 7 Energy components (in kJ/mol) derived from the BLW-ED approach at transition states (TSs) and LMs

Complex	ΔE_{ele}	ΔE_{Pauli}	ΔE_{pol}	ΔE_{CT}	ΔE_{disp}	ΔE_{int}
TS						
1	193.4	11.3	−11.2	−1.1	3.3	195.7
2	124.8	0.1	−9.8	−0.3	5.1	119.9
LM						
1	166.9	47.8	−21.8	−6.9	3.1	189.1
2	57.5	119.9	−52.2	−31.7	−4.7	88.8
3	−135.7	130.5	−33.7	−31.4	−2.0	−72.3
4	−253.8	246.6	−82.5	−84.7	−19.0	−193.4
LM-TS						
1	−26.6	36.5	−10.6	−5.8	−0.2	−6.6
2	−67.3	119.8	−42.4	−31.4	−9.8	−31.1

Electrostatic interaction governs the well depths in both inter-anion hydrogen and halogen bonding cases, with polarization and charge transfer interactions also making positive contributions to metastability. Conversely, the Pauli exchange repulsion favors the transition state over the minimum structure, playing a negative role in the kinetic stability. We conducted the above potential energy surface analysis based on the BLW-ED analyses for a series of counterintuitive non-covalent interactions between ions of like charges, confirming the key role of electrostatic interaction in the metastability [80–82, 86, 99, 223, 230].

8 Concluding Remarks

Non-covalent interactions are ubiquitous in all fields including molecular biology, materials science, and pharmaceutical industry and play a key role in the properties of molecules and materials. The theoretical understanding of their underlying forces or factors can benefit the further rational design of novel molecules and materials with preferable and tunable properties. While numerous energy decomposition approaches have been developed and extensively applied in the past several decades, the BLW method is unique as it can derive the self-consistent wavefunction for the electron-localized (diabatic) state, allowing quantitative estimates of the charge transfer interaction in non-covalent interactions. The BLW-based energy decomposition scheme can, thus, provide a unified solution to probe the nature of various non-covalent interactions, as showcased in this chapter.

Acknowledgements CW acknowledges the support from the National Natural Science Foundation of China (No. 22073060). This work was performed in part at the Joint School of Nanoscience and Nanoengineering, a member of the National Nanotechnology Coordinated Infrastructure (NNCI), which is supported by the National Science Foundation (grant ECCS-2025462).

References

1. Stone A (2013) The theory of intermolecular forces. Oxford University Press, Oxford
2. Beno BR, Yeung K-S, Bartberger MD, Pennington LD, Meanwell NA (2015) A survey of the role of noncovalent sulfur interactions in drug design. J Med Chem 58(11):4383–4438. https://doi.org/10.1021/jm501853m
3. Nepal B, Scheiner S (2015) Long-range behavior of noncovalent bonds. Neutral and charged H-bonds, pnicogen, chalcogen, and halogen bonds. Chem Phys 456:34–40. https://doi.org/10.1016/j.chemphys.2015.04.008
4. Hobza P, Muller-Dethlefs K (2009) Non-covalent interactions: theory and experiment. The Royal Society of Chemistry, Cambridge
5. Alvarez S (2013) A cartography of the van der Waals territories. Dalton Trans 42(24):8617–8636. https://doi.org/10.1039/C3DT50599E
6. Arunan E (2013) Hydrogen bond seen, halogen bond defined and carbon bond proposed: intermolecular bonding, a field that is maturing! Curr Sci 105(7):892–894

7. Ball P (2014) What is a bond? Chem World 11:50–54
8. Guevara-Vela JM, Romero-Montalvo E, Mora Gómez VA, Chávez-Calvillo R, García-Revilla M, Francisco E et al (2016) Hydrogen bond cooperativity and anticooperativity within the water hexamer. Phys Chem Chem Phys 18(29):19557–19566. https://doi.org/10.1039/C6CP00763E
9. Gibb BC (2020) The centenary (maybe) of the hydrogen bond. Nat Chem 12(8):665–667. https://doi.org/10.1038/s41557-020-0524-2
10. Dereka B, Yu Q, Lewis NHC, Carpenter WB, Bowman JM, Tokmakoff A (2021) Crossover from hydrogen to chemical bonding. Science 371(6525):160–164. https://doi.org/10.1126/science.abe1951
11. Hansen AS, Du L, Kjaergaard HG (2014) Positively charged phosphorus as a hydrogen bond acceptor. J Phys Chem Lett 5(23):4225–4231. https://doi.org/10.1021/jz502150d
12. Zhang X, Dai H, Yan H, Zou W, Cremer D (2016) B–H···π interaction: a new type of nonclassical hydrogen bonding. J Am Chem Soc 138(13):4334–4337. https://doi.org/10.1021/jacs.6b01249
13. Lin X, Wu W, Mo Y (2018) A direct proof of the resonance-impaired hydrogen bond (RIHB) concept. Chem Eur J 24(5):1053–1056. https://doi.org/10.1002/chem.201705424
14. Lin X, Wu W, Mo Y (2019) How resonance modulates multiple hydrogen bonding in self-assembled systems. J Org Chem 84(22):14805–14815. https://doi.org/10.1021/acs.joc.9b02381
15. Mata I, Alkorta I, Molins E, Espinosa E (2013) Tracing environment effects that influence the stability of anion–anion complexes: the case of phosphate–phosphate interactions. Chem Phys Lett 555:106–109. https://doi.org/10.1016/j.cplett.2012.10.073
16. Weinhold F, Klein RA (2014) Anti-electrostatic hydrogen bonds. Angew Chem Int Ed 53(42): 11214–11217. https://doi.org/10.1002/anie.201405812
17. Gamrad W, Dreier A, Goddard R, Pörschke K-R (2015) Cation–cation pairing by N–C–H···O hydrogen bonds. Angew Chem Int Ed 54(15):4482–4487. https://doi.org/10.1002/anie.201408278
18. Mata I, Molins E, Alkorta I, Espinosa E (2015) The paradox of hydrogen-bonded anion–anion aggregates in oxoanions: a fundamental electrostatic problem explained in terms of electrophilic···nucleophilic interactions. J Phys Chem A 119(1):183–194. https://doi.org/10.1021/jp510198g
19. MacDonald TSC, Feringa BL, Price WS, Wezenberg SJ, Beves JE (2020) Controlled diffusion of photoswitchable receptors by binding anti-electrostatic hydrogen-bonded phosphate oligomers. J Am Chem Soc 142(47):20014–20020. https://doi.org/10.1021/jacs.0c09072
20. Auffinger P, Hays FA, Westhof E, Ho PS (2004) Halogen bonds in biological molecules. Proc Natl Acad Sci 101(48):16789–16794. https://doi.org/10.1073/pnas.0407607101
21. Mínguez Espallargas G, Zordan F, Arroyo Marín L, Adams H, Shankland K, van de Streek J et al (2009) Rational modification of the hierarchy of intermolecular interactions in molecular crystal structures by using tunable halogen bonds. Chem Eur J 15(31):7554–7568. https://doi.org/10.1002/chem.200900410
22. Legon AC (2010) The halogen bond: an interim perspective. Phys Chem Chem Phys 12(28): 7736–7747. https://doi.org/10.1039/C002129F
23. Bertani R, Sgarbossa P, Venzo A, Lelj F, Amati M, Resnati G et al (2010) Halogen bonding in metal–organic–supramolecular networks. Coord Chem Rev 254(5):677–695. https://doi.org/10.1016/j.ccr.2009.09.035
24. Hardegger LA, Kuhn B, Spinnler B, Anselm L, Ecabert R, Stihle M et al (2011) Systematic investigation of halogen bonding in protein–ligand interactions. Angew Chem Int Ed 50(1): 314–318. https://doi.org/10.1002/anie.201006781
25. Parker AJ, Stewart J, Donald KJ, Parish CA (2012) Halogen bonding in DNA base pairs. J Am Chem Soc 134(11):5165–5172. https://doi.org/10.1021/ja2105027

26. Mani D, Arunan E (2013) The X–C···Y (X = O/F, Y = O/S/F/Cl/Br/N/P) 'carbon bond' and hydrophobic interactions. Phys Chem Chem Phys 15(34):14377–14383. https://doi.org/10.1039/C3CP51658J
27. Meyer F, Dubois P (2013) Halogen bonding at work: recent applications in synthetic chemistry and materials science. CrystEngComm 15(16):3058–3071. https://doi.org/10.1039/C2CE26150B
28. Politzer P, Lane P, Concha MC, Ma Y, Murray JS (2007) An overview of halogen bonding. J Mol Model 13(2):305–311. https://doi.org/10.1007/s00894-006-0154-7
29. Politzer P, Murray JS, Clark T (2013) Halogen bonding and other σ-hole interactions: a perspective. Phys Chem Chem Phys 15(27):11178–11189. https://doi.org/10.1039/C3CP00054K
30. Montaña ÁM (2017) The σ and π holes. The halogen and tetrel bondings: their nature, importance and chemical, biological and medicinal implications. ChemistrySelect 2(28):9094–9112. https://doi.org/10.1002/slct.201701676
31. Angarov V, Kozuch S (2018) On the σ, π and δ hole interactions: a molecular orbital overview. New J Chem 42(2):1413–1422. https://doi.org/10.1039/C7NJ03632A
32. Metrangolo P, Resnati G (2001) Halogen bonding: a paradigm in supramolecular chemistry. Chem Eur J 7(12):2511–2519. https://doi.org/10.1002/1521-3765(20010618)7:12<2511::AID-CHEM25110>3.0.CO;2-T
33. Metrangolo P, Neukirch H, Pilati T, Resnati G (2005) Halogen bonding based recognition processes: a world parallel to hydrogen bonding. Acc Chem Res 38(5):386–395. https://doi.org/10.1021/ar0400995
34. Li Q, Xu X, Liu T, Jing B, Li W, Cheng J et al (2010) Competition between hydrogen bond and halogen bond in complexes of formaldehyde with hypohalous acids. Phys Chem Chem Phys 12(25):6837–6843. https://doi.org/10.1039/B926355A
35. Chen X, Liu S, Liu L, Liu X, Cai Y, Liu N et al (2013) Imaging hydrogen bond in real space. arXiv: Chem Phys
36. Zhang J, Chen P, Yuan B, Ji W, Cheng Z, Qiu X (2013) Real-space identification of intermolecular bonding with atomic force microscopy. Science 342(6158):611–614. https://doi.org/10.1126/science.1242603
37. Johnson ER, Keinan S, Mori-Sánchez P, Contreras-García J, Cohen AJ, Yang W (2010) Revealing noncovalent interactions. J Am Chem Soc 132(18):6498–6506. https://doi.org/10.1021/ja100936w
38. Guerra C, Burgos J, Ayarde-Henríquez L, Chamorro E (2024) Formulating reduced density gradient approaches for noncovalent interactions. J Phys Chem A 128(30):6158–6166. https://doi.org/10.1021/acs.jpca.4c01667
39. Mo Y, Peyerimhoff SD (1998) Theoretical analysis of electronic delocalization. J Chem Phys 109(5):1687–1697. https://doi.org/10.1063/1.476742
40. Mo Y, Song L, Wu W, Cao Z, Zhang Q (2002) Electronic delocalization: a quantitative study from modern ab initio valence bond theory. J Theor Comput Chem 01(01):137–151. https://doi.org/10.1142/s0219633602000099
41. Mo Y (2003) Geometrical optimization for strictly localized structures. J Chem Phys 119(3):1300–1306. https://doi.org/10.1063/1.1580094
42. Mo Y, Song L, Lin Y (2007) Block-localized wavefunction (BLW) method at the density functional theory (DFT) level. J Phys Chem A 111(34):8291–8301. https://doi.org/10.1021/jp0724065
43. Gallup GA (2002) Valence bond methods: theory and applications. Cambridge University Press, Cambridge
44. Gallup GA (2002) Valence bond theory, vol 10. Elsevier, Amsterdam
45. Shaik S, Hiberty PC (2008) A chemist's guide to valence bond theory. Wiley, Hoboken
46. Wu W, Su P, Shaik S, Hiberty PC (2011) Classical valence bond approach by modern methods. Chem Rev 111(11):7557–7593. https://doi.org/10.1021/cr100228r

47. Grimme S, Hansen A, Brandenburg JG, Bannwarth C (2016) Dispersion-corrected mean-field electronic structure methods. Chem Rev 116(9):5105–5154. https://doi.org/10.1021/acs.chemrev.5b00533
48. Bader RFW, Larouche A, Gatti C, Carroll MT, MacDougall PJ, Wiberg KB (1987) Properties of atoms in molecules: dipole moments and transferability of properties. J Chem Phys 87(2): 1142–1152. https://doi.org/10.1063/1.453294
49. Mata I, Molins E, Alkorta I, Espinosa E (2009) Effect of an external electric field on the dissociation energy and the electron density properties: the case of the hydrogen bonded dimer HF···HF. J Chem Phys 130(4):044104. https://doi.org/10.1063/1.3065972
50. Wu P, Chaudret R, Hu X, Yang W (2013) Noncovalent interaction analysis in fluctuating environments. J Chem Theory Comput 9(5):2226–2234. https://doi.org/10.1021/ct4001087
51. Konkoli Z, Cremer D (1998) A new way of analyzing vibrational spectra. III. Characterization of normal vibrational modes in terms of internal vibrational modes. Int J Quant Chem 67(1): 29–40. https://doi.org/10.1002/(SICI)1097-461X(1998)67:1<29::AID-QUA3>3.0.CO;2-0
52. Konkoli Z, Cremer D (1998) A new way of analyzing vibrational spectra. I. Derivation of adiabatic internal modes. Int J Quant Chem 67(1):1–9. https://doi.org/10.1002/(SICI)1097-461X(1998)67:1<1::AID-QUA1>3.0.CO;2-Z
53. Konkoli Z, Larsson JA, Cremer D (1998) A new way of analyzing vibrational spectra. IV. Application and testing of adiabatic modes within the concept of the characterization of normal modes. Int J Quant Chem 67(1):41–55. https://doi.org/10.1002/(SICI)1097-461X(1998)67:1<41::AID-QUA4>3.0.CO;2-Z
54. Konkoli Z, Larsson JA, Cremer D (1998) A new way of analyzing vibrational spectra. II. Comparison of internal mode frequencies. Int J Quant Chem 67(1):11–27. https://doi.org/10.1002/(SICI)1097-461X(1998)67:1<11::AID-QUA2>3.0.CO;2-1
55. Cremer D, Wu A, Larsson A, Kraka E (2000) Some thoughts about bond energies, bond lengths, and force constants. J Mol Model 6(4):396–412. https://doi.org/10.1007/PL00010739
56. Cremer D, Kraka E (2017) Generalization of the Tolman electronic parameter: the metal–ligand electronic parameter and the intrinsic strength of the metal–ligand bond. Dalton Trans 46(26):8323–8338. https://doi.org/10.1039/C7DT00178A
57. Kraka E, Cremer D (2019) Dieter Cremer's contribution to the field of theoretical chemistry. Int J Quantum Chem 119(6):e25849. https://doi.org/10.1002/qua.25849
58. Tao Y, Zou W, Sethio D, Verma N, Qiu Y, Tian C et al (2019) In situ measure of intrinsic bond strength in crystalline structures: local vibrational mode theory for periodic systems. J Chem Theory Comput 15(3):1761–1776. https://doi.org/10.1021/acs.jctc.8b01279
59. Kitaura K, Morokuma K (1976) A new energy decomposition scheme for molecular interactions within the Hartree-Fock approximation. Int J Quantum Chem 10(2):325–340. https://doi.org/10.1002/qua.560100211
60. Morokuma K (1977) Why do molecules interact? The origin of electron donor-acceptor complexes, hydrogen bonding and proton affinity. Acc Chem Res 10(8):294–300. https://doi.org/10.1021/ar50116a004
61. Jeziorski B, Moszynski R, Szalewicz K (1994) Perturbation theory approach to intermolecular potential energy surfaces of van der Waals complexes. Chem Rev 94(7):1887–1930. https://doi.org/10.1021/cr00031a008
62. Sinnokrot MO, Sherrill CD (2004) Substituent effects in $\pi-\pi$ interactions: sandwich and T-shaped configurations. J Am Chem Soc 126(24):7690–7697. https://doi.org/10.1021/ja049434a
63. Geng Y, Takatani T, Hohenstein EG, Sherrill CD (2010) Accurately characterizing the $\pi-\pi$ interaction energies of indole−benzene complexes. J Phys Chem A 114(10):3576–3582. https://doi.org/10.1021/jp9099495
64. Sherrill CD (2013) Energy component analysis of π interactions. Acc Chem Res 46(4): 1020–1028. https://doi.org/10.1021/ar3001124

65. Mitoraj MP, Michalak A, Ziegler T (2009) A combined charge and energy decomposition scheme for bond analysis. J Chem Theory Comput 5(4):962–975. https://doi.org/10.1021/ct800503d
66. Ziegler T, Rauk A, Baerends EJ (1977) On the calculation of multiplet energies by the Hartree-Fock-Slater method. Theor Chim Acta 43(3):261–271. https://doi.org/10.1007/BF00551551
67. Su P, Jiang Z, Chen Z, Wu W (2014) Energy decomposition scheme based on the generalized Kohn–Sham scheme. J Phys Chem A 118(13):2531–2542. https://doi.org/10.1021/jp500405s
68. Su P, Tang Z, Wu W (2020) Generalized Kohn-Sham energy decomposition analysis and its applications. WIREs Comput Mol Sci 10(5):e1460. https://doi.org/10.1002/wcms.1460
69. Grimme S (2003) Improved second-order Møller–Plesset perturbation theory by separate scaling of parallel- and antiparallel-spin pair correlation energies. J Chem Phys 118(20): 9095–9102. https://doi.org/10.1063/1.1569242
70. Takatani T, David Sherrill C (2007) Performance of spin-component-scaled Møller–Plesset theory (SCS-MP2) for potential energy curves of noncovalent interactions. Phys Chem Chem Phys 9(46):6106–6114. https://doi.org/10.1039/B709669K
71. Kozuch S, Martin JML (2013) Halogen bonds: benchmarks and theoretical analysis. J Chem Theory Comput 9(4):1918–1931. https://doi.org/10.1021/ct301064t
72. Rigby J, Izgorodina EI (2014) New SCS- and SOS-MP2 coefficients fitted to semi-coulombic systems. J Chem Theory Comput 10(8):3111–3122. https://doi.org/10.1021/ct500309x
73. Neese F, Wennmohs F, Hansen A (2009) Efficient and accurate local approximations to coupled-electron pair approaches: an attempt to revive the pair natural orbital method. J Chem Phys 130(11). https://doi.org/10.1063/1.3086717
74. Altun A, Neese F, Bistoni G (2020) Extrapolation to the limit of a complete pair natural orbital space in local coupled-cluster calculations. J Chem Theory Comput 16(10):6142–6149. https://doi.org/10.1021/acs.jctc.0c00344
75. Gray M, Herbert JM (2024) Assessing the domain-based local pair natural orbital (DLPNO) approximation for non-covalent interactions in sizable supramolecular complexes. J Chem Phys 161(5). https://doi.org/10.1063/5.0206533
76. Chen W, Gordon MS (1996) Energy decomposition analyses for many-body interaction and applications to water complexes. J Phys Chem 100(34):14316–14328. https://doi.org/10.1021/jp960694r
77. Tang Z, Song Y, Zhang S, Wang W, Xu Y, Wu D et al (2021) XEDA, a fast and multipurpose energy decomposition analysis program. J Comput Chem 42(32):2341–2351. https://doi.org/10.1002/jcc.26765
78. Lei S-B, Deng K, Yang Y-L, Zeng Q-D, Wang C, Jiang J-Z (2008) Electric driven molecular switching of asymmetric tris(phthalocyaninato) lutetium Triple-Decker complex at the liquid/solid Interface. Nano Lett 8(7):1836–1843. https://doi.org/10.1021/nl0803186
79. Wang C, Mo Y, Wagner JP, Schreiner PR, Jemmis ED, Danovich D et al (2015) The self-association of graphane is driven by London dispersion and enhanced orbital interactions. J Chem Theory Comput 11(4):1621–1630. https://doi.org/10.1021/acs.jctc.5b00075
80. Wang C, Danovich D, Shaik S, Mo Y (2017) Halogen bonds in novel polyhalogen monoanions. Chem Eur J 23(36):8719–8728. https://doi.org/10.1002/chem.201701116
81. Wang C, Fu Y, Zhang L, Danovich D, Shaik S, Mo Y (2018) Hydrogen- and halogen-bonds between ions of like charges: are they anti-electrostatic in nature? J Comput Chem 39(9): 481–487. https://doi.org/10.1002/jcc.25068
82. Wang C, Danovich D, Shaik S, Wu W, Mo Y (2019) Attraction between electrophilic caps: a counterintuitive case of noncovalent interactions. J Comput Chem 40(9):1015–1022. https://doi.org/10.1002/jcc.25566
83. Mo Y (2014) The block-localized wavefunction (BLW) perspective of chemical bonding. In: Frenking G, Shaik S (eds) The chemical bond: fundamental aspects of chemical bonding. Wiley, pp 199–232

84. Wang C, Guan L, Danovich D, Shaik S, Mo Y (2016) The origins of the directionality of noncovalent intermolecular interactions. J Comput Chem 37(1):34–45. https://doi.org/10.1002/jcc.23946
85. Wang C, Danovich D, Shaik S, Mo Y (2017) A unified theory for the blue- and red-shifting phenomena in hydrogen and halogen bonds. J Chem Theory Comput 13(4):1626–1637. https://doi.org/10.1021/acs.jctc.6b01133
86. Wang C, Aman Y, Ji X, Mo Y (2019) Tetrel bonding interaction: an analysis with the block-localized wavefunction (BLW) approach. Phys Chem Chem Phys 21(22):11776–11784. https://doi.org/10.1039/C9CP01710K
87. Schmidt MW, Baldridge KK, Boatz JA, Elbert ST, Gordon MS, Jensen JH et al (1993) General atomic and molecular electronic structure system. J Comput Chem 14(11):1347–1363. https://doi.org/10.1002/jcc.540141112
88. Song L, Mo Y, Zhang Q, Wu W (2005) XMVB: a program for ab initio nonorthogonal valence bond computations. J Comput Chem 26(5):514–521. https://doi.org/10.1002/jcc.20187
89. Chen Z, Ying F, Chen X, Song J, Su P, Song L et al (2015) XMVB 2.0: a new version of Xiamen valence bond program. Int J Quantum Chem 115(11):731–737. https://doi.org/10.1002/qua.24855
90. Pauling L (1960) The nature of the chemical bond. Cornell University Press, Ithaca
91. Kollman PA, Allen LC (1972) Theory of the hydrogen bond. Chem Rev 72(3):283–303. https://doi.org/10.1021/cr60277a004
92. Liu SY, Dykstra CE (1986) A theory of vibrational transition frequency shifts due to hydrogen bonding. J Phys Chem 90(14):3097–3103. https://doi.org/10.1021/j100405a013
93. Gordon MS, Jensen JH (1996) Understanding the hydrogen bond using quantum chemistry. Acc Chem Res 29(11):536–543. https://doi.org/10.1021/ar9600594
94. Ghanty TK, Staroverov VN, Koren PR, Davidson ER (2000) Is the hydrogen bond in water dimer and ice covalent? J Am Chem Soc 122(6):1210–1214. https://doi.org/10.1021/ja9937019
95. Guerra CF, Bickelhaupt FM, Baerends EJ (2004) Hydrogen bonding in mimics of Watson–Crick base pairs involving C–H proton donor and f proton acceptor groups: a theoretical study. ChemPhysChem 5(4):481–487. https://doi.org/10.1002/cphc.200301069
96. Perrin CL (2010) Are short, low-barrier hydrogen bonds unusually strong? Acc Chem Res 43(12):1550–1557. https://doi.org/10.1021/ar100097j
97. Grabowski SJ (2011) What is the covalency of hydrogen bonding? Chem Rev 111(4): 2597–2625. https://doi.org/10.1021/cr800346f
98. Mulliken RS (1950) Structures of complexes formed by halogen molecules with aromatic and with oxygenated solvents. J Am Chem Soc 72(1):600–608. https://doi.org/10.1021/ja01157a151
99. Wang C, Danovich D, Mo Y, Shaik S (2014) On the nature of the halogen bond. J Chem Theory Comput 10(9):3726–3737. https://doi.org/10.1021/ct500422t
100. Bader RFW (1964) A model for the hydrogen bond and proton transfer reaction. Can J Chem 42(8):1822–1834. https://doi.org/10.1139/v64-274
101. Ratajczak H (1972) Charge-transfer properties of the hydrogen bond. I. Theory of the enhancement of dipole moment of hydrogen-bonded systems. J Phys Chem 76(21): 3000–3004. https://doi.org/10.1021/j100665a013
102. Thompson WH, Hynes JT (2000) Frequency shifts in the hydrogen-bonded OH stretch in halide−water clusters. The importance of charge transfer. J Am Chem Soc 122(26): 6278–6286. https://doi.org/10.1021/ja993058q
103. Trudeau G, Dumas J-M, Dupuis P, Guérin M, Sandorfy C (1980) Intermolecular interactions and anesthesia: infrared spectroscopic studies. In: Van der Waals systems. Springer, Berlin, pp 91–125
104. Buděšínský M, Fiedler P, Arnold Z (1989) Triformylmethane: an efficient preparation, some derivatives, and spectra. Synthesis 1989(11):858–860. https://doi.org/10.1055/s-1989-27411

105. Hobza P, Špirko V, Selzle HL, Schlag EW (1998) Anti-hydrogen bond in the benzene dimer and other carbon proton donor complexes. J Phys Chem A 102(15):2501–2504. https://doi.org/10.1021/jp973374w
106. Hobza P, Špirko VR, Havlas Z, Buchhold K, Reimann B, Barth H-D et al (1999) Anti-hydrogen bond between chloroform and fluorobenzene. Chem Phys Lett 299(2):180–186. https://doi.org/10.1016/S0009-2614(98)01264-0
107. Hobza P, Havlas Z (2000) Blue-shifting hydrogen bonds. Chem Rev 100(11):4253–4264. https://doi.org/10.1021/cr990050q
108. van der Veken BJ, Herrebout WA, Szostak R, Shchepkin DN, Havlas Z, Hobza P (2001) The nature of improper, blue-shifting hydrogen bonding verified experimentally. J Am Chem Soc 123(49):12290–12293. https://doi.org/10.1021/ja010915t
109. Egli M, Sarkhel S (2007) Lone pair−aromatic interactions: to stabilize or not to stabilize. Acc Chem Res 40(3):197–205. https://doi.org/10.1021/ar068174u
110. Wright AM, Howard AA, Howard JC, Tschumper GS, Hammer NI (2013) Charge transfer and blue shifting of vibrational frequencies in a hydrogen bond acceptor. J Phys Chem A 117(26): 5435–5446. https://doi.org/10.1021/jp401642b
111. Alabugin IV, Manoharan M, Peabody S, Weinhold F (2003) Electronic basis of improper hydrogen bonding: a subtle balance of hyperconjugation and rehybridization. J Am Chem Soc 125(19):5973–5987. https://doi.org/10.1021/ja034656e
112. Wang W, Wong N-B, Zheng W, Tian A (2004) Theoretical study on the blueshifting halogen bond. J Phys Chem A 108(10):1799–1805. https://doi.org/10.1021/jp036769q
113. Wang W, Hobza P (2008) Origin of the X−Hal (Hal = Cl, Br) bond-length change in the halogen-bonded complexes. J Phys Chem A 112(17):4114–4119. https://doi.org/10.1021/jp710992h
114. Wang W, Zhang Y, Ji B (2010) On the difference of the properties between the blue-shifting halogen bond and the blue-shifting hydrogen bond. J Phys Chem A 114(26):7257–7260. https://doi.org/10.1021/jp103457u
115. Joy J, Jemmis ED, Vidya K (2015) Negative hyperconjugation and red-, blue- or zero-shift in X–Z···Y complexes. Faraday Discuss 177:33–50. https://doi.org/10.1039/C4FD00183D
116. Joy J, Jose A, Jemmis ED (2016) Continuum in the X–Z···Y weak bonds: Z = main group elements. J Comput Chem 37(2):270–279. https://doi.org/10.1002/jcc.24036
117. Hermansson K (2002) Blue-shifting hydrogen bonds. J Phys Chem A 106(18):4695–4702. https://doi.org/10.1021/jp0143948
118. Qian W, Krimm S (2002) Vibrational spectroscopy of hydrogen bonding: origin of the different behavior of the C−H···O hydrogen bond. J Phys Chem A 106(28):6628–6636. https://doi.org/10.1021/jp020438g
119. McDowell SAC, Buckingham AD (2005) On the correlation between bond-length change and vibrational frequency shift in hydrogen-bonded complexes: a computational study of Y···HCl dimers (Y = N2, CO, BF). J Am Chem Soc 127(44):15515–15520. https://doi.org/10.1021/ja0543651
120. McDowell SAC, Buckingham AD (2005) Comparison of some vibrational features of FArH···Rg and FH···Rg complexes (Rg = He, Ne, Ar, Kr). Spectrochim Acta A Mol Biomol Spectrosc 61(7):1603–1609. https://doi.org/10.1016/j.saa.2004.11.028
121. Buckingham AD, Del Bene JE, McDowell SAC (2008) The hydrogen bond. Chem Phys Lett 463(1):1–10. https://doi.org/10.1016/j.cplett.2008.06.060
122. Joseph J, Jemmis ED (2007) Red-, blue-, or no-shift in hydrogen bonds: a unified explanation. J Am Chem Soc 129(15):4620–4632. https://doi.org/10.1021/ja067545z
123. Kraka E, Setiawan D, Cremer D (2016) Re-evaluation of the bond length–bond strength rule: the stronger bond is not always the shorter bond. J Comput Chem 37(1):130–142. https://doi.org/10.1002/jcc.24207
124. Kaupp M, Danovich D, Shaik S (2017) Chemistry is about energy and its changes: a critique of bond-length/bond-strength correlations. Coord Chem Rev 344:355–362. https://doi.org/10.1016/j.ccr.2017.03.002

125. Huber SM, Scanlon JD, Jimenez-Izal E, Ugalde JM, Infante I (2013) On the directionality of halogen bonding. Phys Chem Chem Phys 15(25):10350–10357. https://doi.org/10.1039/C3CP50892G
126. Murray JS, Riley KE, Politzer P, Clark T (2010) Directional weak intermolecular interactions: σ-hole bonding. Aust J Chem 63(12):1598–1607. https://doi.org/10.1071/CH10259
127. Shields ZP, Murray JS, Politzer P (2010) Directional tendencies of halogen and hydrogen bonds. Int J Quantum Chem 110(15):2823–2832. https://doi.org/10.1002/qua.22787
128. Steiner T, Desiraju R, G. (1998) Distinction between the weak hydrogen bond and the van der Waals interaction. Chem Commun 8:891–892. https://doi.org/10.1039/A708099I
129. Tsuzuki S, Wakisaka A, Ono T, Sonoda T (2012) Magnitude and origin of the attraction and directionality of the halogen bonds of the complexes of C_6F_5X and C_6H_5X (X=I, Br, cl and F) with pyridine. Chem Eur J 18(3):951–960. https://doi.org/10.1002/chem.201102562
130. Adhikari U, Scheiner S (2012) Sensitivity of pnictogen, chalcogen, halogen and H-bonds to angular distortions. Chem Phys Lett 532:31–35. https://doi.org/10.1016/j.cplett.2012.02.064
131. Reed AE, Curtiss LA, Weinhold F (1988) Intermolecular interactions from a natural bond orbital, donor-acceptor viewpoint. Chem Rev 88(6):899–926. https://doi.org/10.1021/cr00088a005
132. Glendening ED, Landis CR, Weinhold F (2012) Natural bond orbital methods. Wiley Interdiscip Rev Comput Mol Sci 2(1):1–42. https://doi.org/10.1002/wcms.51
133. Weinhold F, Landis CR, Glendening ED (2016) What is NBO analysis and how is it useful? Int Rev Phys Chem 35(3):399–440. https://doi.org/10.1080/0144235X.2016.1192262
134. Bauzá A, Alkorta I, Frontera A, Elguero J (2013) On the reliability of pure and hybrid DFT methods for the evaluation of halogen, chalcogen, and pnictogen bonds involving anionic and neutral electron donors. J Chem Theory Comput 9(11):5201–5210. https://doi.org/10.1021/ct400818v
135. Werling KA, Hutchison GR, Lambrecht DS (2013) Piezoelectric effects of applied electric fields on hydrogen-bond interactions: first-principles electronic structure investigation of weak electrostatic interactions. J Phys Chem Lett 4(9):1365–1370. https://doi.org/10.1021/jz400355v
136. Velpula G, Teyssandier J, De Feyter S, Mali KS (2017) Nanoscale control over the mixing behavior of surface-confined bicomponent supramolecular networks using an oriented external electric field. ACS Nano 11(11):10903–10913. https://doi.org/10.1021/acsnano.7b04610
137. Demirörs AF, Alison L (2018) Electric field assembly of colloidal superstructures. J Phys Chem Lett 9(15):4437–4443. https://doi.org/10.1021/acs.jpclett.8b01538
138. Mali KS, Wu D, Feng X, Müllen K, Van der Auweraer M, De Feyter S (2011) Scanning tunneling microscopy-induced reversible phase transformation in the two-dimensional crystal of a positively charged discotic polycyclic aromatic hydrocarbon. J Am Chem Soc 133(15): 5686–5688. https://doi.org/10.1021/ja111699n
139. Cui K, Mali KS, Ivasenko O, Wu D, Feng X, Walter M et al (2014) Squeezing, then stacking: from breathing pores to three-dimensional ionic self-assembly under electrochemical control. Angew Chem Int Ed 53(47):12951–12954. https://doi.org/10.1002/anie.201406246
140. Zheng Q-N, Liu X-H, Liu X-R, Chen T, Yan H-J, Zhong Y-W et al (2014) Bilayer molecular assembly at a solid/liquid interface as triggered by a mild electric field. Angew Chem Int Ed 53(49):13395–13399. https://doi.org/10.1002/anie.201406523
141. Cometto FP, Kern K, Lingenfelder M (2015) Local conformational switching of supramolecular networks at the solid/liquid interface. ACS Nano 9(5):5544–5550. https://doi.org/10.1021/acsnano.5b01658
142. Lee S-L, Fang Y, Velpula G, Cometto FP, Lingenfelder M, Müllen K et al (2015) Reversible local and global switching in multicomponent supramolecular networks: controlled guest release and capture at the solution/solid interface. ACS Nano 9(12):11608–11617. https://doi.org/10.1021/acsnano.5b06081

143. Ramos M, Alkorta I, Elguero J, Golubev NS, Denisov GS, Benedict H et al (1997) Theoretical study of the influence of electric fields on hydrogen-bonded acid−base complexes. J Phys Chem A 101(50):9791–9800. https://doi.org/10.1021/jp972586q
144. Toledo EJL, Custodio R, Ramalho TC, Porto MEG, Magriotis ZM (2009) Electrical field effects on dipole moment, structure and energetic of $(H_2O)n$ (2⩽n⩽15) cluster. J Mol Struct THEOCHEM 915(1):170–177. https://doi.org/10.1016/j.theochem.2009.08.035
145. Arabi AA, Matta CF (2011) Effects of external electric fields on double proton transfer kinetics in the formic acid dimer. Phys Chem Chem Phys 13(30):13738–13748. https://doi.org/10.1039/C1CP20175A
146. Mata I, Alkorta I, Espinosa E, Molins E (2011) Erratum to Relationships between interaction energy, intermolecular distance and electron density properties in hydrogen bonded complexes under external electric fields. Chem Phys Lett 508(4):332. https://doi.org/10.1016/j.cplett.2011.04.053
147. Jissy AK, Datta A (2012) Effect of external electric field on H-bonding and π-stacking interactions in guanine aggregates. ChemPhysChem 13(18):4163–4172. https://doi.org/10.1002/cphc.201200690
148. Zoppi L, Ferretti A, Baldridge KK (2013) Static and field-oriented properties of bowl-shaped polynuclear aromatic hydrocarbon fragments. J Chem Theory Comput 9(11):4797–4804. https://doi.org/10.1021/ct400575p
149. Foroutan-Nejad C, Marek R (2014) Potential energy surface and binding energy in the presence of an external electric field: modulation of anion–π interactions for graphene-based receptors. Phys Chem Chem Phys 16(6):2508–2514. https://doi.org/10.1039/C3CP52671B
150. Muruganathan M, Sun J, Imamura T, Mizuta H (2015) Electrically tunable van der Waals interaction in graphene–molecule complex. Nano Lett 15(12):8176–8180. https://doi.org/10.1021/acs.nanolett.5b03653
151. Fried SD, Boxer SG (2015) Measuring electric fields and noncovalent interactions using the vibrational stark effect. Acc Chem Res 48(4):998–1006. https://doi.org/10.1021/ar500464j
152. Dutta BJ, Bhattacharyya PK (2015) DFT studies on hydrogen-bonding, stacking, and X–H···π-bonded systems in presence of external electric field. Int J Quantum Chem 115(20):1459–1466. https://doi.org/10.1002/qua.24950
153. Novák M, Foroutan-Nejad C, Marek R (2016) Modulating electron sharing in ion-π-receptors via substitution and external electric field: a route toward bond strengthening. J Chem Theory Comput 12(8):3788–3795. https://doi.org/10.1021/acs.jctc.6b00586
154. Sarmah N, Bhattacharyya PK (2016) Behaviour of cation–pi interaction in presence of external electric field. RSC Adv 6(102):100008–100015. https://doi.org/10.1039/C6RA21334K
155. Xu T, Momen R, Azizi A, van Mourik T, Früchtl H, Kirk SR et al (2019) The destabilization of hydrogen bonds in an external E-field for improved switch performance. J Comput Chem 40(21):1881–1891. https://doi.org/10.1002/jcc.25843
156. Tokatlı A, Tunç F, Ucun F (2019) Effect of external electric field on C–X···π halogen bonds. J Mol Model 25(3):57. https://doi.org/10.1007/s00894-019-3938-2
157. Cai Z-F, Zhan G, Daukiya L, Eyley S, Thielemans W, Severin K et al (2019) Electric-field-mediated reversible transformation between supramolecular networks and covalent organic frameworks. J Am Chem Soc 141(29):11404–11408. https://doi.org/10.1021/jacs.9b05265
158. Bader R (1994) Atoms in molecules: a quantum theory. Oxford University Press, New York
159. Chen L, Dang J, Du J, Wang C, Mo Y (2021) Hydrogen and halogen bonding in homogeneous external electric fields: modulating the bond strengths. Chem Eur J 27(56):14042–14050. https://doi.org/10.1002/chem.202102284
160. Shriver DF, Biallas MJ (1967) Observation of the chelate effect with a bidentate Lewis acid, $F_2BCH_2CH_2BF_2$. J Am Chem Soc 89(5):1078–1081. https://doi.org/10.1021/ja00981a007
161. Park CH, Simmons HE (1968) Macrobicyclic amines. III. Encapsulation of halide ions by in,in-1,(k + 2)-diazabicyclo[k.l.m.]alkane ammonium ions. J Am Chem Soc 90(9):2431–2432. https://doi.org/10.1021/ja01011a047

162. Dastineh F, Salehzadeh S, Bayat M, Maghsoud Y (2015) Comparison of the selectivity of [M (12-Crown-4)]$^+$ (M=Li$^+$, Na$^+$, K$^+$) complexes for halide anions and some neutral molecules; a computational study. J Theor Comput Chem 14(08):1550057. https://doi.org/10.1142/s0219633615500571
163. Gale PA, Howe ENW, Wu X (2016) Anion receptor chemistry. Chem 1(3):351–422. https://doi.org/10.1016/j.chempr.2016.08.004
164. Gale PA, Howe ENW, Wu X, Spooner MJ (2018) Anion receptor chemistry: highlights from 2016. Coord Chem Rev 375:333–372. https://doi.org/10.1016/j.ccr.2018.02.005
165. Biagini C, Fielden SDP, Leigh DA, Schaufelberger F, Di Stefano S, Thomas D (2019) Dissipative catalysis with a molecular machine. Angew Chem Int Ed 58(29):9876–9880. https://doi.org/10.1002/anie.201905250
166. Kaabel S, Stein RS, Fomitšenko M, Järving I, Friščić T, Aav R (2019) Size-control by anion templating in mechanochemical synthesis of hemicucurbiturils in the solid state. Angew Chem Int Ed 58(19):6230–6234. https://doi.org/10.1002/anie.201813431
167. Park S-H, Hyun JY, Shin I (2019) A lysosomal chloride ion-selective fluorescent probe for biological applications. Chem Sci 10(1):56–66. https://doi.org/10.1039/C8SC04084B
168. Tapia L, Pérez Y, Bolte M, Casas J, Solà J, Quesada R et al (2019) pH-dependent chloride transport by pseudopeptidic cages for the selective killing of cancer cells in acidic microenvironments. Angew Chem Int Ed 58(36):12465–12468. https://doi.org/10.1002/anie.201905965
169. Wu X, Small JR, Cataldo A, Withecombe AM, Turner P, Gale PA (2019) Voltage-switchable HCl transport enabled by lipid headgroup–transporter interactions. Angew Chem Int Ed 58(42):15142–15147. https://doi.org/10.1002/anie.201907466
170. Chen L, Berry SN, Wu X, Howe ENW, Gale PA (2020) Advances in anion receptor chemistry. Chem 6(1):61–141. https://doi.org/10.1016/j.chempr.2019.12.002
171. Ong WSY, Smaldone RA, Dodani SC (2020) A neutral porous organic polymer host for the recognition of anionic dyes in water. Chem Sci 11(29):7716–7721. https://doi.org/10.1039/D0SC02941F
172. Shinde S, Incel A, Mansour M, Olsson GD, Nicholls IA, Esen C et al (2020) Urea-based imprinted polymer hosts with switchable anion preference. J Am Chem Soc 142(26):11404–11416. https://doi.org/10.1021/jacs.0c00707
173. Al Isawi WA, Salome AZ, Ahmed BM, Zeller M, Mezei G (2021) Selective binding of anions by rigidified nanojars: sulfate vs. carbonate. Org Biomol Chem 19(35):7641–7654. https://doi.org/10.1039/D1OB01318A
174. Williams GT, Haynes CJE, Fares M, Caltagirone C, Hiscock JR, Gale PA (2021) Advances in applied supramolecular technologies. Chem Soc Rev 50(4):2737–2763. https://doi.org/10.1039/D0CS00948B
175. Macreadie LK, Gilchrist AM, McNaughton DA, Ryder WG, Fares M, Gale PA (2022) Progress in anion receptor chemistry. Chem 8(1):46–118. https://doi.org/10.1016/j.chempr.2021.10.029
176. Kang SO, Llinares JM, Powell D, VanderVelde D, Bowman-James K (2003) New polyamide cryptand for anion binding. J Am Chem Soc 125(34):10152–10153. https://doi.org/10.1021/ja034969+
177. Alcalde E, Mesquida N, Pérez-García L, Alvarez-Rúa C, García-Granda S, García-Rodriguez E (1999) Hydrogen bonded driven anion binding by dicationic [14]imidazoliophanes. Chem Commun 3:295–296. https://doi.org/10.1039/A808503J
178. Sato K, Arai S, Yamagishi T (1999) A new tripodal anion receptor with C H···X$^-$ hydrogen bonding. Tetrahedron Lett 40(28):5219–5222. https://doi.org/10.1016/S0040-4039(99)00942-9
179. Li Y, Flood AH (2008) Pure C–H hydrogen bonding to chloride ions: a preorganized and rigid macrocyclic receptor. Angew Chem Int Ed 47(14):2649–2652. https://doi.org/10.1002/anie.200704717

180. Ramabhadran RO, Hua Y, Li Y-J, Flood AH, Raghavachari K (2011) From atomic to molecular anions: a neutral receptor captures cyanide using strong C–H hydrogen bonds. Chem Eur J 17(33):9123–9129. https://doi.org/10.1002/chem.201100793
181. Song NR, Moon JH, Choi J, Jun EJ, Kim Y, Kim S-J et al (2013) Cyclic benzobisimidazolium derivative for the selective fluorescent recognition of HSO_4^- via a combination of C–H hydrogen bonds and charge interactions. Chem Sci 4(4):1765–1771. https://doi.org/10.1039/C3SC22201B
182. Lee S, Chen C-H, Flood AH (2013) A pentagonal cyanostar macrocycle with cyanostilbene CH donors binds anions and forms dialkylphosphate [3]rotaxanes. Nat Chem 5(8):704–710. https://doi.org/10.1038/nchem.1668
183. Lisbjerg M, Valkenier H, Jessen BM, Al-Kerdi H, Davis AP, Pittelkow M (2015) Biotin[6]uril esters: chloride-selective transmembrane anion carriers employing C–H···anion interactions. J Am Chem Soc 137(15):4948–4951. https://doi.org/10.1021/jacs.5b02306
184. Liu Y, Zhao W, Chen C-H, Flood AH (2019) Chloride capture using a C–H hydrogen-bonding cage. Science 365(6449):159–161. https://doi.org/10.1126/science.aaw5145
185. Bonifazi D, Mohnani S, Llanes-Pallas A (2009) Supramolecular chemistry at interfaces: molecular recognition on nanopatterned porous surfaces. Chem Eur J 15(29):7004–7025. https://doi.org/10.1002/chem.200900900
186. Langton MJ, Beer PD (2014) Rotaxane and catenane host structures for sensing charged guest species. Acc Chem Res 47(7):1935–1949. https://doi.org/10.1021/ar500012a
187. Vallavoju N, Sivaguru J (2014) Supramolecular photocatalysis: combining confinement and non-covalent interactions to control light initiated reactions. Chem Soc Rev 43(12): 4084–4101. https://doi.org/10.1039/C3CS60471C
188. Grimme S, Bannwarth C, Shushkov P (2017) A robust and accurate tight-binding quantum chemical method for structures, vibrational frequencies, and noncovalent interactions of large molecular systems parametrized for all spd-block elements ($Z = 1$–86). J Chem Theory Comput 13(5):1989–2009. https://doi.org/10.1021/acs.jctc.7b00118
189. Bannwarth C, Ehlert S, Grimme S (2019) GFN2-xTB – an accurate and broadly parametrized self-consistent tight-binding quantum chemical method with multipole electrostatics and density-dependent dispersion contributions. J Chem Theory Comput 15(3):1652–1671. https://doi.org/10.1021/acs.jctc.8b01176
190. Neese F (2022) Software update: the ORCA program system – version 5.0. WIREs Comput Mol Sci 12(5):e1606. https://doi.org/10.1002/wcms.1606
191. Li J, Wang C, Mo Y (2023) Selectivity rule of cryptands for anions: molecular rigidity and bonding site. Chem Eur J 29(16):e202203558. https://doi.org/10.1002/chem.202203558
192. Fatila EM, Twum EB, Sengupta A, Pink M, Karty JA, Raghavachari K et al (2016) Anions stabilize each other inside macrocyclic hosts. Angew Chem Int Ed 55(45):14057–14062. https://doi.org/10.1002/anie.201608118
193. Fatila EM, Twum EB, Karty JA, Flood AH (2017) Ion pairing and co-facial stacking drive high-fidelity bisulfate assembly with cyanostar macrocyclic hosts. Chem Eur J 23(44): 10652–10662. https://doi.org/10.1002/chem.201701763
194. McNally JS, Wang XP, Hoffmann C, Wilson AD (2017) Self-assembly of molecular ions via like-charge ion interactions and through-space defined organic domains. Chem Commun 53(79):10934–10937. https://doi.org/10.1039/C7CC06401B
195. White NG (2019) Antielectrostatically hydrogen bonded anion dimers: counter-intuitive, common and consistent. CrystEngComm 21(33):4855–4858. https://doi.org/10.1039/C9CE01118H
196. Zhao W, Flood AH, White NG (2020) Recognition and applications of anion–anion dimers based on anti-electrostatic hydrogen bonds (AEHBs). Chem Soc Rev 49(22):7893–7906. https://doi.org/10.1039/D0CS00486C
197. Quiñonero D, Alkorta I, Elguero J (2016) Cation–cation and anion–anion complexes stabilized by halogen bonds. Phys Chem Chem Phys 18(40):27939–27950. https://doi.org/10.1039/C6CP03662G

198. Holthoff JM, Engelage E, Weiss R, Huber SM (2020) "Anti-electrostatic" halogen bonding. Angew Chem Int Ed 59(27):11150–11157. https://doi.org/10.1002/anie.202003083
199. Loy C, Holthoff JM, Weiss R, Huber SM, Rosokha SV (2021) "Anti-electrostatic" halogen bonding in solution. Chem Sci 12(23):8246–8251. https://doi.org/10.1039/D1SC01863A
200. Wang G, Chen Z, Xu Z, Wang J, Yang Y, Cai T et al (2016) Stability and characteristics of the halogen bonding interaction in an anion–anion complex: a computational chemistry study. J Phys Chem B 120(4):610–620. https://doi.org/10.1021/acs.jpcb.5b08139
201. Knorr A, Stange P, Fumino K, Weinhold F, Ludwig R (2016) Spectroscopic evidence for clusters of like-charged ions in ionic liquids stabilized by cooperative hydrogen bonding. ChemPhysChem 17(4):458–462. https://doi.org/10.1002/cphc.201501134
202. Strate A, Niemann T, Michalik D, Ludwig R (2017) When like charged ions attract in ionic liquids: controlling the formation of cationic clusters by the interaction strength of the counterions. Angew Chem Int Ed 56(2):496–500. https://doi.org/10.1002/anie.201609799
203. Weinhold F (2018) Theoretical prediction of robust second-row oxyanion clusters in the metastable domain of antielectrostatic hydrogen bonding. Inorg Chem 57(4):2035–2044. https://doi.org/10.1021/acs.inorgchem.7b02943
204. Prohens R, Portell A, Font-Bardia M, Bauzá A, Frontera A (2018) H-bonded anion–anion complex trapped in a squaramido-based receptor. Chem Commun 54(15):1841–1844. https://doi.org/10.1039/C7CC09241E
205. Barbas R, Prohens R, Bauzá A, Franconetti A, Frontera A (2019) H-bonded anion–anion complexes in fentanyl citrate polymorphs and solvates. Chem Commun 55(1):115–118. https://doi.org/10.1039/C8CC09028A
206. Cullen DA, Gardiner MG, White NG (2019) A three dimensional hydrogen bonded organic framework assembled through antielectrostatic hydrogen bonds. Chem Commun 55(80): 12020–12023. https://doi.org/10.1039/C9CC06707H
207. Gogoi A, Dutta D, Verma AK, Nath H, Frontera A, Guha AK et al (2019) Energetically favorable anti-electrostatic hydrogen bonded cationic clusters in Ni(II) 3,5-dimethylpyrazole complexes: anticancer evaluation and theoretical studies. Polyhedron 168:113–126. https://doi.org/10.1016/j.poly.2019.04.043
208. Niemann T, Stange P, Strate A, Ludwig R (2019) When hydrogen bonding overcomes Coulomb repulsion: from kinetic to thermodynamic stability of cationic dimers. Phys Chem Chem Phys 21(16):8215–8220. https://doi.org/10.1039/C8CP06417B
209. Azofra LM, Elguero J, Alkorta I (2020) A conceptual DFT study of phosphonate dimers: dianions supported by H-bonds. J Phys Chem A 124(11):2207–2214. https://doi.org/10.1021/acs.jpca.9b10681
210. Chen J, Qian K, Xiao K, Luo J, Li H, Ma T et al (2020) Co-ion effects in the self-assembly of macroions: from Co-ions to co-macroions and to the unique feature of self-recognition. Langmuir 36(35):10519–10527. https://doi.org/10.1021/acs.langmuir.0c01797
211. Khorrami F, Kowsari MH (2020) Tracing local nanostructure of the aqueous solutions of the biocompatible [Cho][Gly] ionic liquid: importance of hydrogen bond attraction between like-charged ions. J Phys Chem B 124(18):3770–3783. https://doi.org/10.1021/acs.jpcb.0c01796
212. Philipp JK, Ludwig R (2020) Clusters of hydroxyl-functionalized cations stabilized by cooperative hydrogen bonds: the role of polarizability and alkyl chain length. Molecules 25(21):4972. https://doi.org/10.3390/molecules25214972
213. Miranda MO, Duarte DJR, Alkorta I (2020) Anion-anion complexes established between aspartate dimers. ChemPhysChem 21(10):1052–1059. https://doi.org/10.1002/cphc.201901200
214. Quiñonero D, Alkorta I, Elguero J (2020) Metastable dianions and dications. ChemPhysChem 21(14):1597–1607. https://doi.org/10.1002/cphc.202000172
215. Scheiner S, Wysokiński R, Michalczyk M, Zierkiewicz W (2020) Pnicogen bonds pairing anionic Lewis acid with neutral and anionic bases. J Phys Chem A 124(24):4998–5006. https://doi.org/10.1021/acs.jpca.0c03881

216. Zierkiewicz W, Wysokiński R, Michalczyk M, Scheiner S (2020) On the stability of interactions between pairs of anions – complexes of MCl_3^- (M=Be, Mg, Ca, Sr, Ba) with pyridine and CN^-. ChemPhysChem 21(9):870–877. https://doi.org/10.1002/cphc.202000098
217. Zapata F, González L, Bastida A, Bautista D, Caballero A (2020) Formation of self-assembled supramolecular polymers by anti-electrostatic anion–anion and halogen bonding interactions. Chem Commun 56(52):7084–7087. https://doi.org/10.1039/D0CC02831B
218. Zhao W, Tropp J, Qiao B, Pink M, Azoulay JD, Flood AH (2020) Tunable adhesion from stoichiometry-controlled and sequence-defined supramolecular polymers emerges hierarchically from cyanostar-stabilized anion–anion linkages. J Am Chem Soc 142(5):2579–2591. https://doi.org/10.1021/jacs.9b12645
219. dos Passos Gomes G, Xu G, Zhu X, Chamoreau L-M, Zhang Y, Bistri-Aslanoff O et al (2021) Mapping C−H···M interactions in confined spaces: (α-ICyDMe)Au, Ag, Cu complexes reveal "contra-electrostatic H bonds" masquerading as anagostic interactions. Chem Eur J 27(31): 8127–8142. https://doi.org/10.1002/chem.202100263
220. Holthoff JM, Weiss R, Rosokha SV, Huber SM (2021) "Anti-electrostatic" halogen bonding between ions of like charge. Chem Eur J 27(67):16530–16542. https://doi.org/10.1002/chem.202102549
221. Grabarz A, Michalczyk M, Zierkiewicz W, Scheiner S (2021) Anion–anion interactions in aerogen-bonded complexes. Influence of solvent environment. Molecules 26(8):2116. https://doi.org/10.3390/molecules26082116
222. Wysokiński R, Zierkiewicz W, Michalczyk M, Scheiner S (2021) Anion···anion $(MX_3^-)_2$ dimers (M = Zn, Cd, Hg; X = Cl, Br, I) in different environments. Phys Chem Chem Phys 23(25):13853–13861. https://doi.org/10.1039/D1CP01502H
223. Fan D, Chen L, Wang C, Yin S, Mo Y (2021) Inter-anion chalcogen bonds: are they anti-electrostatic in nature? J Chem Phys 155(23):234302. https://doi.org/10.1063/5.0076872
224. Andreo L, Gomila RM, Priola E, Giordana A, Pantaleone S, Diana E et al (2022) Anion···anion $[AuI_4]^-\cdots[AuI_2]^-$ complex trapped in the solid state by tetramethylammonium cations. Cryst Growth Des 22(11):6539–6544. https://doi.org/10.1021/acs.cgd.2c00749
225. Mata I, Alkorta I, Molins E, Espinosa E (2012) Electrostatics at the origin of the stability of phosphate-phosphate complexes locked by hydrogen bonds. ChemPhysChem 13(6): 1421–1424. https://doi.org/10.1002/cphc.201200068
226. Chen L, Feng Q, Wang C, Yin S, Mo Y (2021) Classical electrostatics remains the driving force for interanion hydrogen and halogen bonding. J Phys Chem A 125(48):10428–10438. https://doi.org/10.1021/acs.jpca.1c09250
227. Shaik S, Danovich D, Wu W, Hiberty PC (2009) Charge-shift bonding and its manifestations in chemistry. Nat Chem 1(6):443–449. https://doi.org/10.1038/nchem.327
228. Frenking G, Caramori GF (2015) No need for a re-examination of the electrostatic notation of the hydrogen bonding: a comment. Angew Chem Int Ed 54(9):2596–2599. https://doi.org/10.1002/anie.201411374
229. Martín-Fernández C, Montero-Campillo MM, Alkorta I (2024) Hydrogen bonds are never of an "anti-electrostatic" nature: a brief tour of a misleading nomenclature. J Phys Chem Lett 15(15):4105–4110. https://doi.org/10.1021/acs.jpclett.4c00779
230. Li J, Feng Q, Wang C, Mo Y (2023) On the nature of inter-anion coinage bonds. Phys Chem Chem Phys 25(22):15371–15381. https://doi.org/10.1039/D3CP00978E
231. Li J, Peng X, Yin S, Wang C, Mo Y (2023) On the metastability of π-π stacking between closed-shell ions of like charges. Chin J Struct Chem 43(5):100213. https://doi.org/10.1016/j.cjsc.2023.100213

Struct Bond (2026) 190: 47–70
https://doi.org/10.1007/430_2025_103

Published online: 31 July 2025

Local Energy Decomposition of Coupled Cluster Energies: Principles and Applications

Martina Colucci, Gianluca Regni, Isaac F. Leach, and Giovanni Bistoni

Contents

Abstract Intermolecular interactions shape chemistry and biology, influencing everything from drug binding to material stability. But how do we quantify and understand these subtle yet powerful effects? This chapter dives into the local energy decomposition (LED) method, a state-of-the-art computational tool that breaks down complex interactions into chemically intuitive components – such as electrostatics, exchange, and London dispersion – at the local coupled cluster level of theory. LED reveals the role these forces play in asymmetric catalysis, protein-ligand binding, and even solid-state materials. We introduce the theory behind local coupled cluster computations and the framework of LED and its recent extensions and discuss several illustrative case studies.

M. Colucci, G. Regni, I. F. Leach, and G. Bistoni (✉)
Department of Chemistry, Biology and Biotechnology, University of Perugia, Perugia, Italy
e-mail: giovanni.bistoni@unipg.it

Keywords Energy decomposition analysis (EDA) · Intermolecular interactions · Local correlation · Local energy decomposition (LED) · London dispersion

1 Introduction

Intermolecular interactions play a fundamental role in chemistry and biology, governing processes such as the formation and stability of reaction intermediates, the selectivity of organic transformations, and the binding of drugs to protein targets. In particular, the selectivity and sensitivity of drug binding are dictated by a complex interplay of intermolecular forces. Understanding and harnessing these interactions for chemical and pharmacological applications remain a major challenge, requiring accurate quantification and characterization of their nature and strength.

Computational approaches play a crucial role in this context by providing a means to isolate and quantify individual inter- and intramolecular interactions. This chapter presents the development and application of efficient and accurate computational methods for the analysis and quantification of molecular interactions. These methods are based on a local variant of the coupled cluster singles and doubles method with perturbative triples, CCSD(T), widely regarded as the "gold standard" of quantum chemistry. By exploiting the local nature of electron correlation, these approaches enable the study of systems with hundreds of atoms and thousands of basis functions, offering invaluable insights into molecular chemistry, material sciences, and biology (Fig. 1).

A wide range of quantum chemical methods exist for the accurate quantification and analysis of intermolecular interactions. In recent reviews, a critical comparison of these approaches was provided [2, 3], with particular emphasis on the coupled cluster-based techniques explored in this chapter. One of the most widely used methods for studying weak intermolecular interactions is symmetry-adapted perturbation theory (SAPT) [4], which decomposes the interaction energy into physically meaningful components, such as electrostatics, exchange-repulsion, induction, and London dispersion (LD). However, because SAPT is based on perturbation theory, it has limitations in its canonical implementation when dealing with strong interactions or systems with significant covalent character, such as metal-ligand bonding in organometallic chemistry.

For such challenging interactions, a supermolecular approach is often preferred, wherein the interaction energy is obtained as the energy difference between a molecular complex and its isolated monomers. To further dissect the interaction energy, energy decomposition analysis (EDA) schemes have been developed, inspired by the pioneering work of Morokuma [5]. Most EDA schemes rely on density functional theory (DFT), but the accuracy of these approaches depends on the chosen exchange-correlation functional. A key limitation of DFT-based EDAs is the treatment of London dispersion interactions, which are poorly captured by

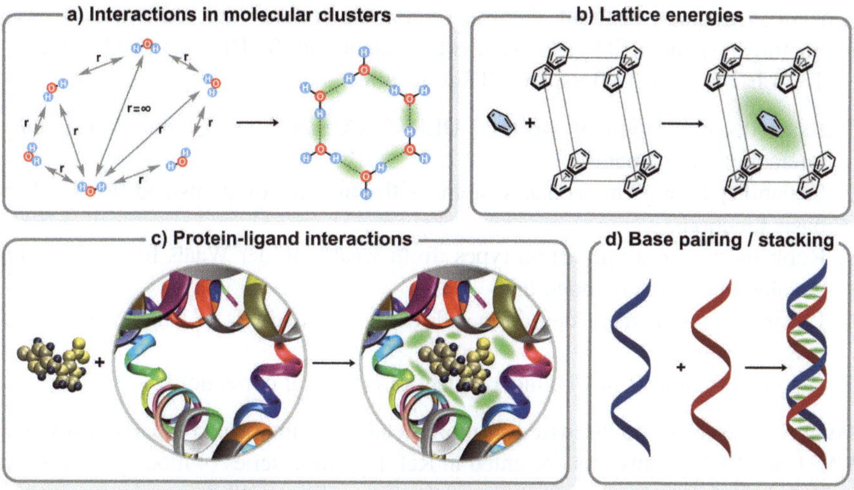

Fig. 1 Illustrative examples of inter-fragment interactions that influence binding strength in different chemical systems: (**a**) small molecules within a cluster, (**b**) a molecule and its adjacent units in a crystal, (**c**) a ligand interacting with a protein, and (**d**) two complementary strands in a DNA duplex. The primary interactions formed during binding are represented in green. (Adapted from Ref. [1] under a CC-BY 4.0 license, https://creativecommons.org/licenses/by/4.0/)

standard functionals. Various corrections have been proposed [6], including the addition of empirical dispersion terms.

Post-Hartree-Fock methods such as second-order Møller-Plesset perturbation theory (MP2) and CCSD(T) naturally incorporate all components of intermolecular interactions, including London dispersion, without empirical corrections. However, EDA schemes for correlated wavefunction methods remain limited. Some approaches utilize local representations of the occupied and virtual orbitals to assign correlation effects to specific monomers, enabling a SAPT-like decomposition of the interaction energy [7, 8]. Unfortunately, MP2-based approaches tend to overestimate London dispersion interactions [9, 10] and perform poorly for open-shell systems [11–14].

To address these challenges, coupled cluster-based EDA schemes have been developed [15]. Among them, the domain-based local pair natural orbital (DLPNO) variant of CCSD(T) (DLPNO-CCSD(T)) has emerged as a highly accurate and computationally efficient method for large systems. Implemented in the freely available ORCA quantum chemistry code [16–19], the local energy decomposition (LED) method [20], which was developed in the context of DLPNO-CCSD (T), provides a clear-cut decomposition of the interaction energy into electrostatics, exchange, London dispersion, and electronic preparation terms. Unlike mean field-based approaches, LED defines London dispersion energy in an electronic structure-dependent manner, without empirical parametrization. Importantly, in weakly interacting systems, LED yields results consistent with SAPT regimes [20, 21].

The LED approach offers several key features:

1. Exact decomposition of the total DLPNO-CCSD(T) energy into fundamental interaction components
2. Scalability to large molecular systems with hundreds of atoms and thousands of basis functions
3. Robustness across interaction types, from weak van der Waals forces to strong covalent metal-ligand bonding
4. Flexibility in fragment definition, allowing for hundreds of fragments without loss of accuracy
5. Rigorous treatment of LD, independent of empirical corrections

Point 5 deserves to be discussed in more detail. A numerical comparison between SAPT and LED results was presented in Ref. [20] for a series of model systems and in Ref. [22] for various conformers of the water dimer. While a strict one-to-one comparison between the energy components obtained with these schemes is not possible due to the different philosophies (and mathematical definitions) underlying these approaches, a thorough discussion of the similarities and differences between LED and wavefunction-based SAPT can be found in Ref. [21]. In general, it was found that they provide similar trends for the LD energy.

The LED approach has found a wide range of applications in many fields of chemical research, as discussed in the following sections. For instance, the first application of the scheme focused on the interaction of Lewis acids and bases for a series of classical Lewis adducts and frustrated Lewis pairs (FLPs) [23]. It was found that LD dominates the interaction energy in FLPs and is also remarkably strong in the Lewis adducts. In the latter systems, its magnitude can be easily modulated by tuning the size (and therefore the polarizability) of the substituents on the basic center. Hence, LD forces may be the factor that tips the scale on whether Lewis pairs form adducts or eventually become "frustrated," thus determining their reactivity.

The success of the LED method stimulated us to develop an open-shell variant of this scheme to extend its range of applicability. Its theory, implementation, and benchmarks are reported in Ref. [24]. The resulting approach allows for an exact decomposition of UHF-DLPNO-CCSD(T) energies that are entirely consistent with the closed-shell approach. As a proof-of-concept case study, the open-shell LED scheme was used to elucidate the mechanism that governs the change in the singlet-triplet energy gap of heme upon its interaction with a CO ligand [24]. In a subsequent study [25], the same approach was used to investigate the energetic components that govern the interaction of carbenes with hydrogen and halogen bond donors and the physics behind the differential stabilization of singlet and triplet spin states upon adduct formation. The study revealed that electrostatic and LD effects often influence different spin states in opposing ways, with one preferentially stabilizing the singlet state while the other favoring the triplet. This seemingly counterintuitive effect arises because, in these systems, electrostatics is highly directional in the singlet state, enforcing structural rigidity in the adduct through a hydrogen or halogen bond with the carbenic lone pair. In the triplet state, this interaction weakens as the electrons delocalize across two carbon orbitals rather than remaining confined to the lone pair. Consequently, the molecules reorient to maximize secondary dispersion interactions.

Having established the usefulness of our scheme for various applications, we analyzed in more detail the relation between LD and correlation energy in Ref. [26]. To do so, we applied our scheme to four model systems with very different interaction natures, namely Ar_2, $(HF)_2$, Be_2, and $ArLi^+$. Although LD is a purely dynamic correlation effect, we found that electron correlation contains several different physical effects besides LD, such as the correlation correction to the electrostatic interaction, which is typically overestimated at the HF level of theory. Nevertheless, LD is typically the dominant contribution to the correlation binding energy.

This finding prompted us to explore the possibility of using LED to develop cost-effective methods for the calculation of intermolecular interactions within the DLPNO-CCSD(T)/LED framework. The key idea was to focus on the intermolecular part of the correlation energy, which contains the LD energy. This led to the development of the Hartree-Fock plus LD (HFLD) approach [27, 28]. Essentially, with the HFLD scheme, the coupled cluster equations are solved while neglecting intramolecular correlation. Surprisingly, HFLD was found to be more accurate than DLPNO-CCSD on various benchmark sets for intermolecular interactions while retaining essentially the same efficiency as the HF method.

This chapter describes our recent efforts in developing computational techniques for the quantification and analysis of intermolecular interactions in large systems, as well as relevant case studies:

- Section 2 provides a theoretical overview of the main local correlation methods integrated within the LED scheme, with a particular focus on the DLPNO-CCSD (T) method.
- Section 3 presents a detailed description of the fundamental principles of the LED scheme, including its most recent developments and extensions [1].
- Section 4 explores the application of the LED scheme to various areas of chemical research, particularly in the study of noncovalent interactions (NCIs) in large systems. Special attention is given to asymmetric catalysis, protein-ligand interactions, conformational equilibria, and solid-state materials.

2 Theoretical Background: Local Correlation Methods

The accurate treatment of electron correlation is essential for computational chemistry. The application of traditional highly correlated post-Hartree-Fock methods, such as CCSD(T), is often limited to relatively small systems due to the steep scaling of the computational cost with the system size. To overcome this, local correlation methods exploit the nearsightedness of electron correlation [29–31]. By restricting the treatment of electron correlation to spatially close orbitals, these methods achieve significant computational savings while maintaining high accuracy [32]. Among them, the DLPNO-CCSD(T) method has proven particularly effective [20, 33–35], offering near-linear scaling of the computational cost with system size while retaining an accuracy comparable to that of canonical CCSD(T) [36].

In this section, we outline the fundamental concepts underlying the main local correlation methods integrated with the LED scheme, including DLPNO-CCSD(T), multilevel DLPNO approaches [37], and HFLD. Emphasis is placed on the DLPNO-CCSD(T) method, detailing key aspects such as orbital localization, pair correlation approximations, and role of pair natural orbitals (PNOs) in reducing computational cost while maintaining high accuracy.

2.1 The DLPNO-CCSD(T) Approach

According to Nesbet's theorem [38], the correlation energy E_C of any post-HF scheme can be expressed as the sum of pair correlation energies, ϵ_{ij}:

$$E_C = \frac{1}{2}\sum_{ij}\epsilon_{ij} \tag{1}$$

In this expression, i and j are the occupied orbitals. In DLPNO-based schemes, a local representation of the molecular orbital (MO) space is used for both occupied and virtual orbitals. The occupied orbitals, as obtained from an HF calculation, are typically localized using schemes developed by Foster and Boys [39] or Pipek and Mezey [40]. The value of the pair correlation energy is strongly influenced by the distance between the electrons: the closer the electrons, the greater their contribution to the total correlation energy.

A key approximation that enhances the efficiency of DLPNO-CCSD(T) is the "strong pair approximation". In this scheme, only a subset of electron pairs (the strong pairs) are treated at the coupled cluster level, while the remaining pairs (the weak pairs) are handled using more approximate methods – typically some (local) variant of MP2. The energy threshold T_{CutPairs} controls the number of electron pairs included in the coupled cluster treatment. When T_{CutPairs} is set to 0, all electron pairs are considered to be strong (i.e., within the algorithm), and the error associated with the strong pair approximation vanishes.

To further enhance efficiency, a local representation of the virtual space is constructed using pair natural orbitals (PNOs) [41, 42]. For each ij pair, a set of PNOs is constructed (labeled as $\tilde{a}_{ij}, \tilde{b}_{ij}, \tilde{c}_{ij}, \ldots$). As previously discussed (Eq. 1), the correlation energy is expressed as a sum of pair correlation energies ϵ_{ij}, each of which can be further decomposed as a sum of double excitations from the occupied orbitals i and j to the PNOs $\tilde{a}_{ij}$ and $\tilde{b}_{ij}$. The pair correlation energy of the electron pair ij is expressed as

$$\epsilon_{ij} = \sum_{\tilde{a}_{ij}\tilde{b}_{ij}} \left(i\tilde{a}_{ij}|j\tilde{b}_{ij}\right) \tilde{\tau}^{ij}_{\tilde{a}_{ij}\tilde{b}_{ij}} \quad (2)$$

where $\tilde{\tau}^{ij}_{\tilde{a}_{ij}\tilde{b}_{ij}}$ are the cluster amplitudes in the PNO basis. In DLPNO calculations, the size of the virtual orbital space is controlled by the tunable T_{CutPNO} threshold. The DLPNO machinery ensures that each electron pair features a specifically tailored virtual space. Typically, a compact virtual space is sufficient to retain canonical coupled cluster accuracy for each electron pair, dramatically increasing the efficiency of DLPNO-CCSD(T) compared to its canonical counterpart. Another key tunable parameter within the DLPNO scheme is the T_{CutPairs} threshold, which determines which electron pairs are considered strong (see above). The recommended values of T_{CutPairs} and T_{CutPNO} have been established through extensive benchmarking. Three sets of thresholds – *LoosePNO*, *NormalPNO*, and *TightPNO* – are defined, with increasing accuracy and computational cost. Their performance has been rigorously evaluated across various chemical systems, demonstrating an excellent balance between accuracy and efficiency when using standard DLPNO settings [36, 43, 44]. A comprehensive 2020 study by Liakos et al. [36] assessed DLPNO accuracy on the GMTKN55 superset of structures [45], which comprises 55 datasets, 1,505 relative energies, and 2,462 single-point calculations. The benchmark results showed that with *TightPNO* settings, mean absolute deviations (MADs) with respect to canonical CCSD(T) are typically below 0.5 kcal/mol.

To further reduce the DLPNO error for challenging systems or when higher accuracy is required, extrapolation schemes can be employed, where results obtained with two T_{CutPNO} thresholds, e.g., 10^{-X} and 10^{-Y}, are extrapolated to the complete PNO space [43]. Among these, $10^{-6}/10^{-7}$ extrapolations (which correspond to *LoosePNO* and *TightPNO*, respectively) significantly enhance the accuracy of DLPNO-CCSD(T) calculations, often achieving sub-kJ/mol precision.

In summary, the DLPNO-CCSD(T) method provides a good balance between accuracy and computational efficiency, making it a practical alternative to canonical CCSD(T) for large systems. Benchmark studies show that *TightPNO* settings yield MADs typically below 0.5 kcal/mol, with further accuracy improvements achievable through PNO extrapolation. These features make DLPNO-CCSD(T) a reliable approach for high-accuracy quantum chemical calculations at a significantly reduced computational cost.

For larger systems, for which the DLPNO-CCSD(T) is not directly applicable, LED can be used in conjunction with a multilevel DLPNO-CCSD(T) approach. These schemes effectively balance accuracy and computational cost by selectively applying high-accuracy treatments to the chemically relevant regions of a system. By fragmenting the system and assigning different levels of theory or truncation thresholds to each region, computational demands are significantly reduced, while accuracy is maintained in key areas. A particularly successful variant of multilevel approaches is the HFLD scheme, which has found widespread applications in the study of NCIs. This approach leverages LED to treat only "dispersion-like" double excitations at the coupled cluster level, greatly reducing the computational cost of the correlation treatment while retaining high accuracy. Further details are provided in Sect. 3.3, following the introduction of the LED scheme.

3 Local Energy Decomposition

The LED scheme has been developed to facilitate the interpretation of local coupled cluster results [20, 24]. Specifically, the LED scheme allows the total energy and binding energy of any molecular system to be decomposed into fragment-dependent contributions. This enables the identification of the parts of the system that provide the greatest energy contribution and allows for the breakdown of the overall energy into physically meaningful components. This section is dedicated to the description of the LED scheme and its fragment-pairwise variant (fp-LED). In addition, we also show how LED can be used as a basis for the development of a novel electronic structure method, HFLD, for the accurate quantification of NCIs in large and complex systems.

3.1 Basic Principles of LED

The local energy decomposition (LED) scheme provides a chemically meaningful breakdown of the total and binding energies of a system into physically interpretable contributions. This approach allows the system to be partitioned into different fragments, which can be functional groups, molecules, or even individual atoms. The total DLPNO-CCSD(T) energy is expressed as the sum of the reference (typically HF) energy and the correlation energy. Both components contain terms that depend on one- and/or two-particle interactions. Since the occupied and virtual orbital spaces are both localized, it is possible to assign each particle in the system – nuclei and electrons – to specific fragments, allowing for the decomposition of total and binding energies into fragment-dependent contributions. Based on this assignment, the DLPNO-CCSD(T) energy can be expressed as the following sum:

$$E = \sum_{x} E_x + \sum_{x>y} E_{xy} \tag{3}$$

where x and y refer to different fragments. The intra-fragment contributions E_x contain all the energy terms associated with particles belonging to the same fragment, while the inter-fragment contributions E_{xy} account for two-particle terms of the HF and correlation energy, where the particles belong to different fragments.

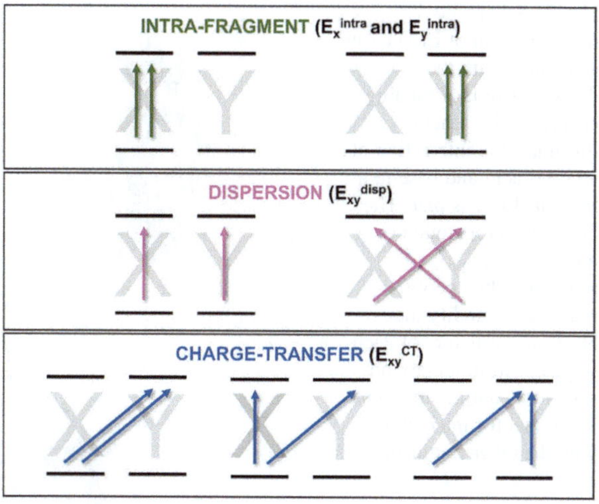

Fig. 2 Schematic representation of double-electronic contributions to the correlation energy. (1) Intra-fragment terms, deriving from the double excitation of two electrons in the same fragment. (2) Dispersion inter-fragment contributions, leading to the formation of one hole and one particle on each fragment. (3) Non-dispersion inter-fragment contributions, leading to a different number of holes and particles on each fragment

When a more detailed understanding of the physical nature of the interactions is required, the inter-fragment terms can be further decomposed to provide insights into the nature of the intermolecular interactions. An electrostatic (Coulomb) interaction term between the fragments (E_{xy}^{els}) and a Slater exchange contribution (E_{xy}^{exch}) can be defined directly from the reference determinant:

$$E_{xy}^{\text{els}} = P_{x\leftrightarrow y}\left[\sum_{A_x > B_y} \frac{Z_{A_x} Z_{B_y}}{\left|R_{A_x} - R_{B_y}\right|} - 2\sum_{i_x, A_y} \left\langle i_x \left| \frac{Z_{A_y}}{\left|r_{i_x} - R_{A_y}\right|} \right| i_x \right\rangle + \sum_{i_x > j_y} 4\left(i_x i_x | j_y j_y\right)\right] \tag{4}$$

$$E_{xy}^{\text{exch}} = P_{x\leftrightarrow y}\left[-2\sum_{i_x > j_y}\left(i_x j_y | j_y i_x\right)\right] \tag{5}$$

By exploiting the localization of both the occupied orbitals and PNOs, three families of double excitations constituting the correlation energy can be identified: the dispersion (E_{xy}^{disp}), charge transfer (E_{xy}^{CT}), and intra-fragment excitations ($E_x^{\text{intra}}, E_y^{\text{intra}}$), represented in Fig. 2.

To summarize, the total electronic energy of the system can be expressed as

$$E = \sum_x E_x + \sum_{x>y}\left[E_{xy}^{\text{els}} + E_{xy}^{\text{exch}} + E_{xy}^{\text{disp}} + E_{xy}^{\text{CT}}\right] \tag{6}$$

This equation highlights how, within the LED scheme, the total energy of a chemical system is decomposed into contributions associated with individual fragments and their interactions. Performing this analysis requires only single-point energy

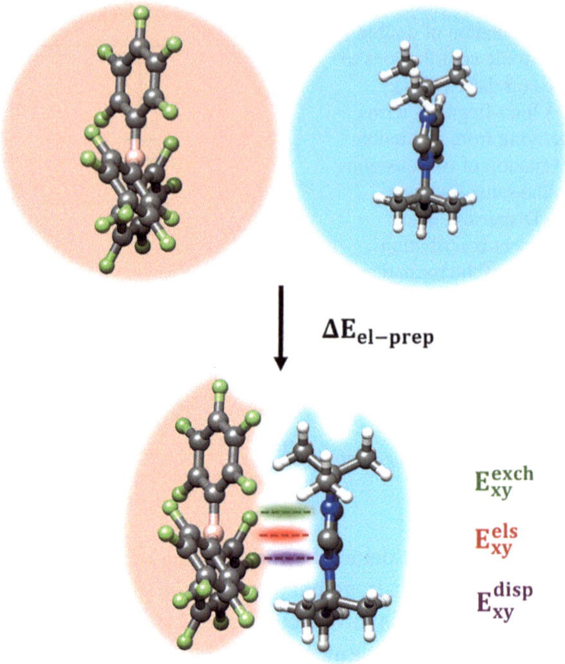

Fig. 3 Schematic representation of LED terms associated with the decomposition of binding energies. The scheme illustrates the interaction of a Lewis acid and base in a frustrated Lewis pair. The electronic preparation energy, $\Delta E_{\text{el - prep}}$, represents the energy required to modify the electronic structure of each fragment from its ground state to a configuration that is optimal for interaction with the other fragment

calculations using a local correlation approach and the partitioning of the system into fragments, with negligible computational overhead. The fragment definition is fully flexible, allowing for tailored insights depending on the specific application.

The LED scheme proves particularly useful for the decomposition of binding energies, offering a detailed breakdown into *electronic preparation* ($\Delta E_{\text{el - prep},\, x}$) and inter-fragment interaction terms ($E_{xy}^{\text{els}}, E_{xy}^{\text{exch}}, E_{xy}^{\text{disp}}$), as depicted schematically in Fig. 3. The contribution from the charge transfer excitations E_{xy}^{CT} is typically not discussed in standard applications and incorporated into the correlation contribution of E_x and hence of $\Delta E_{\text{el - prep},\, x}$.

For the sake of simplicity, consider the association reaction of the monomers X and Y forming the dimer XY:

$$X + Y \rightarrow XY$$

Within a supermolecular approach, the association (or binding) energy is computed as the difference between the energy of the adduct, $E(XY)$, and the energies of the separated monomers in their relaxed geometries, $E^0(X)$ and $E^0(Y)$:

$$\Delta E_{\text{bind}} = E(XY) - E^0(X) - E^0(Y) \tag{7}$$

By applying LED, the total energy of the dimer is expressed in terms of fragment contributions:

$$\Delta E_{\text{bind}} = E_x + E_y + E_{xy} - E^0(X) - E^0(Y) \tag{8}$$

Adding and subtracting the energy of each isolated fragment in the dimer geometry ($E(X)$ and $E(Y)$), we obtain

$$\begin{aligned}\Delta E_{\text{bind}} &= E_x - E(X) + E_y - E(Y) + E(X) - E^0(X) + E(Y) - E^0(Y) + E_{xy} \\ &= \Delta E_{\text{el-prep},x} + \Delta E_{\text{el-prep},y} + \Delta E_{\text{geo-prep},x} + \Delta E_{\text{geo-prep},y} + E_{xy}\end{aligned} \tag{9}$$

This expression can be generalized for an arbitrary number of fragments:

$$\Delta E_{\text{bind}} = \sum_x \Delta E_{\text{el-prep},x} + \sum_x \Delta E_{\text{geo-prep},x} + \sum_{xy} E_{xy} \tag{10}$$

where the electronic preparation energy $\Delta E_{\text{el - prep}}$ represents the energy needed to change the electronic structure of each fragment from its ground state to a state that is optimal for the interaction with the other fragments. The geometric preparation term $\Delta E_{\text{geo - prep}}$ denotes the energy required to distort the monomers from their isolated geometry to the geometry they adopt in the dimer. Both the electronic and geometric preparation terms are positive by construction. It is often useful to define the interaction energy as the difference between the binding energy and the geometric preparation terms:

$$\Delta E_{\text{int}} = \sum_x \Delta E_{\text{el-prep},x} + \sum_{xy} E_{xy} \tag{11}$$

Hence, the final expression for the interaction energy decomposition in the LED framework is

$$\Delta E_{\text{int}} = \sum_x \Delta E_{\text{el-prep},x} + \sum_{xy} \left(E_{xy}^{\text{els}} + E_{xy}^{\text{exch}} + E_{xy}^{\text{disp}} + E_{xy}^{\text{CT}} \right) \tag{12}$$

For systems held together purely by noncovalent interactions, it has been found useful in some cases to group several terms in Eq. (12) and decompose ΔE_{int} into dispersion and non-dispersion contributions. Under these conditions, the latter accounts for the combined effects of electrostatics and steric interactions.

3.2 *The Fragment-Pairwise Local Energy Decomposition Scheme*

In systems consisting of only two fragments, the electronic preparation energy of each fragment quantifies the extent to which its electronic structure is influenced by the presence of the other fragment. However, when the system involves more than two fragments, the electronic structure of each fragment is affected by the presence of all the other ($N-1$) fragments, making the electronic preparation energy the only term that is not pairwise additive in the interaction energy decomposition. The fragment-pairwise LED (fp-LED) method was developed to fully decompose the total binding energy between complex chemical systems into pairwise additive terms [1]. This is accomplished by distributing the electronic preparation energy of each fragment across all the interaction energies in which it participates. This distributed electronic preparation term captures how the interaction between each pair of fragments influences their respective electronic structures. Given two fragments, X and Y, the distributed electronic preparation is defined as:

$$\Delta E_{\text{el-prep}}^{(x,y)} = \omega_x^{(x,y)} \Delta E_{\text{el-prep},x} + \omega_y^{(x,y)} \Delta E_{\text{el-prep},y} \tag{13}$$

In this expression, the ω terms represent the contributions of the pairwise interaction between X and Y to the total interaction of the two fragments:

$$\omega_x^{(x,y)} = \frac{|E_{xy}|}{\sum_{\substack{k=1 \\ k \neq x}}^{N} |E_{xy}|} \tag{14}$$

$$\omega_y^{(x,y)} = \frac{|E_{xy}|}{\sum_{\substack{k=1 \\ k \neq y}}^{N} |E_{xy}|} \tag{15}$$

Note that absolute values are typically used for the E_{xy} terms in the calculations of $\omega_x^{(x,y)}$and $\omega_y^{(x,y)}$ when both positive and negative interactions significantly contribute to the perturbation of the electronic structure of the fragment. The total interaction energy can be written in terms of fragment-pairwise contributions:

$$\Delta E_{\text{int}} = \sum_{x>y} \left(\Delta E_{\text{el-prep}}^{(x,y)} + E_{xy} \right) = \sum_{x>y} \Delta E_{\text{int}}^{xy} \tag{16}$$

As an illustrative example, the decomposition of the interaction energy in the water trimer is shown in Fig. 4. This decomposition is represented in the standard LED map, where the diagonal elements correspond to the electronic preparation energy of each fragment, while the off-diagonal elements represent the interaction energies between each pair of fragments.

Comparing the standard LED map (Fig. 4, left) to its fp-LED counterpart (Fig. 4, right), it is evident that while the standard LED scheme decomposes the total interaction energy into large positive and negative energy terms, the fp-LED scheme yields results that are closer to the DLPNO-CCSD(T) value for an isolated dimer, i.e., −5.1 kcal/mol. The difference arises from the cooperative effects of various NCIs. This approach allows for a deeper understanding of how each fragment pair

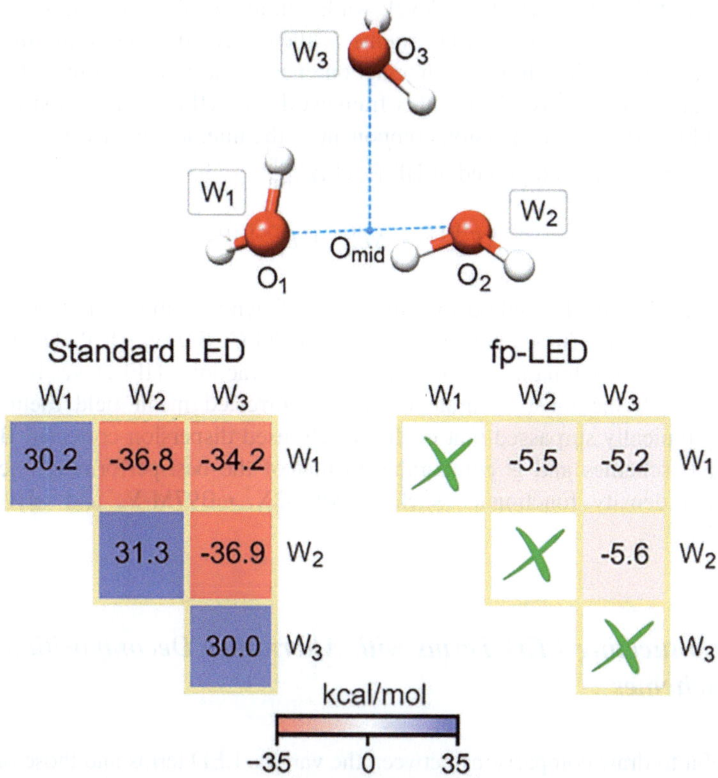

Fig. 4 Structure of the water trimer with the definition of its fragments. Comparison between standard and fp-LED maps. (Adapted from Ref. [1] under a CC-BY 4.0 license, https://creativecommons.org/licenses/by/4.0/)

influences the overall binding energy, emphasizing the individual contributions to the system's energy.

3.3 *The HFLD Method*

To efficiently quantify dispersion interactions in complex systems with high accuracy, the HFLD method has been developed [27, 28]. This approach is based on the multilevel implementation of the DLPNO-CCSD(T) method in conjunction with the LED scheme. As detailed above, the multilevel DLPNO-CCSD(T) framework allows for the computation of both fragment energies and their interaction energies at different levels of accuracy. In this context, HFLD is performed by excluding all electron pairs localized on the same fragment from the coupled cluster treatment. The key idea behind HFLD is to focus on intermolecular interaction energies, which are treated at the DLPNO-CCSD level, while intramolecular interactions are computed at the HF level. The LED scheme is then used to identify the dispersion contributions to the interfragment interaction energy and to correct the HF interaction energies accordingly. Within this framework, the HFLD interaction energy is obtained by adding the dispersion component of the interaction energy, $E_{\mathrm{disp}}^{\mathrm{C\text{-}CCSD}}$, to the interaction energy computed at HF level $\Delta E_{\mathrm{int}}^{\mathrm{HF}}$:

$$\Delta E_{\mathrm{int}}^{\mathrm{HFLD}} = \Delta E_{\mathrm{int}}^{\mathrm{HF}} + E_{\mathrm{disp}}^{\mathrm{C\text{-}CCSD}} \tag{17}$$

This allows the HFLD method to achieve an efficiency comparable to that of HF while retaining an accuracy between that of DLPNO-CCSD and DLPNO-CCSD(T). On challenging benchmark sets for noncovalent interactions, HFLD was found to be on a par with the most accurate dispersion-corrected mean field methods. Its accuracy typically surpassed that of the widely used dispersion-corrected B3LYP-D3/D4/NL schemes and is comparable to that of the best-performing exchange-correlation density functionals, such as M06-2X, ωB97M-V, and ωB97M-D3 (BJ) [27].

3.4 Connecting LED Terms with Alternative Decomposition Schemes

It is useful to draw comparisons between the various LED terms and those originating from alternative energy decomposition approaches. While a strict one-to-one correspondence is not possible, several LED terms are conceptually related to those found in other partitioning schemes. For example, the electronic preparation energy in LED is inherently positive and represents the primary repulsive contribution to the interaction. This arises from the perturbation of the electronic structure of the monomers as they adjust to maximize interaction with each other. As such, it plays a role similar to the so-called Pauli repulsion in certain Morokuma-like EDA schemes. The latter, formulated within a framework that begins with non-orthogonal fragment wavefunctions, introduces an energy penalty for bringing the fragments together while enforcing antisymmetry in the initial "noninteracting" state.

Another key term to consider is exchange. In LED, the exchange interaction naturally arises from the antisymmetry requirement of the wavefunction, reducing Coulomb repulsion between electrons of the same spin. This effect is fundamentally linked to Fermi correlation, as discussed in standard Hartree-Fock theory textbooks. The exchange interaction plays a crucial role in many contexts, particularly in determining the relative energy of different spin states. However, this term has no direct counterpart in the positive exchange-repulsion terms found in schemes like SAPT. The latter originate from the necessity of introducing antisymmetry corrections in Rayleigh-Schrödinger perturbation theory, which starts from non-antisymmetric zeroth-order solutions. Consequently, SAPT exchange-repulsion is more closely related to the concept of Pauli repulsion mentioned earlier.

Electrostatics is another term that warrants discussion. In LED, the electrostatic interaction is conceptually similar to that found in alternative partitioning schemes. However, in LED, this term accounts for interactions between the perturbed one-electron densities of the interacting fragments, whereas in EDA and SAPT, electrostatics is typically evaluated using "frozen" electron densities of non-perturbed monomers. Charge transfer excitations in LED, on the other hand, belong to the realm of dynamic correlation and should not be confused with the conventional notion of charge transfer used in coordination chemistry. As a result, charge transfer excitations are rarely discussed in the LED framework. Finally, the definition of dispersion in LED is fully analogous to that used in other decomposition schemes, as previously discussed. The key aspects of LED's treatment of dispersion are its reliance on a high level of theory (coupled cluster) and the use of orthogonal, localized orbitals.

4 Key Applications

With the theoretical framework of LED established, we now turn our attention to its real-world applications in various branches of chemical research. This section outlines how the method has been applied to a wide range of systems, including biological structures and molecular crystals, to provide valuable insights to understand the role of NCIs in diverse research contexts.

4.1 Asymmetric Catalysis

Asymmetric catalysis represents a pillar of modern chemistry and, among all of the NCIs, London dispersion (LD) plays a pivotal role in this area [46], especially because small energy differences in the stereoselective transition states lead to significant changes in the selectivity. In addition, transition states are in general more polarizable than reactants, due to weakened bonds, and are therefore more sensitive to dispersion interactions [47]. In this context, the LED scheme has been widely applied [48–50]. It can be used to identify and quantify the key (non)covalent interactions that drive the selectivity of chemical reactions, as well as contributions of catalyst-substrate dispersion forces, opening up new opportunities for the design of novel catalysts with tailored properties.

For example, LED has been employed to identify and quantify key catalyst-substrate interactions in enantiodetermining transition states, providing insights into the stereocontrolling factors in asymmetric Diels-Alder reactions catalyzed by a chiral Brønsted acid [48]. This study revealed that the observed enantioselectivity arises from a delicate balance between steric repulsion and LD attraction (Fig. 5). In

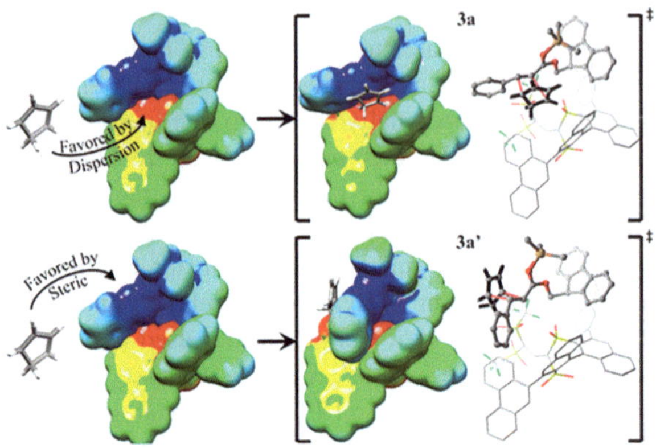

	$\Delta\Delta G^{\ddagger}$	e.r.
Experimental	1.1	86.5
DLPNO-CCSD(T)	2.6	98.8:1.2

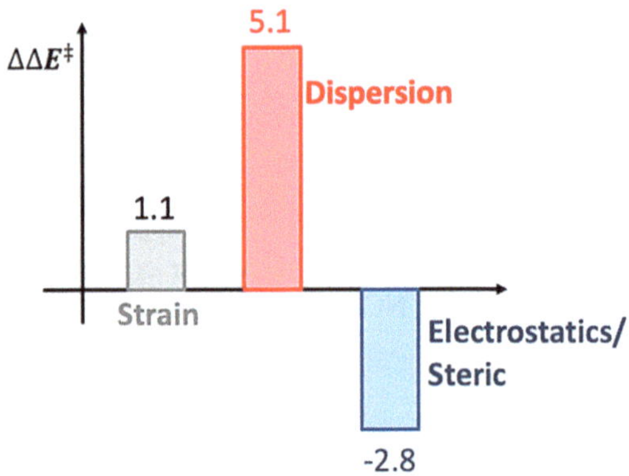

Fig. 5 Competing transition states for cyclopentadiene attack at a chiral ion pair. Although the attack of cyclopentadiene to the less hindered face (transition state 3a′) is sterically favored, LD interactions favor the attack to the more crowded face (transition state 3a). (Adapted from Ref. [3] under a CC-BY 4.0 license, https://creativecommons.org/licenses/by/4.0/)

particular, the scheme was used to decompose the relative energy between the competing low-energy transition states – corresponding to the relative activation barrier ($\Delta\Delta E^{+}$) that governs enantioselectivity – into distinct physical effects, including catalyst strain, steric (non-dispersion), and LD contributions. While non-dispersion interactions favor the attack of the reagent on the less crowded face of the substrate, dispersion interactions stabilize the transition state associated with the attack on the more congested face. In this case, dispersion interactions are stronger than steric effects, ultimately determining the observed enantioselectivity.

LED can also be used to unravel individual energy contributions associated with the relative strain of both the catalyst and substrate in transition states of competitive reaction channels, influencing for example the selectivity of enantioselective transformations. For example, the LED scheme can quantify relative changes in intra-catalyst dispersion interactions resulting from modifications to specific functional groups. This can be achieved through a series of systematic geometry modifications of the catalyst, followed by energy decomposition analysis in the corresponding transition states. Thus, LED enables the monitoring of how different contributions to strain energy evolve as the catalyst is modified, aiding in the identification of key interactions responsible for selectivity, as discussed, for example, in Ref. [49].

Recently, LED was also employed to analyze van der Waals interactions between transition states and an organic solvent in asymmetric catalytic transformations [50]. This analysis confirmed the presence of strong interactions between the transition state and the solvent. Additionally, the atomic decomposition of the London dispersion energy scheme [51] revealed that these interactions are significant only for specific groups of the catalyst, highlighting the potential of LED in catalyst design.

4.2 Protein-Ligand Interactions

The DLPNO-CCSD(T)/LED approach can be applied to study the association of biomolecular systems in order to identify the contribution of each fragment of the system to the overall interaction [52–55].

In this context, LED and its fragment-pairwise variant have been employed to investigate the intermolecular interactions governing biomolecular stability and recognition, including DNA stabilization [1, 56] and protein-ligand binding [1, 52, 57]. It has provided insights into the dominant role of Watson-Crick pairing in DNA stability, alongside the contributions of stacking interactions and London dispersion forces. Furthermore, in the context of protein-ligand interactions, LED has been used to quantify the impact of specific amino acid residues on ligand binding, enhancing the understanding of molecular selectivity and informing rational drug design. In this section, an illustrative example of the application of the LED scheme to a protein-ligand system is discussed [57].

Nicotinic acetylcholine receptors (nAChRs) are ligand-gated ion channels, essential for rapid neuronal signal transduction in animal nervous systems [58, 59]. Here, we investigate the binding of two agonists of the nAChR, namely imidacloprid and nicotine. These ligands have different selectivity patterns: while nicotine is highly toxic for mammals and has limited efficacy as an insecticide, imidacloprid is a synthetic insecticide, with high selectivity for insects. Some insects are resistant to imidacloprid, as a consequence of a mutation in the binding site of the receptor (R55T mutation), consisting of the exchange of an arginine (R) residue with a threonine (T) residue. This mutation, present in resistant insects and vertebrates, is responsible for insect resistance to imidacloprid. In this example, only the resistant form of the protein is considered, binding to the nicotine and imidacloprid ligands. By defining as fragments the ligand and the amino acid residues of the protein pocket, the LED scheme is used to analyze the key interactions between the ligands and the active site of the protein, providing detailed insights into the selectivity of the ligands. The decomposition of the total interaction energy, presented in the LED

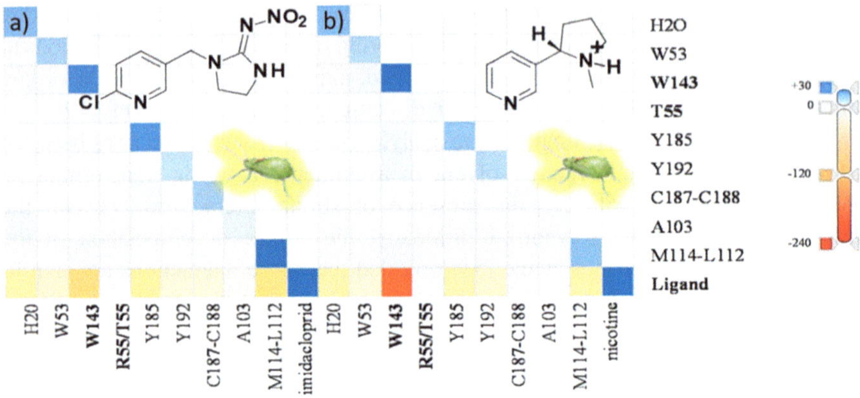

Fig. 6 LED interaction maps. Interaction energy (in kcal/mol) decomposition of imidacloprid (**a**) and nicotine (**b**) to the receptor with R55T mutation. (Adapted from Ref. [57] under a CC-BY 4.0 license, https://creativecommons.org/licenses/by/4.0/)

kcal/mol (-60 -30 0 30 60)	H_2O	W53	T55	A103	L112-M114	S142-T144	Y185	C187-C188	Y192	rest
Imidacloprid	-10.2	-2.7	-0.3	-1.0	-11.6	-18.3	-4.8	1.2	-4.4	-0.5
Nicotine	-11.2	-3.5	0.2	-0.2	-11.3	-41.7	-9.6	-0.9	-4.9	0.7
Δ	1.0	0.7	-0.5	-0.8	-0.4	23.4	4.7	2.1	0.5	-1.2

Fig. 7 fp-LED map showing the interactions of the two ligands with the residues of the protein and the difference (Δ) in the interaction between each residue and the ligands. (Adapted from Ref. [1] under the CC-BY 4.0 license, https://creativecommons.org/licenses/by/4.0/)

maps in Fig. 6, shows the key interactions of the two ligands in the resistant form of the protein.

The LED map reveals that, in the resistant form, the loss of the positively charged arginine residue leads to a decrease of the overall interaction energy with the imidacloprid. Indeed, comparison between the LED maps demonstrates that the resistant form of the receptor preferentially binds to nicotine, through the interaction with the W143 residue. Furthermore, the fp-LED map [1] shown in Fig. 7 enables the identification of the interaction strength of each amino acid residue with two ligands. The map confirms a significantly stronger binding energy to the protein for nicotine, as compared to imidacloprid.

4.3 Conformational Equilibria

In the study of conformational equilibria, accurately identifying and quantifying the specific intramolecular interactions within conformers are essential for understanding their relative stability [60]. In this regard, the LED scheme has proved to be highly valuable, allowing for the decomposition of noncovalent interaction energies driving the conformational equilibria into physically meaningful components, all while maintaining accuracy at the gold standard coupled cluster level. This scheme has been employed to determine the contribution of LD forces to the stabilization of different conformers [61], providing crucial insight into the case of sterically hindered conformers [62].

As an illustrative example, the effectiveness of the scheme was demonstrated by analyzing the unusual conformational equilibrium between the 1,6-folded and 1,4-unfolded conformers of di-*tert*-butyl-cyclooctatetraene (COT) [63], initially observed experimentally by Streitwieser et al. (Fig. 8) [64]. The study focuses on calculating the Gibbs free energy conformational equilibrium, conducted in 16 different solvents. While the entropic factor, driven by solvent reorganization, favors 1,4-COT, experimental results indicate that 1,6-COT is enthalpically preferred in all tested solvents. This preference arises from LD interactions between the *tert*-butyl groups. SAPT calculations on model dimers obtained by removing the COT ring, and saturating the resulting radicals with hydrogens, found that dispersion forces are the primary stabilizing factor for 1,6-COT. LED analysis confirmed this result, showing that CH-CH dispersion interactions (E_{xy}^{disp}) between the *tert*-butyl groups contribute 0.95 kcal/mol more in 1,6-COT, due to the decreased distance between these groups in 1,6-COT. It is worth noting that these dispersion interactions are often overlooked in solution, as it is commonly assumed that they cancel out by solvophobic effects in the net interaction balance [65].

4.4 Solids and Materials

The LED scheme has proven to be a valuable tool for analyzing the factors influencing the formation and stability of polymorphs in crystalline materials [66]. By decomposing noncovalent interactions, it provides insight into the forces

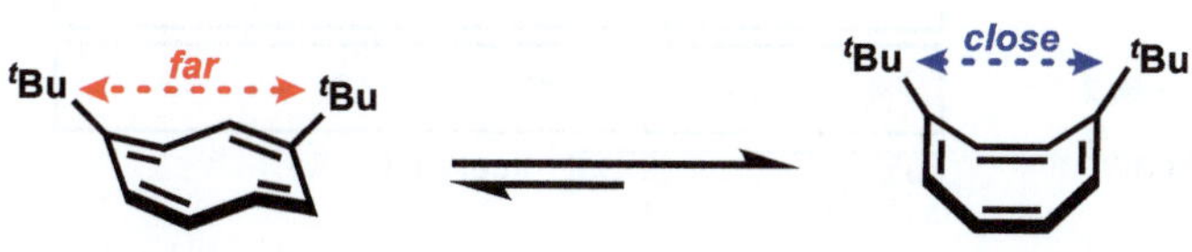

(unfolded) 1,4-COT

(folded) 1,6-COT

Fig. 8 Conformational equilibrium of 1,4- (left) and 1,6- (right) di-*tert*-butyl-cyclooctatetraene (COT) governing molecular packing and aggregation in the solid state. In this example, the LED scheme was used to investigate the properties of the 1,4-di-adamantyl-cyclooctatetraene (DIAD-COT) crystal, demonstrating its effectiveness in quantifying intermolecular interactions and understanding the stability of its crystalline forms [1].

A schematic representation of the fragmentation of the system is shown in Fig. 9a. The LED scheme is first used to estimate the lattice energy of the DIAD-COT crystal. By computing the energy associated with the removal of the central DIAD-COT molecule from the molecular cluster, a lattice energy of −48.9 kcal/mol was estimated. The decomposition of the lattice energy conducted with the LED scheme (Fig. 9b) highlights the crucial role of LD forces in stabilizing the crystal structure, with a contribution of −68.2 kcal/mol. Moreover, the fp-LED map reveals that the strongest interaction occurs when monomers (1–6) are aligned parallel to the central monomer (0), whereas head-to-head interactions between monomer 0 and monomers 7–10 are significantly weaker.

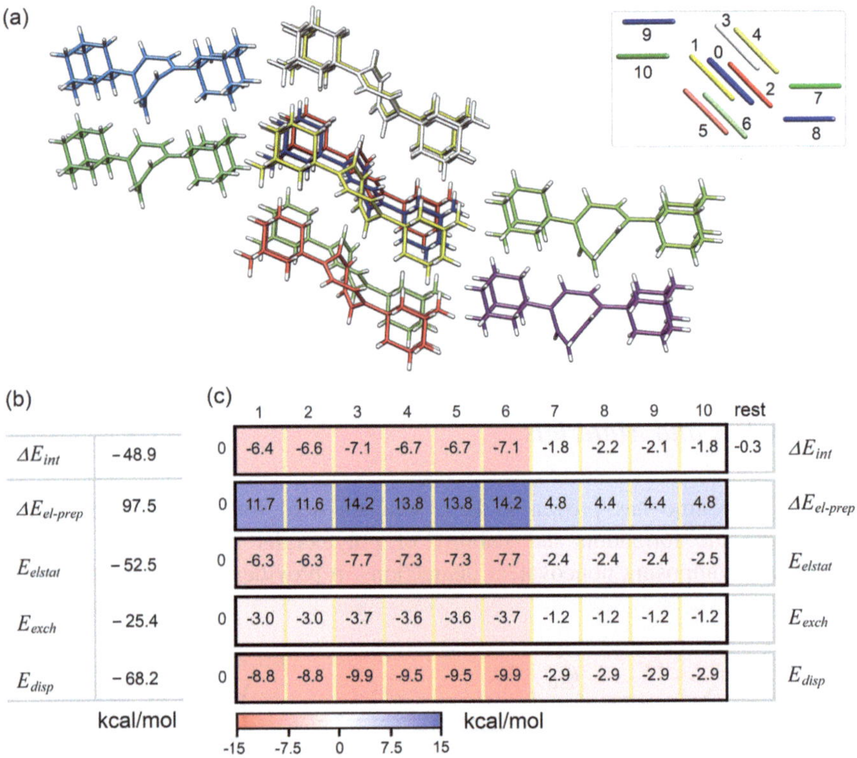

Fig. 9 (**a**) Structure of the DIAD-COT crystal. (**b**) Decomposition of the interaction energy. (**c**) fp-LED map showing the key interactions between the fragments. (Adapted from Ref. [1] under a CC-BY 4.0 license, https://creativecommons.org/licenses/by/4.0/)

5 Conclusions

Intermolecular interactions shape chemical and biological processes, influencing reaction mechanisms, selectivity, and molecular recognition. Their accurate quantification requires computational methods balancing accuracy and efficiency. This chapter presents local correlation approaches, focusing on the LED scheme in the DLPNO-CCSD(T) framework. LED decomposes interaction energies into fundamental components – electrostatics, exchange, dispersion, and electronic preparation – ensuring exact energy partitioning at the coupled cluster level. Its scalability and rigorous treatment of dispersion make it broadly applicable across different interaction regimes.

LED has been instrumental in studying asymmetric catalysis, protein-ligand interactions, and solid-state materials. Extensions to open-shell systems provide insights into spin-dependent interactions, while cost-effective approximations such as HFLD enhance computational efficiency. These methodologies offer powerful tools for analyzing intermolecular interactions in large systems, advancing computational chemistry, molecular design, and materials science. Ongoing work focuses on extending LED to the study of covalent bonds and introduces atomic decomposition of interaction components, further broadening its applicability.

References

1. Altun A, Leach IF, Neese F, Bistoni G (2024) A generally applicable method for disentangling the effect of individual noncovalent interactions on the binding energy. Angew Chem Int Ed 64: e202421922
2. Bistoni G (2020) Finding chemical concepts in the Hilbert space: coupled cluster analyses of noncovalent interactions. WIREs Comput Mol Sci 10:e1442
3. Bistoni G, Altun A, Wang Z, Neese F (2024) Local energy decomposition analysis of London dispersion effects: from simple model dimers to complex biomolecular assemblies. Acc Chem Res 57:1411–1420
4. Jeziorski B, Moszynski R, Szalewicz K (1994) Perturbation theory approach to intermolecular potential energy surfaces of van der Waals complexes. Chem Rev 94:1887–1930
5. Kitaura K, Morokuma K (1976) A new energy decomposition scheme for molecular interactions within the Hartree-Fock approximation. Int J Quantum Chem 10:325–340
6. Grimme S, Hansen A, Brandenburg JG, Bannwarth C (2016) Dispersion-corrected mean-field electronic structure methods. Chem Rev 116:5105–5154
7. Schütz M, Rauhut G, Werner H-J (1998) Local treatment of electron correlation in molecular clusters: structures and stabilities of $(H_2O)_n$, $n = 2–4$. J Phys Chem A 102:5997–6003
8. Thirman J, Head-Gordon M (2015) An energy decomposition analysis for second-order Møller–Plesset perturbation theory based on absolutely localized molecular orbitals. J Chem Phys 143:084124
9. Jurečka P, Šponer J, Černý J, Hobza P (2006) Benchmark database of accurate (MP2 and CCSD (T) complete basis set limit) interaction energies of small model complexes, DNA base pairs, and amino acid pairs. Phys Chem Chem Phys 8:1985–1993
10. Cybulski SM, Lytle ML (2007) The origin of deficiency of the supermolecule second-order Møller-Plesset approach for evaluating interaction energies. J Chem Phys 127:141102

11. Byrd EFC, Sherrill CD, Head-Gordon M (2001) The theoretical prediction of molecular radical species: a systematic study of equilibrium geometries and harmonic vibrational frequencies. J Phys Chem A 105:9736–9747
12. Farnell L, Pople JA, Radom L (1983) Structural predictions for open-shell systems: a comparative assessment of ab initio procedures. J Phys Chem 87:79–82
13. Tozer DJ et al (1993) Theory and applications of spin-restricted open-shell Møller-Plesset theory. Mol Phys 79:777–793
14. Wong MW, Radom L (1998) Radical addition to alkenes: further assessment of theoretical procedures. J Phys Chem A 102:2237–2245
15. Azar RJ, Head-Gordon M (2012) An energy decomposition analysis for intermolecular interactions from an absolutely localized molecular orbital reference at the coupled-cluster singles and doubles level. J Chem Phys 136:024103
16. Neese F (2012) The ORCA program system. WIREs Comput Mol Sci 2:73–78
17. Neese F (2018) Software update: the ORCA program system, version 4.0. WIREs Comput Mol Sci 8:e1327
18. Neese F, Wennmohs F, Becker U, Riplinger C (2020) The ORCA quantum chemistry program package. J Chem Phys 152:224108
19. Neese F (2022) Software update: the ORCA program system-version 5.0. WIREs Comput Mol Sci 12:e1606
20. Schneider WB et al (2016) Decomposition of intermolecular interaction energies within the local pair natural orbital coupled cluster framework. J Chem Theory Comput 12:4778–4792
21. Altun A, Izsák R, Bistoni G (2021) Local energy decomposition of coupled-cluster interaction energies: Interpretation, benchmarks, and comparison with symmetry-adapted perturbation theory. Int J Quantum Chem 121:e26339
22. Altun A, Neese F, Bistoni G (2018) Local energy decomposition analysis of hydrogen-bonded dimers within a domain-based pair natural orbital coupled cluster study. Beilstein J Org Chem 14:919–929
23. Bistoni G, Auer AA, Neese F (2017) Understanding the role of dispersion in frustrated Lewis pairs and classical Lewis adducts: a domain-based local pair natural orbital coupled cluster study. Chem Eur J 23:865–873
24. Altun A, Saitow M, Neese F, Bistoni G (2019) Local energy decomposition of open-shell molecular systems in the domain-based local pair natural orbital coupled cluster framework. J Chem Theory Comput 15:1616–1632
25. Ghafarian Shirazi R, Neese F, Pantazis DA, Bistoni G (2019) Physical nature of differential spin-state stabilization of carbenes by hydrogen and halogen bonding: a domain-based pair natural orbital coupled cluster study. J Phys Chem A 123:5081–5090
26. Altun A, Neese F, Bistoni G (2019) Effect of electron correlation on intermolecular interactions: a pair natural orbitals coupled cluster based local energy decomposition study. J Chem Theory Comput 15:215–228
27. Altun A, Neese F, Bistoni G (2022) Open-shell variant of the London dispersion-corrected Hartree–Fock method (HFLD) for the quantification and analysis of noncovalent interaction energies. J Chem Theory Comput 18:2292–2307
28. Altun A, Neese F, Bistoni G (2019) HFLD: a nonempirical London dispersion-corrected Hartree-Fock method for the quantification and analysis of noncovalent interaction energies of large molecular systems. J Chem Theory Comput 15:5894–5907
29. Sæbø S, Pulay P (1985) Local configuration interaction: an efficient approach for larger molecules. Chem Phys Lett 113:13–18
30. Sæbø S, Pulay P (1987) Fourth-order Møller–Plessett perturbation theory in the local correlation treatment. I. Method. J Chem Phys 86:914–922
31. Sæbø S, Pulay P (1988) The local correlation treatment. II. Implementation and tests. J Chem Phys 88:1884–1890
32. Nagy PR (2024) State-of-the-art local correlation methods enable affordable gold standard quantum chemistry for up to hundreds of atoms. Chem Sci 15:14556–14584

33. Riplinger C, Neese F (2013) An efficient and near linear scaling pair natural orbital based local coupled cluster method. J Chem Phys 138:034106
34. Neese F, Hansen A, Wennmohs F, Grimme S (2009) Accurate theoretical chemistry with coupled pair models. Acc Chem Res 42:641–648
35. Neese F, Wennmohs F, Hansen A (2009) Efficient and accurate local approximations to coupled-electron pair approaches: an attempt to revive the pair natural orbital method. J Chem Phys 130:114108
36. Liakos DG, Guo Y, Neese F (2020) Comprehensive benchmark results for the domain based local pair natural orbital coupled cluster method (DLPNO-CCSD(T)) for closed- and open-shell systems. J Phys Chem A 124:90–100
37. Sparta M et al (2017) Multilevel approaches within the local pair natural orbital framework. J Chem Theory Comput 13:3198–3207
38. Nesbet RK (1965) Electronic correlation in atoms and molecules. In: Prigogine I (ed) Advances in chemical physics, vol 9. Wiley, pp 321–363
39. Foster JM, Boys SF (1960) Canonical configurational interaction procedure. Rev Mod Phys 32: 300–302
40. Pipek J, Mezey PG (1989) A fast intrinsic localization procedure applicable for ab initio and semiempirical linear combination of atomic orbital wave functions. J Chem Phys 90:4916–4926
41. Edmiston C, Krauss M (1965) Configuration-interaction calculation of H_3 and H_2. J Chem Phys 42:1119–1120
42. Edmiston C, Krauss M (1966) Pseudonatural orbitals as a basis for the superposition of configurations. I. He_2^+. J Chem Phys 45:1833–1839
43. Altun A, Neese F, Bistoni G (2020) Extrapolation to the limit of a complete pair natural orbital space in local coupled-cluster calculations. J Chem Theory Comput 16:6142–6149
44. Minenkov Y et al (2017) Pair natural orbital and canonical coupled cluster reaction enthalpies involving light to heavy alkali and alkaline earth metals: the importance of sub-valence correlation. Phys Chem Chem Phys 19:9374–9391
45. Goerigk L et al (2017) A look at the density functional theory zoo with the advanced GMTKN55 database for general main group thermochemistry, kinetics and noncovalent interactions. Phys Chem Chem Phys 19:32184–32215
46. Knowles RR, Jacobsen EN (2010) Attractive noncovalent interactions in asymmetric catalysis: links between enzymes and small molecule catalysts. Proc Natl Acad Sci 107:20678–20685
47. Wagner JP, Schreiner PR (2015) London dispersion in molecular chemistry – reconsidering steric effects. Angew Chem Int Ed 54:12274–12296
48. Yepes D, Neese F, List B, Bistoni G (2020) Unveiling the delicate balance of steric and dispersion interactions in organocatalysis using high-level computational methods. J Am Chem Soc 142:3613–3625
49. Harden I, Neese F, Bistoni G (2022) An induced-fit model for asymmetric organocatalytic reactions: a case study of the activation of olefins *via* chiral Brønsted acid catalysts. Chem Sci 13:8848–8859
50. Kayal R, Baldinelli L, Harden I, Neese F, Bistoni G (2025) Understanding and quantifying the impact of solute–solvent van der Waals interactions on the selectivity of asymmetric catalytic transformations. Chem Sci 16:2700–2709
51. Baldinelli L, De Angelis F, Bistoni G (2024) Unraveling atomic contributions to the London dispersion energy: insights into molecular recognition and reactivity. J Chem Theory Comput 20:1923–1931
52. Park G et al (2024) Pentaphosphorylation via the anhydride of dihydrogen pentametaphosphate: access to nucleoside hexa- and heptaphosphates and study of their interaction with ribonuclease A. ACS Cent Sci 10:1415–1422
53. Elangovan N et al (2023) Influence of the solvents through molecular structure, wavefunction studies, biological activity prediction and molecular docking studies of 4,4-((1,2-phenylenebis (azaneylylidene)) bis (methaneylylidene)) bis(2-methoxyphenol). J Mol Liq 391:123334

54. Kojasoy V, Tantillo DJ (2023) Impacts of noncovalent interactions involving sulfur atoms on protein stability, structure, folding, and bioactivity. Org Biomol Chem 21:11–23
55. Sylvetsky N (2020) Toward simple, predictive understanding of protein-ligand interactions: electronic structure calculations on Torpedo californica acetylcholinesterase join forces with the chemist's intuition. Sci Rep 10:9218
56. Altun A, Garcia-Ratés M, Neese F, Bistoni G (2021) Unveiling the complex pattern of intermolecular interactions responsible for the stability of the DNA duplex. Chem Sci 12: 12785–12793
57. Beck ME, Riplinger C, Neese F, Bistoni G (2021) Unraveling individual host–guest interactions in molecular recognition from first principles quantum mechanics: insights into the nature of nicotinic acetylcholine receptor agonist binding. J Comput Chem 42:293–302
58. Unwin N (2005) Refined structure of the nicotinic acetylcholine receptor at 4Å resolution. J Mol Biol 346:967–989
59. Jeschke P, Nauen R, Beck ME (2013) Nicotinic acetylcholine receptor agonists: a milestone for modern crop protection. Angew Chem Int Ed 52:9464–9485
60. Strauss MA, Wegner HA (2019) Molecular systems for the quantification of London dispersion interactions. Eur J Org Chem 2019:295–302
61. Žabka M, Naviri L, Gschwind RM (2021) Noncovalent CH–π and π–π interactions in phosphoramidite palladium(II) complexes with strong conformational preference. Angew Chem Int Ed 60:25832–25838
62. Rummel L, Domanski MHJ, Hausmann H, Becker J, Schreiner PR (2022) London dispersion favors sterically hindered diarylthiourea conformers in solution. Angew Chem Int Ed 61: e202204393
63. Schümann JM, Wagner JP, Eckhardt AK, Quanz H, Schreiner PR (2021) Intramolecular London dispersion interactions do not cancel in solution. J Am Chem Soc 143:41–45
64. Lyttle MH, Streitwieser A, Kluttz RQ (1981) Unusual equilibrium between 1,4- and 1,6-di-tert-butylcyclooctatetraenes. J Am Chem Soc 103:3232–3233
65. Yang L, Adam C, Nichol GS, Cockroft SL (2013) How much do van der Waals dispersion forces contribute to molecular recognition in solution? Nat Chem 5:1006–1010
66. Altun A et al (2024) Rationalizing polymorphism with local correlation-based methods: a case study of pnictogen molecular crystals. Phys Chem Chem Phys 26:28733–28745

Struct Bond (2026) 190: 71–108
https://doi.org/10.1007/430_2025_96

Published online: 10 September 2025

Electron Density-Based Energy Decomposition Analysis: From QM to QM/MM Calculations

Marcos Mandado

Contents

Abstract Weak interactions involving closed-shell molecules, such as hydrogen bonding, van der Waals forces, and π–π stacking, are pivotal to understanding a wide array of phenomena in chemistry, physics, and biology. These interactions dictate the behavior and properties of molecular assemblies, from protein folding and DNA stability to material properties and reaction mechanisms. Computational studies have become essential tools for investigating these interactions, providing detailed insights into their nature, strength, and effects. Modern computational methods enable researchers to model complex systems with high precision, decompose intermolecular energies into meaningful components, and design innovative materials and drugs. By bridging theory and experiment, computational approaches advance our understanding of intermolecular forces, driving progress across scientific disciplines. This chapter introduces a relatively new scheme for decomposing intermolecular interaction energy into distinct, physically meaningful terms – such

M. Mandado
Department of Physical Chemistry, University of Vigo, Vigo, Spain
e-mail: mandado@uvigo.es

as electrostatic, exchange, spin repulsion, and polarization – using the fragmentation of the complex's electron density. The primary advantages of this method are its simplicity and low computational cost, making it highly practical compared to other techniques. Following an introductory section that reviews widely used methods in the theoretical community, three subsequent sections delve into the theoretical framework of this electron density-based energy decomposition analysis and evaluate its performance through previously reported results. These sections focus on the original scheme's application to SCF and post-SCF levels; its extension to decompose polarization energy into induction and dispersion terms, and its use with large complexes in the context of DFT; and, finally, its generalization for application with QM/MM hybrid methods.

Keywords Computational chemistry · Energy decomposition analysis · Interaction energy · Intermolecular interactions · Non-covalent interactions

1 Introduction

Intermolecular interactions are fundamental to understanding the behavior and properties of matter across physics, chemistry, and biology [1, 2]. In physics, they govern phenomena such as phase transitions, determining whether substances exist as solids, liquids, or gases under specific conditions. In chemistry, these interactions play a crucial role in dictating reaction mechanisms, molecular structures, and stability of compounds, influencing processes from crystallization to catalysis. In biology, intermolecular forces are indispensable for life, enabling the formation of complex biomolecular structures like DNA, proteins, and membranes as well as facilitating critical processes such as enzyme-substrate binding and cell signaling. Together, these forces shape the physical world, drive chemical reactivity, and sustain biological function, highlighting their universal importance.

In the theory of intermolecular interactions [3–5], interaction energy refers to the total energy resulting from the forces acting between two or more molecules. It encompasses contributions from various components, such as van der Waals forces, electrostatic interactions, hydrogen bonding, and dispersion forces. This energy characterizes the potential energy landscape that governs how molecules approach, orient, and influence each other. On the other hand, binding energy quantifies the energy required to separate two molecules from their bound state to an infinitely separated state where the molecules are relaxed to their equilibrium configuration. The binding energy accounts for the interaction energy plus the relaxation energy of the interacting molecules from their geometry in the complex to their equilibrium geometries. While interaction energy provides a comprehensive view of the forces at play, binding energy is a specific measure of the stability and strength of the molecular association.

The most commonly employed method for calculating intermolecular interaction energies is the supermolecular ansatz [6, 7], where the interaction energy is determined as the difference between the total energies of the complex and its constituent molecules. Due to its simplicity, this approach is widely utilized; indeed, any ab initio or density functional theory (DFT) software package can readily provide the energies of both the complex and the monomers, with the latter corrected for the undesirable basis set superposition error (BSSE) [8–10]. However, the supermolecular approach offers limited insight into the nature of intermolecular interactions, as it only yields the value of the interaction energy without elucidating the underlying physical mechanisms, intrinsically related to the type of forces involved.

Energy decomposition analysis (EDA) has emerged as a powerful tool in quantum chemistry for understanding the nature of intermolecular interactions. The development of EDA methods began with the pioneering work of Morokuma in the early 1970s [11, 12], which laid the groundwork for subsequent advancements in the field. The Morokuma method introduced a systematic approach to decompose the interaction energy between two molecular fragments into the following physically meaningful components: electrostatic, arising from the classical Coulomb attraction between charge distributions; exchange-repulsion, also known as Pauli repulsion, which accounts for the destabilizing overlap of electron clouds due to the Pauli exclusion principle; charge transfer, due to the electron charge migration between interacting molecules; and polarization, resulting from the intramolecular electron charge reorganization. This method allowed chemists to gain insights into the contributions of different forces to the overall interaction energy. The main limitation of Morokuma's method was its restriction to the Hartree-Fock (HF) level of theory, which excludes electron correlation, a critical factor in the stability of many intermolecular complexes, particularly those dominated by dispersion forces.

In 1977, Ziegler and Rauk introduced an EDA method within the HF framework [13], which was subsequently extended to Kohn-Sham density functional theory (KS-DFT) [14]. This approach divides the interaction energy into three primary components: electrostatic interaction, Pauli repulsion, and orbital interaction energy. The orbital interaction term encompasses stabilizing effects such as charge transfer, polarization, and covalent bonding between molecular fragments. As the method operates within the KS-DFT framework, the extent and type of electron correlation included in the calculations depend heavily on the chosen functional. However, long-range electron correlation remains inadequately described in all DFT functionals, with varying performance depending on the functional used [15]. To accurately account for dispersion energy, empirical dispersion corrections are often necessary, with Grimme et al.'s widely adopted corrections [16, 17] being among the most commonly employed.

Building upon the works of Morokuma and Ziegler and Rauk, several EDA methods have been developed over the past few decades. Some of these methods are based on valence bond theory, such as the generalized product function (GPF) EDA by Nascimento et al. [18, 19] and the block-localized wavefunction (BLW)

EDA by Mo et al. [20–22]. Others, like the interacting quantum atom (IQA) method by Pendás et al. [23, 24], are grounded in the fragmentation of real space into atomic domains. In the IQA method, the energy components included in the Hamiltonian – namely, kinetic energy, core attraction, electron repulsion, and nuclear repulsion – are decomposed into intra-atomic and interatomic contributions. Additionally, several widely used and versatile methods deserve mention, such as the absolute localized molecular orbital (ALMO) EDA by Head-Gordon et al. [25, 26], the natural energy decomposition analysis (NEDA) by Weinhold et al. [27, 28], and the EDA-natural orbitals for chemical valence (NOCV) [29]. The latter combines the EDA approach of Ziegler and Rauk [13, 14] with the NOCV method developed by Mitoraj and Michalak [30].

The methods mentioned so far can be classified as variational methods. An alternative approach to studying intermolecular interactions is through perturbation theory [2–4, 31], which, unlike the supermolecular approach, does not inherently require the computation of the total energy of the combined system. Perturbation theory has been a cornerstone in the study of intermolecular forces since the early days of quantum chemistry, serving as a foundational framework for their understanding, modeling, and computation. Perturbation methods provide valuable interpretative insights and have played a crucial role in the development of accurate ab initio intermolecular force fields [3, 32], advancing our ability to predict and analyze complex molecular interactions. In the perturbation theory of intermolecular forces, the unperturbed system consists of isolated monomers and their interaction is treated as a perturbation. The intermolecular interactions are decomposed into four fundamental physical contributions: electrostatic, exchange-repulsion, induction, and dispersion. The exchange-repulsion term accounts for both the spin repulsion governed by the Pauli exclusion principle and the exchange energy resulting from electron tunneling between interacting systems. Induction and dispersion energies, often grouped under the polarization term, describe the mutual influence of polarized charge distributions within the interacting molecules.

The simplest perturbative method to studying intermolecular interactions is the so-called polarization theory [2, 33, 34], which entails the standard application of Rayleigh-Schrödinger perturbation theory (RSPT) to an intermolecular complex. While this method effectively describes long-range interactions between molecules, it falls short in accurately capturing short-range interactions, where exchange-repulsion plays an important role. To address this limitation, the symmetry-adapted perturbation theory (SAPT) was developed by Jeziorski et al. [31, 35]. SAPT not only provides a more comprehensive treatment of intermolecular forces but also offers a computationally efficient framework for evaluating the perturbative contributions, including the effects of electron exchange and spin repulsion between interacting molecules.

A significant limitation of SAPT within wavefunction-based methods is its high computational cost, particularly when introducing intramonomer correlation effects. Originally, SAPT was based on a double-perturbation theory expansion, considering two perturbations: the intermolecular interaction operator and the intramonomer correlation operator. This approach requires a large number of terms to accurately describe each specific interaction energy component. This restricts its applicability to smaller complexes, as the computational demands become prohibitive for larger systems. Advanced versions of SAPT, such as SAPT(DFT) [36] or DFT-SAPT [37], circumvent the double-perturbation expansion by introducing intramonomer correlation effects through DFT. In SAPT(DFT), the monomers' wavefunctions and response properties are described using KS-DFT rather than traditional wavefunction-based methods [38]. This reduces the computational cost and extends the method's applicability to larger and more complex systems. The balance between computational efficiency and accuracy makes SAPT(DFT) an attractive tool for studying intricate intermolecular phenomena. However, despite its advantages, the computational cost of SAPT(DFT) remains significantly higher than that of variational methods, such as those based on Morokuma's and Ziegler and Rauk's schemes.

In quest for an EDA scheme that offers accuracy comparable to SAPT or variational methods like that of Ziegler and Rauk, yet with lower computational cost and broad applicability across self-consistent field (SCF) methods – including KS-DFT – and post-SCF levels, Mandado et al. proposed a deformation density-based EDA scheme in 2011 [39], hereafter referred to as DD-EDA. In the DD-EDA scheme, the one-electron density and exchange-correlation density of the complex are fragmented into contributions from the isolated molecules, along with deformation densities associated with intermolecular electron exchange, spin repulsion, and electron polarization induced by the intermolecular interaction. This fragmentation is incorporated into the supermolecular framework to decompose the total interaction energy into electrostatic, exchange, repulsion, and polarization components. The DD-EDA scheme can be applied at any computational level, provided that the one-electron and exchange-correlation densities are available for those levels. In practice, at post-SCF levels, approximated expressions based on natural orbitals are used for the intermolecular exchange density to circumvent the challenging task of calculating the exact matrices associated with this density [39].

The DD-EDA scheme maintains strong connections to previously proposed EDA approaches. It provides first-order contributions, namely, electrostatic and exchange-repulsion energies, that align closely with those of SAPT method, making the results directly comparable [39]. Additionally, the DD-EDA method introduces a refinement by separating the short-range exchange-repulsion term into distinct exchange and repulsion components. This distinction arises from the use of the exchange density to calculate the exchange energy and the one-electron density to determine the repulsion energy. On the other hand, the polarization energy is analogous to the orbital interaction energy in the Ziegler and Rauk's scheme. However, in DD-EDA, the polarization energy is composed of various terms linked to the deformation of either the one-electron density or the exchange-correlation density. This structure enables the analysis of the polarization energy in terms of contributions from distinct physical origins.

Subsequently, second-order Rayleigh-Schrödinger perturbation theory was introduced into the DD-EDA scheme to extract the induction term from the polarization energy [40]. This converted DD-EDA into a hybrid scheme, combining variationally obtained electron densities with an expression for the induction energy derived from a perturbational treatment. Once the induction term is obtained, dispersion corresponds to the remaining polarization energy at the second-order perturbation level. To date, this scheme has been applied exclusively using densities obtained at the KS-DFT level, with explicit corrections for dispersion included [40–44]. The primary reason for this is that dispersion energy becomes more significant as the size of the interacting molecules increases, making the computation of such systems and their analysis more challenging at post-SCF levels.

More recently, the DD-EDA scheme has been extended to calculations using hybrid quantum mechanics/molecular mechanics (QM/MM) methods [45–47]. This allows the study of intermolecular interactions in solution [47] and biological environments [45, 46], among other situations involving very large systems. Other EDA schemes based on QM/MM calculations can be found in the literature, but they involve either a fragmentation of the complex system into a QM fragment and an MM fragment [48, 49] or a fragmentation into two QM fragments polarized by an MM environment [50]. In the first case, part of the interaction center is described classically, so that the accuracy of the results obtained is limited since the interaction center in QM/MM calculation should be fully described at QM level. In the second case, the interaction center is fully described at QM level, but the classical potential only serves to polarize the QM region and it is not included explicitly in the calculation of the different energy terms. Only very recently has a QM/MM EDA scheme based on the generalized Kohn-Sham (GKS) EDA been proposed [51], where the interaction center is fully described at QM level and the classical potential is explicitly included in the calculation of the different energy terms as in the QM/MM DD-EDA scheme.

In the following sections, the theoretical framework of the DD-EDA is reviewed in detail. This includes the original energy decomposition at QM level, the hybrid version that allows decomposing the polarization energy into induction and dispersion, and its extension to QM/MM calculations. The performance of the DD-EDA is also discussed in each section by including results obtained for different complexes of interest in chemistry, biology, and materials science.

2 DD-EDA

2.1 Theoretical Framework

Herein, the derivation of the equations for the different interaction energy terms is introduced for a purely QM treatment. All the equations are given in atomic units. The extension to hybrid QM/MM treatments is summarized in a further section.

The primary objective of the DD-EDA scheme is to extend the insights offered by the supermolecular approach beyond merely quantifying the energy involved in a complexation process. It aims to provide additional information about the nature of the interaction. Therefore, the mathematical foundation for deriving the various interaction energy terms begins with the total energy of the supermolecule, expressed in terms of densities and density matrices. For an interacting quantum system formed by two closed-shell molecules A and B, the total nonrelativistic energy can be expressed exactly as a functional of the one-electron density (1-D), the one-electron reduced density matrix (1-RDM), $\rho(\boldsymbol{r})$ and $\rho(\boldsymbol{r},\boldsymbol{r}')_{\boldsymbol{r}=\boldsymbol{r}'}$, and the exchange-correlation density (XCD), $\rho_{XC}(\boldsymbol{r}_1,\boldsymbol{r}_2)$:

$$\begin{aligned} E_{AB} = & -\frac{1}{2}\int \nabla^2\rho(\boldsymbol{r},\boldsymbol{r}')_{\boldsymbol{r}=\boldsymbol{r}'}d\boldsymbol{r} + \int \widehat{v}_N\rho(\boldsymbol{r})d\boldsymbol{r} + \frac{1}{2}\iint \frac{\rho(\boldsymbol{r}_1)\rho(\boldsymbol{r}_2)}{|\boldsymbol{r}_2-\boldsymbol{r}_1|}d\boldsymbol{r}_1 d\boldsymbol{r}_2 \\ & + \frac{1}{2}\iint \frac{\rho_{XC}(\boldsymbol{r}_1,\boldsymbol{r}_2)}{|\boldsymbol{r}_2-\boldsymbol{r}_1|}d\boldsymbol{r}_1 d\boldsymbol{r}_2 + \sum_{I=1}^{N-1}\sum_{J>I}^{N}\frac{Z_I Z_J}{|\boldsymbol{R}_I-\boldsymbol{R}_J|} \end{aligned} \tag{1}$$

In the above equation, $\boldsymbol{r}_1$, $\boldsymbol{r}_2$ represent the electron positions; Z_I, Z_J and $\boldsymbol{R}_I$, $\boldsymbol{R}_J$ represent, respectively, nuclei atomic numbers and positions; and $\widehat{v}_N$ represents the potential operator of the nuclei, which correspond to the sum of the potentials of the nuclei of A and B:

$$\widehat{v}_N = \widehat{v}_{N_A} + \widehat{v}_{N_B} \tag{2}$$

Equivalent equations can be written for the isolated molecules:

$$\begin{aligned} E_{A/B} = & -\frac{1}{2}\int \nabla^2\rho_{A/B}(\boldsymbol{r},\boldsymbol{r}')_{\boldsymbol{r}=\boldsymbol{r}'}d\boldsymbol{r} + \int \widehat{v}_{N_{A/B}}\rho_{A/B}(\boldsymbol{r})d\boldsymbol{r} \\ & + \frac{1}{2}\iint \frac{\rho_{A/B}(\boldsymbol{r}_1)\rho_{A/B}(\boldsymbol{r}_2)}{|\boldsymbol{r}_2-\boldsymbol{r}_1|}d\boldsymbol{r}_1 d\boldsymbol{r}_2 + \frac{1}{2}\iint \frac{\rho_{XC,A/B}(\boldsymbol{r}_1,\boldsymbol{r}_2)}{|\boldsymbol{r}_2-\boldsymbol{r}_1|}d\boldsymbol{r}_1 d\boldsymbol{r}_2 \\ & + \sum_{I=1}^{N_{A/B}-1}\sum_{J>I}^{N_{A/B}}\frac{Z_I Z_J}{|\boldsymbol{R}_I-\boldsymbol{R}_J|} \end{aligned} \tag{3}$$

The terms represented in Eq. (3) for A and B are implicitly accounted for in Eq. (1). These terms can be excluded from the total energy of the complex, isolating only the contributions associated with intermolecular interactions. This forms the core concept of the supermolecular approach, wherein the interaction energy is determined as

$$E_{int} = E_{AB}(AB) - [E_A(AB) + E_B(AB)] \tag{4}$$

Two key considerations must be mentioned regarding Eq. (4), which must be taken into account for the DD-EDA scheme. First, the geometries of the molecules correspond to those adopted within the complex. As a result, the energy associated with the relaxation from the complex geometry to that of the isolated molecules is excluded from the interaction energy. Incorporating the geometry relaxation energy is straightforward and, as mentioned in the previous section, leads to what is commonly referred to as binding energy. Second, the energies of the complex and the isolated molecules must be calculated using a sufficiently large basis set to eliminate the well-known basis set superposition error (BSSE). Since the use of very large basis sets is impractical for large complexes, the counterpoise (CP) correction is often employed [10]. This involves using the basis set of the complex for the calculation of the energies of the isolated molecules instead of their own basis sets. While this method is not without criticism, mainly due to its tendency to overestimate BSSE [52, 53], recent studies have shown that its application is essential and provides accurate results for calculations using single-determinant wavefunctions [54], such as the KS-DFT approach.

The basic idea behind the DD-EDA scheme is the introduction of fragmentations for the complex 1-D, 1-RDM, and XCD that enable the subtraction of the energies of the isolated molecules from the total complex energy and yield intermolecular terms that can be grouped into physically meaningful components. In its initial formulation [39], these components corresponded to the traditional concepts of electrostatic, Pauli repulsion, and polarization energies, introduced in the context of the theory of intermolecular forces. Such fragmentation is given by

$$\rho(\boldsymbol{r}) = \rho_A(\boldsymbol{r}) + \rho_B(\boldsymbol{r}) + \Delta\rho_{Pau}(\boldsymbol{r}) + \Delta\rho_{Pol}(\boldsymbol{r}) \quad (5)$$

$$\rho(\boldsymbol{r},\boldsymbol{r}')_{\boldsymbol{r}=\boldsymbol{r}'} = \rho_A(\boldsymbol{r},\boldsymbol{r}')_{\boldsymbol{r}=\boldsymbol{r}'} + \rho_B(\boldsymbol{r},\boldsymbol{r}')_{\boldsymbol{r}=\boldsymbol{r}'} + \Delta\rho_{Pau}(\boldsymbol{r},\boldsymbol{r}')_{\boldsymbol{r}=\boldsymbol{r}'} + \Delta\rho_{Pol}(\boldsymbol{r},\boldsymbol{r}')_{\boldsymbol{r}=\boldsymbol{r}'} \quad (6)$$

$$\rho_{XC}(\boldsymbol{r}_1,\boldsymbol{r}_2) = \rho_{XC,A}(\boldsymbol{r}_1,\boldsymbol{r}_2) + \rho_{XC,B}(\boldsymbol{r}_1,\boldsymbol{r}_2) + \rho_{X,AB}(\boldsymbol{r}_1,\boldsymbol{r}_2) + \Delta\rho_{XC}(\boldsymbol{r}_1,\boldsymbol{r}_2) \quad (7)$$

where the 1-D and the 1-RDM of the complex (as described in Eqs. 5 and 6) are expressed as the sum of those corresponding to the isolated molecules, along with their deformations resulting from Pauli repulsion (indicated by the subscript "*Pau*") and electron polarization (indicated by the subscript "*Pol*"). Furthermore, in Eq. (7), the XCD of the complex is broken down into the sum of the XCDs of the isolated molecules; the intermolecular exchange density (XD), $\rho_{X,\ AB}(\boldsymbol{r}_1,\boldsymbol{r}_2)$, which arises from the antisymmetrization of the product of the isolated molecules' wavefunctions; and the deformation of the XCD, $\Delta\rho_{XC}(\boldsymbol{r}_1,\boldsymbol{r}_2)$, caused by electron polarization.

Introducing the densities and density matrices of Eqs. (5)–(7), as well as the partition of the nuclear potential operator of Eq. (2), in the expression of E_{AB}, the following expanded equation is obtained:

$$
\begin{aligned}
E_{AB} = & -\frac{1}{2}\int \nabla^2\rho_A(\boldsymbol{r},\boldsymbol{r}')_{r=r'}d\boldsymbol{r} - \frac{1}{2}\int \nabla^2\rho_B(\boldsymbol{r},\boldsymbol{r}')_{r=r'}d\boldsymbol{r} - \frac{1}{2}\int \nabla^2\Delta\rho_{Pau}(\boldsymbol{r},\boldsymbol{r}')_{r=r'}d\boldsymbol{r} \\
& -\frac{1}{2}\int \nabla^2\Delta\rho_{Pol}(\boldsymbol{r},\boldsymbol{r}')_{r=r'}d\boldsymbol{r} + \int \hat{v}_{N_A}\rho_A(\boldsymbol{r})d\boldsymbol{r} + \int \hat{v}_{N_B}\rho_B(\boldsymbol{r})d\boldsymbol{r} \\
& + \int \hat{v}_{N_A}\rho_B(\boldsymbol{r})d\boldsymbol{r} + \int \hat{v}_{N_B}\rho_A(\boldsymbol{r})d\boldsymbol{r} + \int \hat{v}_{N_A}\Delta\rho_{Pau}(\boldsymbol{r})d\boldsymbol{r} + \int \hat{v}_{N_B}\Delta\rho_{Pau}(\boldsymbol{r})d\boldsymbol{r} \\
& + \int \hat{v}_{N_A}\Delta\rho_{Pol}(\boldsymbol{r})d\boldsymbol{r} + \int \hat{v}_{N_B}\Delta\rho_{Pol}(\boldsymbol{r})d\boldsymbol{r} + \frac{1}{2}\iint \frac{\rho_A(\boldsymbol{r}_1)\rho_A(\boldsymbol{r}_2)}{|\boldsymbol{r}_2-\boldsymbol{r}_1|}d\boldsymbol{r}_1 d\boldsymbol{r}_2 \\
& + \frac{1}{2}\iint \frac{\rho_B(\boldsymbol{r}_1)\rho_B(\boldsymbol{r}_2)}{|\boldsymbol{r}_2-\boldsymbol{r}_1|}d\boldsymbol{r}_1 d\boldsymbol{r}_2 + \iint \frac{\rho_A(\boldsymbol{r}_1)\rho_B(\boldsymbol{r}_2)}{|\boldsymbol{r}_2-\boldsymbol{r}_1|}d\boldsymbol{r}_1 d\boldsymbol{r}_2 \\
& + \iint \frac{\rho_A(\boldsymbol{r}_1)\Delta\rho_{Pau}(\boldsymbol{r}_2)}{|\boldsymbol{r}_2-\boldsymbol{r}_1|}d\boldsymbol{r}_1 d\boldsymbol{r}_2 + \iint \frac{\rho_B(\boldsymbol{r}_1)\Delta\rho_{Pau}(\boldsymbol{r}_2)}{|\boldsymbol{r}_2-\boldsymbol{r}_1|}d\boldsymbol{r}_1 d\boldsymbol{r}_2 \\
& + \iint \frac{\rho_A(\boldsymbol{r}_1)\Delta\rho_{Pol}(\boldsymbol{r}_2)}{|\boldsymbol{r}_2-\boldsymbol{r}_1|}d\boldsymbol{r}_1 d\boldsymbol{r}_2 + \iint \frac{\rho_B(\boldsymbol{r}_1)\Delta\rho_{Pol}(\boldsymbol{r}_2)}{|\boldsymbol{r}_2-\boldsymbol{r}_1|}d\boldsymbol{r}_1 d\boldsymbol{r}_2 \\
& + \iint \frac{\Delta\rho_{Pau}(\boldsymbol{r}_1)\Delta\rho_{Pol}(\boldsymbol{r}_2)}{|\boldsymbol{r}_2-\boldsymbol{r}_1|}d\boldsymbol{r}_1 d\boldsymbol{r}_2 + \frac{1}{2}\iint \frac{\Delta\rho_{Pau}(\boldsymbol{r}_1)\Delta\rho_{Pau}(\boldsymbol{r}_2)}{|\boldsymbol{r}_2-\boldsymbol{r}_1|}d\boldsymbol{r}_1 d\boldsymbol{r}_2 \\
& + \frac{1}{2}\iint \frac{\Delta\rho_{Pol}(\boldsymbol{r}_1)\Delta\rho_{Pol}(\boldsymbol{r}_2)}{|\boldsymbol{r}_2-\boldsymbol{r}_1|}d\boldsymbol{r}_1 d\boldsymbol{r}_2 + \frac{1}{2}\iint \frac{\rho_{XC,A}(\boldsymbol{r}_1,\boldsymbol{r}_2)}{|\boldsymbol{r}_2-\boldsymbol{r}_1|}d\boldsymbol{r}_1 d\boldsymbol{r}_2 \\
& + \frac{1}{2}\iint \frac{\rho_{XC,B}(\boldsymbol{r}_1,\boldsymbol{r}_2)}{|\boldsymbol{r}_2-\boldsymbol{r}_1|}d\boldsymbol{r}_1 d\boldsymbol{r}_2 + \frac{1}{2}\iint \frac{\rho_{X,AB}(\boldsymbol{r}_1,\boldsymbol{r}_2)}{|\boldsymbol{r}_2-\boldsymbol{r}_1|}d\boldsymbol{r}_1 d\boldsymbol{r}_2 \\
& + \frac{1}{2}\iint \frac{\Delta\rho_{XC}(\boldsymbol{r}_1,\boldsymbol{r}_2)}{|\boldsymbol{r}_2-\boldsymbol{r}_1|}d\boldsymbol{r}_1 d\boldsymbol{r}_2 + \sum_{I=1}^{N_A-1}\sum_{J>I}^{N_A}\frac{Z_I Z_J}{|\boldsymbol{R}_I-\boldsymbol{R}_J|} + \sum_{I=N_A+1}^{N-1}\sum_{J>I}^{N}\frac{Z_I Z_J}{|\boldsymbol{R}_I-\boldsymbol{R}_J|} \\
& + \sum_{I=1}^{N_A}\sum_{J=N_A+1}^{N}\frac{Z_I Z_J}{|\boldsymbol{R}_I-\boldsymbol{R}_J|}
\end{aligned}
\tag{8}
$$

from which the interaction energy is extracted by removing the terms corresponding to the energies of the isolated molecules (Eq. 3). Thus, the interaction energy remains, in an expanded form, as

$$
\begin{aligned}
E_{int} = & -\frac{1}{2}\int \nabla^2 \Delta\rho_{Pau}(\mathbf{r},\mathbf{r}')_{\mathbf{r}=\mathbf{r}'} d\mathbf{r} - \frac{1}{2}\int \nabla^2 \Delta\rho_{Pol}(\mathbf{r},\mathbf{r}')_{\mathbf{r}=\mathbf{r}'} d\mathbf{r} + \int \hat{v}_{N_A}\rho_B(\mathbf{r})d\mathbf{r} \\
& + \int \hat{v}_{N_B}\rho_A(\mathbf{r})d\mathbf{r} + \int \hat{v}_{N_A}\Delta\rho_{Pau}(\mathbf{r})d\mathbf{r} + \int \hat{v}_{N_B}\Delta\rho_{Pau}(\mathbf{r})d\mathbf{r} \\
& + \int \hat{v}_{N_A}\Delta\rho_{Pol}(\mathbf{r})d\mathbf{r} + \int \hat{v}_{N_B}\Delta\rho_{Pol}(\mathbf{r})d\mathbf{r} + \iint \frac{\rho_A(\mathbf{r}_1)\rho_B(\mathbf{r}_2)}{|\mathbf{r}_2-\mathbf{r}_1|} d\mathbf{r}_1 d\mathbf{r}_2 \\
& + \iint \frac{\rho_A(\mathbf{r}_1)\Delta\rho_{Pau}(\mathbf{r}_2)}{|\mathbf{r}_2-\mathbf{r}_1|} d\mathbf{r}_1 d\mathbf{r}_2 + \iint \frac{\rho_B(\mathbf{r}_1)\Delta\rho_{Pau}(\mathbf{r}_2)}{|\mathbf{r}_2-\mathbf{r}_1|} d\mathbf{r}_1 d\mathbf{r}_2 \\
& + \iint \frac{\rho_A(\mathbf{r}_1)\Delta\rho_{Pol}(\mathbf{r}_2)}{|\mathbf{r}_2-\mathbf{r}_1|} d\mathbf{r}_1 d\mathbf{r}_2 + \iint \frac{\rho_B(\mathbf{r}_1)\Delta\rho_{Pol}(\mathbf{r}_2)}{|\mathbf{r}_2-\mathbf{r}_1|} d\mathbf{r}_1 d\mathbf{r}_2 \\
& + \iint \frac{\Delta\rho_{Pau}(\mathbf{r}_1)\Delta\rho_{Pol}(\mathbf{r}_2)}{|\mathbf{r}_2-\mathbf{r}_1|} d\mathbf{r}_1 d\mathbf{r}_2 + \frac{1}{2}\iint \frac{\Delta\rho_{Pau}(\mathbf{r}_1)\Delta\rho_{Pau}(\mathbf{r}_2)}{|\mathbf{r}_2-\mathbf{r}_1|} d\mathbf{r}_1 d\mathbf{r}_2 \\
& + \frac{1}{2}\iint \frac{\Delta\rho_{Pol}(\mathbf{r}_1)\Delta\rho_{Pol}(\mathbf{r}_2)}{|\mathbf{r}_2-\mathbf{r}_1|} d\mathbf{r}_1 d\mathbf{r}_2 + \frac{1}{2}\iint \frac{\rho_{X,AB}(\mathbf{r}_1,\mathbf{r}_2)}{|\mathbf{r}_2-\mathbf{r}_1|} d\mathbf{r}_1 d\mathbf{r}_2 \\
& + \frac{1}{2}\iint \frac{\Delta\rho_{XC}(\mathbf{r}_1,\mathbf{r}_2)}{|\mathbf{r}_2-\mathbf{r}_1|} d\mathbf{r}_1 d\mathbf{r}_2 + \sum_{I=1}^{N_A}\sum_{J=N_A+1}^{N} \frac{Z_I Z_J}{|\mathbf{R}_I-\mathbf{R}_J|}
\end{aligned}
\tag{9}
$$

Equation (9) is appealing because it includes only terms that are unambiguously associated with electrostatic, Pauli, or polarization energies. By grouping the terms related to the same type of energy, we can derive the following expressions:

$$
E_{Elec} = \int \widehat{v}_{N_A}\rho_B(\mathbf{r})d\mathbf{r} + \int \widehat{v}_{N_B}\rho_A(\mathbf{r})d\mathbf{r} + \iint \frac{\rho_A(\mathbf{r}_1)\rho_B(\mathbf{r}_2)}{|\mathbf{r}_2-\mathbf{r}_1|} d\mathbf{r}_1 d\mathbf{r}_2 + \sum_{I=1}^{N_A}\sum_{J=N_A+1}^{N} \frac{Z_I Z_J}{|\mathbf{R}_{IJ}|}
\tag{10}
$$

$$
\begin{aligned}
E_{Rep} = & -\frac{1}{2}\int \nabla^2 \Delta\rho_{Pau}(\mathbf{r},\mathbf{r}')_{\mathbf{r}_1=\mathbf{r}'_1} d\mathbf{r} + \int \widehat{v}_A\Delta\rho_{Pau}(\mathbf{r})d\mathbf{r} + \int \widehat{v}_B\Delta\rho_{Pau}(\mathbf{r})d\mathbf{r} \\
& + \frac{1}{2}\iint \frac{\Delta\rho_{Pau}(\mathbf{r}_1)\Delta\rho_{Pau}(\mathbf{r}_2)}{|\mathbf{r}_2-\mathbf{r}_1|} d\mathbf{r}_1 d\mathbf{r}_2
\end{aligned}
\tag{11}
$$

$$
E_{Exch} = \frac{1}{2}\iint \frac{\rho_{X,AB}(\mathbf{r}_1,\mathbf{r}_2)}{|\mathbf{r}_2-\mathbf{r}_1|} d\mathbf{r}_1 d\mathbf{r}_2
\tag{12}
$$

$$E_{Pol} = -\frac{1}{2}\int \nabla^2 \Delta\rho_{Pol}(\boldsymbol{r}_1,\boldsymbol{r}'_1)_{\boldsymbol{r}_1=\boldsymbol{r}'_1} d\boldsymbol{r} + \int \widehat{v}_A \Delta\rho_{Pol}(\boldsymbol{r})d\boldsymbol{r} + \int \widehat{v}_B \Delta\rho_{Pol}(\boldsymbol{r})d\boldsymbol{r}$$
$$+\frac{1}{2}\iint \frac{\Delta\rho_{Pol}(\boldsymbol{r}_1)\Delta\rho_{Pol}(\boldsymbol{r}_2)}{|\boldsymbol{r}_2-\boldsymbol{r}_1|} d\boldsymbol{r}_1 d\boldsymbol{r}_2 + \iint \frac{\Delta\rho_{Pau}(\boldsymbol{r}_1)\Delta\rho_{Pol}(\boldsymbol{r}_2)}{|\boldsymbol{r}_2-\boldsymbol{r}_1|} d\boldsymbol{r}_1 d\boldsymbol{r}_2$$
$$+\iint \frac{\Delta\rho_{XC}(\boldsymbol{r}_1,\boldsymbol{r}_2)}{|\boldsymbol{r}_2-\boldsymbol{r}_1|} d\boldsymbol{r}_1 d\boldsymbol{r}_2 \tag{13}$$

where E_{Rep} and E_{Exch} are, respectively, the repulsion and the intermolecular exchange parts of the Pauli energy. The separation of intermolecular exchange energy and repulsion is an advantage of using electron densities over other EDA schemes. It must be noticed that in Eqs. (11) and (13), we have merged the nuclear potentials with the electron potentials of the isolated molecules to get more compact expressions:

$$\widehat{v}_{A/B}(\boldsymbol{r}) = \widehat{v}_{N_{A/B}}(\boldsymbol{r}) + \int \frac{\rho_{A/B}(\boldsymbol{r}_2)}{|\boldsymbol{r}_2-\boldsymbol{r}_1|} d\boldsymbol{r}_2 \tag{14}$$

Equations (10)–(13) offer a straightforward and versatile approach to understanding the nature of any intermolecular interaction, serving as a valuable complement to the supermolecular method. The approach is user-friendly, requiring only three variational calculations, one for the complex and two for the isolated molecules. Alternatively, many-body perturbation theory methods, such as Møller-Plesset, can be employed. Its versatility lies in the fact that it can be employed with any electronic structure method capable of providing the 1-D and the corresponding 1-RDM, as will be shown further. This includes methods such as HF, post-HF, and KS-DFT.

The key point in the implementation of the DD-EDA scheme is the calculation of $\rho_{X,\ AB}(\boldsymbol{r}_1,\boldsymbol{r}_2)$ and $\Delta\rho_{Pau}(\boldsymbol{r})$. As mentioned before, both arise from the antisymmetrization of the product of the wavefunctions of the isolated molecules. Regarding the Hartree product, the antisymmetrized wavefunction incorporates both intermolecular electron exchange and intermolecular repulsion due to the overlap between electrons of the same spin, which is a physical manifestation of the Pauli exclusion principle. At SCF level, the XD can be expressed in terms of spin orbitals. In the case of a restricted HF calculation and an orthogonal spin orbital basis, it takes the following form:

$$\rho_X^{SCF}(\boldsymbol{r}_1,\boldsymbol{r}_2) = -2\sum_{i=1}^{NOO}\sum_{j=1}^{NOO} \varphi_i^O(\boldsymbol{r}_1)\varphi_j^O(\boldsymbol{r}_2)\varphi_j^O(\boldsymbol{r}_1)\varphi_i^O(\boldsymbol{r}_2) \tag{15}$$

where NOO refers to the number of occupied orbitals, and the superscript "O" on the orbitals denotes an orthogonal basis. For the case of a wavefunction constructed as an antisymmetrized product of the isolated molecules' wavefunctions, the XD can be expressed in terms of their molecular orbitals and partitioned into three different terms:

$$\rho_X^{SCF}(\boldsymbol{r}_1,\boldsymbol{r}_2) = -2\sum_{i=1}^{NOO_A}\sum_{j=1}^{NOO_A}\varphi_i(\boldsymbol{r}_1)\varphi_j(\boldsymbol{r}_2)\varphi_j(\boldsymbol{r}_1)\varphi_i(\boldsymbol{r}_2)$$
$$-2\sum_{i=1}^{NOO_B}\sum_{j=1}^{NOO_B}\varphi_i(\boldsymbol{r}_1)\varphi_j(\boldsymbol{r}_2)\varphi_j(\boldsymbol{r}_1)\varphi_i(\boldsymbol{r}_2) + \rho_{X,AB}^{SCF}(\boldsymbol{r}_1,\boldsymbol{r}_2) \tag{16}$$

The first two terms correspond to the intramolecular XDs, and the third term is the intermolecular XD. The calculation of the third term using the orbitals of the isolated molecules is not straightforward as they do not form a mutually orthogonal basis. Therefore, it is necessary to orthogonalize first these orbitals to represent the XD of the unrelaxed complex as in Eq. (15). In the current implementation of the intermolecular exchange energy, the Lowdin's symmetric orthogonalization [55] is employed for this task. Thus, denoting again the new set of orthogonal orbitals with the superscript "O" and isolating the intermolecular XD from Eq. (16), we obtain

$$\rho_{X,AB}^{SCF}(\boldsymbol{r}_1,\boldsymbol{r}_2) = -2\sum_{i=1}^{NOO}\sum_{j=1}^{NOO}\varphi_i^O(\boldsymbol{r}_1)\varphi_j^O(\boldsymbol{r}_2)\varphi_j^O(\boldsymbol{r}_1)\varphi_i^O(\boldsymbol{r}_2)$$
$$+2\sum_{i=1}^{NOO_A}\sum_{j=1}^{NOO_A}\varphi_i(\boldsymbol{r}_1)\varphi_j(\boldsymbol{r}_2)\varphi_j(\boldsymbol{r}_1)\varphi_i(\boldsymbol{r}_2) + 2\sum_{i=1}^{NOO_B}\sum_{j=1}^{NOO_B}\varphi_i(\boldsymbol{r}_1)\varphi_j(\boldsymbol{r}_2)\varphi_j(\boldsymbol{r}_1)\varphi_i(\boldsymbol{r}_2) \tag{17}$$

In practice, the calculations are done on the basis of atomic orbitals, which are represented by basis functions located on the corresponding atoms. The molecular orbitals are then constructed as linear combinations of these basis functions, χ_μ:

$$\varphi_i = \sum_{\mu=1}^{NBF} C_{\mu,i}\chi_\mu \tag{18}$$

where NBF refers to the number of basis functions and $C_{\mu,\, i}$ is the coefficient of the function μ in the orbital i. Defining the elements of the 1-RDM as

$$D_{\mu\nu} = \sum_{i=1}^{NOO} C_{\mu,i}C_{\nu,i} \tag{19}$$

the intermolecular XD can be more compactly represented in terms of the unique set of basis functions of the complex (Eq. 20), instead of three different sets of orbitals (Eq. 17):

$$\rho_{X,AB}^{SCF}(\boldsymbol{r}_1,\boldsymbol{r}_2)$$
$$= -2\sum_{\mu=1}^{NBF}\sum_{\nu=1}^{NBF}\sum_{\lambda=1}^{NBF}\sum_{\sigma=1}^{NBF}\left(D_{\mu\nu}^O D_{\lambda\sigma}^O - D_{\mu\nu}^A D_{\lambda\sigma}^A - D_{\mu\nu}^B D_{\lambda\sigma}^B\right)\chi_\mu(\boldsymbol{r}_1)\chi_\nu(\boldsymbol{r}_2)\chi_\lambda(\boldsymbol{r}_1)\chi_\sigma(\boldsymbol{r}_2) \tag{20}$$

In Eq. (20), the superscript "*O*" in the 1-RDM elements means that they were obtained from the set of orthogonalized orbitals.

In the same way, the Pauli deformation density can be expressed in terms of basis functions and the 1-RDMs of the unrelaxed complex and isolated molecules:

$$\Delta\rho_{Pau}^{SCF}(\mathbf{r}) = 2\sum_{\mu=1}^{NBF}\sum_{\nu=1}^{NBF}\left(D_{\mu\nu}^{O} - D_{\mu\nu}^{A} - D_{\mu\nu}^{B}\right)\chi_{\mu}(\mathbf{r})\chi_{\nu}(\mathbf{r}) \quad (21)$$

For multideterminant post-SCF wavefunctions, the "exact" calculation of the intermolecular XD is a much more complicated task. It requires the handling of the two-electron reduced density matrix (2-RDM), which is impracticable for large molecules. As an alternative to the computation of 2-RDMs, the XCD can be approximated in terms of natural orbitals and their occupation numbers. In the current implementation of the DD-EDA, the approximate form of the XCD derived from the Pauli exclusion principle (Eq. 22) [56] is employed, which is exact for two-electron systems and was found to provide accurate results in atomic energy calculations [57]:

$$\rho_{X}^{SCF}(\mathbf{r}_1,\mathbf{r}_2) = -2\sum_{i=1}^{NNO}\sum_{j=1}^{NNO}\left(n_i^O\right)^{1/2}\left(n_j^O\right)^{1/2}\varphi_i^O(\mathbf{r}_1)\varphi_j^O(\mathbf{r}_2)\varphi_j^O(\mathbf{r}_1)\varphi_i^O(\mathbf{r}_2) \quad (22)$$

In Eq. 22, NNO refers to the number of natural orbitals. This approximate form of the XCD incorporates exchange and correlation effects that are implicitly included in the natural orbitals and their weights (occupation numbers) while also ensuring the *N*-representability of the XCD. Other forms are possible, such as those derived from the natural orbital functional theory (NOFT), although they have not been tested to date.

Similarly to the case of single-determinant wavefunctions, the intermolecular XD for multideterminant wavefunctions may be written as

$$\begin{aligned}\rho_{X,AB}^{SCF}(\mathbf{r}_1,\mathbf{r}_2) = & -2\sum_{i=1}^{NNO}\sum_{j=1}^{NNO}\left(n_i^O\right)^{1/2}\left(n_j^O\right)^{1/2}\varphi_i^O(\mathbf{r}_1)\varphi_j^O(\mathbf{r}_2)\varphi_j^O(\mathbf{r}_1)\varphi_i^O(\mathbf{r}_2) \\ & +2\sum_{i=1}^{NNO_A}\sum_{j=1}^{NNO_A} n_i^{1/2}n_j^{1/2}\varphi_i(\mathbf{r}_1)\varphi_j(\mathbf{r}_2)\varphi_j(\mathbf{r}_1)\varphi_i(\mathbf{r}_2) \\ & +2\sum_{i=1}^{NNO_B}\sum_{j=1}^{NNO_B} n_i^{1/2}n_j^{1/2}\varphi_i(\mathbf{r}_1)\varphi_j(\mathbf{r}_2)\varphi_j(\mathbf{r}_1)\varphi_i(\mathbf{r}_2)\end{aligned} \quad (23)$$

In the multideterminant scenario, the orthogonalization procedure is applied to natural orbitals with non-integer occupations, which necessitates the transformation of the occupation numbers as well. This transformation can be accomplished using a weighted orthogonalization method. The weighted orthogonalization currently utilized in the DD-EDA is similar to the approach developed by Weinhold et al. [58, 59]. It involves an additional diagonalization of the occupation matrix to obtain the final set of orthogonalized natural orbitals.

The intermolecular XD and Pauli deformation density can also be represented more compactly using basis functions along with the corresponding 1-RDMs. Equations (20) and (21) can be extended to post-SCF calculations, but with the matrix elements of Eqs. (24) and (25) for $\rho_{X,\,AB}(\boldsymbol{r}_1,\boldsymbol{r}_2)$ and $\Delta\rho_{Pau}(\boldsymbol{r})$, respectively:

$$D_{\mu\nu} = \sum_{i=1}^{NNO} n_i^{1/2} C_{\mu,i} C_{\nu,i} \tag{24}$$

$$D_{\mu\nu} = \sum_{i=1}^{NNO} n_i C_{\mu,i} C_{\nu,i} \tag{25}$$

The calculation of the terms corresponding to polarization energy requires, additionally, the knowledge of $\Delta\rho_{Pol}(\boldsymbol{r})$ and $\Delta\rho_{XC}(\boldsymbol{r}_1,\boldsymbol{r}_2)$. The first one, the deformation density associated with electron polarization, is easily obtained by subtraction of the Pauli deformation density and the 1-Ds of the isolated molecules from the 1-D of the complex:

$$\Delta\rho_{Pol}(\boldsymbol{r}) = \rho(\boldsymbol{r}) - \rho_A(\boldsymbol{r}) - \rho_B(\boldsymbol{r}) - \Delta\rho_{Pau}(\boldsymbol{r}) \tag{26}$$

On the contrary, the calculation of $\Delta\rho_{XC}(\boldsymbol{r}_1,\boldsymbol{r}_2)$ for post-SCF methods presents the same challenge as the calculation of $\rho_{X,\,AB}(\boldsymbol{r}_1,\boldsymbol{r}_2)$, the requirement for 2-RDMs. To address this issue, we can introduce once again approximated expressions for the exchange-correlation density (XCD) based on natural orbitals and occupation numbers. Alternatively, we can derive the last term of Eq. (13) as the difference between the total interaction energy and the sum of the other terms. The latter approach is the one utilized in the current DD-EDA implementation due to its simplicity and anticipated accuracy, especially when working with large systems.

2.2 *Illustrative Examples*

Table 1 highlights the performance of the DD-EDA by comparing the first-order interaction energy components – electrostatic and exchange-repulsion – obtained using SAPT(0) [60, 61] with those calculated via DD-EDA [62] across a series of small complexes. The comparison is conducted at both SCF and coupled-cluster singles and doubles (CCSD) levels of theory, employing a triple-zeta basis set augmented with double-d and double-p polarization functions, as well as diffuse functions in all atoms for both the SAPT(0) and DD-EDA calculations. Geometries were taken from the database HB6/04 [63–65] for $(H_2O)_2$ and $(HF)_2$, from the database CT7/04 [63–65] for NH_3-Cl_2 and H_2O-ClF, from the database DI6/04

Table 1 Comparison of electrostatic and exchange-repulsion (Pauli) energies for some intermolecular complexes obtained at SCF and CCSD levels using the DD-EDA scheme and the SAPT(0) method. The basis set employed for the calculations was the 6-311++G(2d,2p), except for benzene dimer 6-311++G(d,p). All values in kJ/mol

		E_{Elec}			$E_{Exch} + E_{Rep}$		
		DD-EDA	SAPT	%Diff	DD-EDA	SAPT	%Diff
$(H_2O)_2$	SCF	−33.51	−33.51	0.00	26.69	26.40	1.06
	CCSD	−32.64	−32.68	0.14	33.72	31.46	7.20
$(HF)_2$	SCF	−26.43	−26.42	0.03	18.33	18.20	0.70
	CCSD	−25.19	−25.19	0.01	24.60	22.84	7.87
$(CH_4)_2$	SCF	−0.84	−0.84	0.01	3.22	3.22	0.63
	CCSD	−0.92	−0.96	3.93	3.51	3.72	5.16
$(HCl)_2$	SCF	−10.70	−10.70	0.00	15.34	15.24	0.64
	CCSD	−10.35	−10.40	0.48	16.96	16.59	2.19
$HCl\text{-}H_2S$	SCF	−22.51	−22.51	0.01	30.06	29.75	1.05
	CCSD	−21.52	−22.66	0.67	30.29	31.41	3.56
$H_2O\text{-}ClF$	SCF	−43.98	−43.99	0.02	47.23	46.67	1.19
	CCSD	−43.52	−42.89	1.47	58.28	56.29	3.49
$NH_3\text{-}Cl_2$	SCF	−48.20	−48.21	0.00	63.34	62.39	1.51
	CCSD	−48.80	−48.61	0.48	75.09	69.28	8.39
$(C_6H_6)_2$	SCF	−3.91	−3.91	0.09	20.04	19.86	0.88
	CCSD	−4.33	−4.39	1.39	16.88	18.74	9.93

[63–65] for $(HCl)_2$ and $HCl\text{-}H_2S$, from the database WI7/05 [64–66] for $(CH_4)_2$, and from the database PPS5/05 [64–66] for $(C_6H_6)_2$ (parallel-displaced structure).

The results presented in Table 1 demonstrate a strong correspondence between DD-EDA and SAPT(0) for electrostatic and exchange-repulsion energies. In the DD-EDA scheme, the latter is calculated as the sum of exchange and repulsion terms. As anticipated, the agreement is most accurate at the SCF level, with maximum differences of 0.09% for electrostatic energy and 1.51% for exchange-repulsion energy.

At the CCSD level, the differences in electrostatic energy remain negligible, except for the methane dimer, which shows a discrepancy of 3.93%. Larger deviations are observed for exchange-repulsion energies at the CCSD level, with the largest difference, 9.93%, occurring in the benzene dimer. This discrepancy likely stems from the use of approximate exchange density matrices in the DD-EDA expressions at post-SCF levels. Nevertheless, the relative differences in CCSD exchange-repulsion energies between SAPT(0) and DD-EDA remain small, which justifies the significant computational efficiency gained by using approximated exchange density expressions based on natural orbitals.

In addition to evaluating the ability of DD-EDA to reproduce SAPT(0) first-order energy terms with negligible computational cost, this section examines the other energy components derived using the DD-EDA scheme. Table 2 summarizes the electrostatic energy, the separated exchange and repulsion contributions to the Pauli

Table 2 Interaction energy components for some intermolecular complexes obtained at SCF and CCSD levels using the DD-EDA scheme. For the exchange and repulsion energies, the percentage of contribution to the Pauli energy is shown in parenthesis. The basis set employed for the calculations was the 6-311++G(2d,2p), except for benzene dimer 6-311++G(d,p). All values in kJ/mol

		E_{Elec}	E_{Exch}	E_{Rep}	E_{Pol}	E_{Int}
$(H_2O)_2$	SCF	−33.51	−44.90 (38.6)	71.54 (61.4)	−8.83	−15.71
	CCSD	−32.64	−56.96 (38.6)	90.76 (61.4)	−19.18	−18.00
$(HF)_2$	SCF	−26.43	−33.39 (39.2)	51.71 (60.8)	−7.87	−15.98
	CCSD	−25.19	−46.65 (39.6)	71.30 (60.4)	−15.69	−16.23
$(CH_4)_2$	SCF	−0.84	−7.84 (41.5)	11.06 (58.5)	−0.17	2.20
	CCSD	−0.92	−12.58 (44.1)	15.93 (55.9)	−3.21	−0.82
$(HCl)_2$	SCF	−10.70	−25.86 (38.6)	41.19 (61.4)	−5.35	−0.71
	CCSD	−10.35	−35.40 (40.3)	52.36 (59.7)	−10.85	−4.24
$HCl\text{-}H_2S$	SCF	−22.51	−41.84 (36.8)	71.90 (63.2)	−11.95	−4.40
	CCSD	−21.52	−58.42 (39.7)	88.71 (60.3)	−18.15	−9.38
$H_2O\text{-}ClF$	SCF	−43.98	−64.68 (37.3)	108.70 (62.7)	−16.50	−9.45
	CCSD	−43.52	−94.96 (38.6)	151.18 (61.4)	−32.16	−13.56
$NH_3\text{-}Cl_2$	SCF	−48.20	−80.69 (35.9)	144.02 (64.1)	−21.97	−6.84
	CCSD	−48.80	−100.48 (36.4)	175.58 (63.6)	−39.79	−13.50
$(C_6H_6)_2$	SCF	−3.91	−40.24 (40.0)	60.28 (60.0)	−2.49	13.64
	CCSD	−4.33	−54.70 (43.3)	71.58 (56.7)	−15.26	−2.71

energy, and the polarization energy for the same complexes listed in Table 1, computed at both the SCF and CCSD levels.

The similar values observed for the electrostatic energy at the SCF and CCSD levels suggest that intramolecular electron correlation has minimal impact on this term. In contrast, the inclusion of intramolecular electron correlation significantly affects the magnitudes of the other first-order energy terms, namely exchange and repulsion energies. It is important to note that, in closed-shell intermolecular complexes, intermolecular exchange is always an attractive term, resulting in a negative value, whereas intermolecular spin repulsion is repulsive and therefore positive. Their sum, referred to as the exchange-repulsion or Pauli energy, is positive and, as shown in Table 1, increases when intramolecular electron correlation is accounted for in the calculations with the exception of benzene dimer. Furthermore, the values in Table 2 indicate that the exchange energy decreases, while the repulsion energy increases upon incorporating electron correlation effects.

The separate analysis of the exchange and repulsion energies provides additional insight into the role of Pauli energy in the stability of a given complex. As shown in Table 2, the exchange energy is the most negative term, making it the dominant stabilizing factor compared to electrostatic and polarization energies. Additionally, notable differences in the relative contributions of exchange and repulsion within the Pauli energy can be observed in Table 2. Thus, the exchange energy percentage increases in methane and benzene dimers with respect to the rest, mainly at correlated level, underscoring the significance of electron correlation in the stability of dispersion-bound complexes. The substantial increase in the relative contribution of exchange energy from the SCF to post-SCF levels explains the decrease in Pauli energy observed in the benzene dimer when intramolecular electron correlation is taken into account in the calculation.

Polarization energy is even more influenced by the inclusion of electron correlation in the calculations. This outcome is expected, as dispersion energy, which is entirely absent at the SCF level, relies on long-range electron correlation effects. Consequently, polarization energies computed at the CCSD level are consistently much lower than their SCF counterparts. This trend is particularly notable for complexes involving nonpolar molecules, such as methane and benzene dimers. In these cases, SCF polarization energies are negligible, resulting in globally unstable complexes, as indicated by the total interaction energy values. Conversely, at the CCSD level, polarization energy plays a crucial role in stabilizing these complexes, exhibiting large negative values in comparison to the electrostatic energy. However, it is important to note that the total interaction energy for the benzene dimer, calculated at the CCSD level, is -2.71 kJ/mol. This value significantly deviates from the reported value of -10.38 kJ/mol, highlighting the need for a more accurate computational approach for this complex, incorporating triple excitations in the CC expansion and employing a larger basis set.

Qualitative insights can be obtained from the spatial representations of the spin repulsion and polarization densities. Figure 1 displays these density plots for the previously discussed complexes at both the SCF and CCSD levels. A general trend reveals that negative spin repulsion density is concentrated in the overlapping region between the two molecules.

Notably, this overlapping region expands at the CCSD level, which can be attributed to the influence of intramolecular electron correlation. The incorporation of electron correlation causes the electron densities of each molecule to spread further, occupying a larger spatial region. This effect arises from the increase in the mean electron-electron distance. As a result, the electron overlap between the molecules also increases, thereby enhancing intermolecular spin repulsion and exchange interactions.

Significant differences in polarization densities are also observed between the SCF and CCSD levels. At the SCF level, complexes formed by polar molecules exhibit a pattern of alternating positive and negative polarization regions along each molecule. In the intermolecular region, this pattern includes a positive and a negative lobe, each positioned near the tip of a molecule, reflecting the influence of inductive and charge transfer effects. Conversely, in complexes formed by nonpolar molecules, such as methane and benzene dimers, where inductive and charge transfer effects are negligible, two negative lobes are observed in the intermolecular region, each located near the tip of a molecule.

At the CCSD level, the polarization density pattern for highly polar complexes, such as water or hydrogen fluoride dimers, remains largely similar to that observed at the SCF level. However, as the polarity of the molecules decreases and dispersion forces become more significant, the positive lobe in the intermolecular region grows in size and shifts slightly from the tip of the molecule toward the center of the

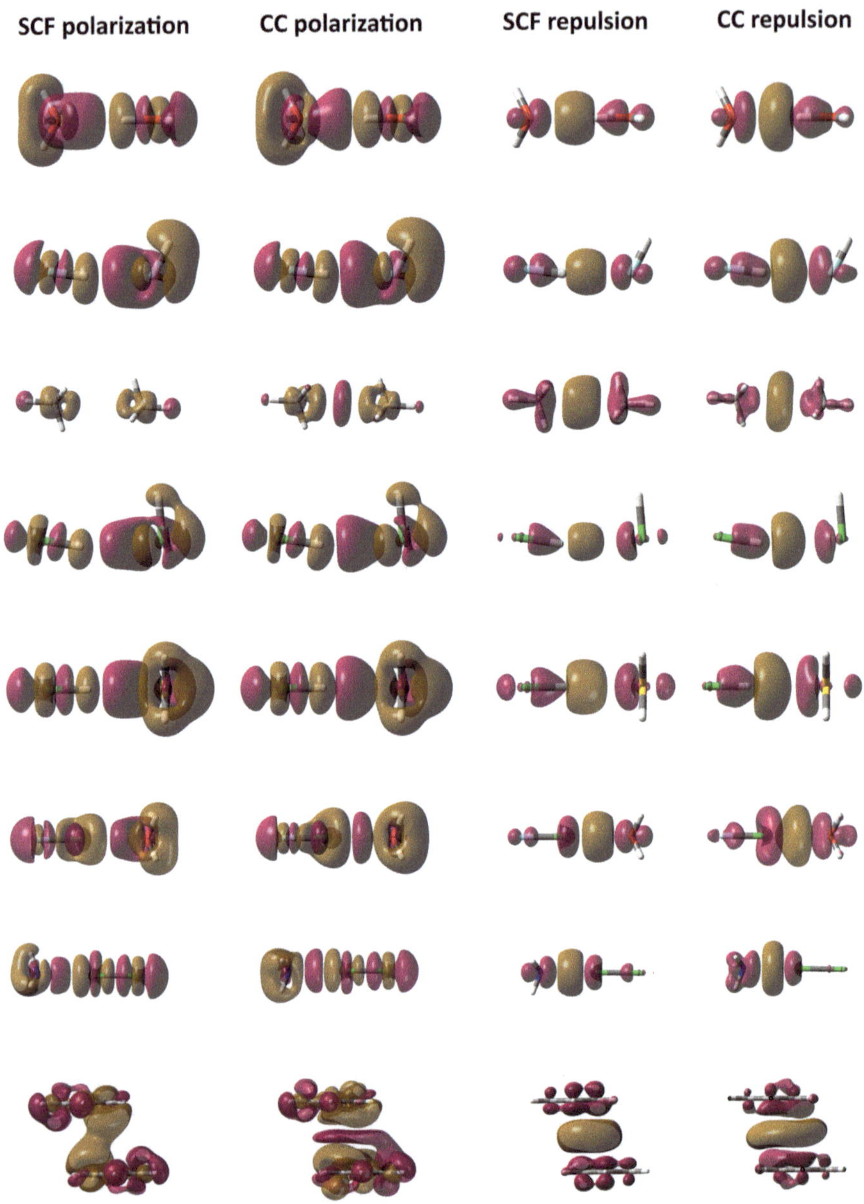

Fig. 1 Plots of polarization and spin repulsion densities obtained at the SCF and CC levels for the intermolecular complexes of Tables 1 and 2. From left to right: SCF polarization, CC polarization, SCF spin repulsion, and CC spin repulsion densities. From top to bottom: $(H_2O)_2$, $(HF)_2$, $(CH_4)_2$, $(HCl)_2$, HCl-H_2S, H_2O-ClF, NH_3-Cl_2, and $(C_6H_6)_2$. The densities are shown for an isosurface value of 5×10^{-4} a.u., with the exception of $(CH_4)_2$ where an isosurface value of 5×10^{-5} is used. Magenta and brown colors represent positive and negative densities, respectively

complex. In nonpolar complexes, where no positive lobes are present in the intermolecular region at the SCF level, the CCSD level reveals a prominent positive polarization region at the center of the complex, flanked by two negative regions at the tips of the molecules. This sandwich-like polarization density pattern is indicative of dispersion interactions. As discussed in the following section, it also offers a valuable tool for assessing the reliability of DFT functionals in accurately accounting for dispersion forces in intermolecular complexes.

3 Hybrid DD-EDA

3.1 Theoretical Framework

One significant limitation of the original DD-EDA scheme is its inability to quantify the contribution of dispersion to the polarization energy. As a result, this partitioning scheme becomes less interesting for studying some systems of great interest, such as weak dispersion complexes formed by small, low-polarity molecules, or complexes involving large systems where dispersion, in general, plays a crucial role. In this second category of complexes, we can mention, for instance, adsorption complexes and those that involve interactions with biological macromolecules, including proteins, lipids, or DNA strands.

To enhance the applicability of the DD-EDA to dispersion complexes, a perturbational scheme was designed to extract the induction energy from the polarization energy. At the second-order perturbation theory (PT) level, polarization can be divided into pure induction and dispersion energies [2]. By integrating our EDA scheme with second-order PT, the induction energy can be separated from the remaining components of polarization [40]. The starting point is the expression of the induction energy obtained by the Rayleigh-Schrödinger PT [3]:

$$E_{ind} = \sum_{m \neq 0} \frac{\left[\int \widehat{v}_A \rho_B^{m0}(\boldsymbol{r}) d\boldsymbol{r} \right]^2}{E_B^m - E_B^0} + \sum_{n \neq 0} \frac{\left[\int \widehat{v}_B \rho_A^{n0}(\boldsymbol{r}) d\boldsymbol{r} \right]^2}{E_A^n - E_A^0} \tag{27}$$

where the notations ρ_A^{n0} and ρ_B^{m0} have been introduced to denote induced transition densities within molecules A and B from ground-state configuration 0 to excited-state configurations m or n. In the equations presented above, the denominators include the energies associated with these transitions. The induction energy represents a second-order term in the perturbative expansion of the interaction energy. The first-order correction to the electron density, which is the only term necessary for calculating the second-order energy, is given by [67]

$$\Delta\rho(\boldsymbol{r}) = 2\sum_{m\neq 0} \frac{\int \widehat{v}_B \rho_A^{n0}(\boldsymbol{r})d\boldsymbol{r}}{E_A^n - E_A^0} \rho_A^{n0} + 2\sum_{n\neq 0} \frac{\int \widehat{v}_A \rho_B^{m0}(\boldsymbol{r})d\boldsymbol{r}}{E_B^m - E_B^0} \rho_B^{m0} \tag{28}$$

This first-order density is unambiguously split into contributions from A and B and, since it represents the electron polarization, it is the equivalent to $\Delta\rho_{Pol}(\boldsymbol{r})$ in the DD-EDA scheme. Thus, replacing $\Delta\rho_{Pol}(\boldsymbol{r})$ by $\Delta\rho(\boldsymbol{r})$, we can straightforwardly extract the induction energy from the second and third terms of Eq. (13). These two terms have been denoted as charge induction energy in Ref. [40]. Following a few simple steps, consisting of identifying the induction energy given by Eq. (27) within the charge induction energy and replacing back the definition of first-order density given by Eq. (28), the following expression is obtained for the induction energy:

$$E_{Ind} = \frac{1}{2}\left[\int \widehat{v}_A \Delta\rho_B(\boldsymbol{r})d\boldsymbol{r} + \int \widehat{v}_B \Delta\rho_A(\boldsymbol{r})d\boldsymbol{r}\right] \tag{29}$$

It should be mentioned that the induction energy given by Eq. (29) resembles the classical definition of the induction energy, where half of the energy is lost during the charge induction process [68]. On the other hand, it must be recalled that a “pure” induction energy term is only obtained at second-order level in the perturbation expansion of the intermolecular interaction energy. Third- and higher-order levels also include induction energy terms and other terms where induction and dispersion cannot be separated. The induction energy of Eq. (29) somehow incorporates higher-order effects as the polarization density in the DD-EDA is not necessarily obtained perturbationally but using variational methods that incorporate terms up to infinite order, such as configuration interaction or coupled-cluster methods.

Since Eq. (29) gives us the induction energy, the rest of polarization energy obtained from Eq. (13) corresponds to dispersion, at least at second-order perturbation level. For higher-order levels or for electron polarizations derived from variational methods, the induction and dispersion energies serve only as approximations of these quantities, which cannot be precisely separated. Thus, the energy decomposition using the hybrid variational-perturbational DD-EDA scheme remains as

$$E_{int} = E_{Elec} + E_{Rep} + E_{Exch} + E_{Ind} + E_{Res-Pol} \tag{30}$$

where the term $E_{Res-Pol}$ includes the second-order dispersion energy along with the less significant higher-order polarization energy terms.

The calculation of induction and dispersion in the DD-EDA requires the definition of molecular domains for the electron density. This definition can be established through an atomic partition of either real space or the Hilbert space of the basis functions. While real space partitions offer advantages over Hilbert space partitions, such as reduced dependence on the basis set, they are significantly more computationally demanding, which limits their applicability in systems. In the current implementation of induction energy within the DD-EDA, a Mulliken-like approach is utilized to define the molecular space. Consequently, it is crucial to carefully select the basis set when separating polarization into induction and dispersion. It is advisable to avoid using diffuse functions for this purpose or, at the very least, to verify that any differences compared to basis sets that do not include diffuse functions are small.

3.2 Illustrative Examples

The decomposition of polarization density into induction and dispersion components has been applied to large complexes where dispersion plays a critical role in stability. Due to the significant computational demands, these large complexes can only be effectively studied in conjunction with DFT. In this context, adsorption complexes on graphene-like surfaces are employed as examples to demonstrate the performance of the hybrid DD-EDA scheme.

The first step will be to show that the hybrid DD-EDA method can be employed to evaluate the capability of DFT functionals to implicitly account for dispersion effects, *i.e.*, without relying on empirical dispersion corrections, which correct the energy but leave the electron density unaffected. To conduct this analysis, complexes formed by a pyridine molecule and different graphene models of increasing size were investigated in Ref. [40]. These graphene models consist of carbon nanodisks with D_{6h} symmetry, saturated with hydrogens. Each model is identified by the number of carbon atoms in the nanodisk (**C24**, **C54**, **C96**, and **C150**).

The energy terms from the hybrid DD-EDA scheme, obtained through HF calculations and calculations using various DFT functionals, are summarized in Table 3. All calculations incorporate the D3 empirical dispersion correction developed by Grimme et al. [16], and its contribution to the total interaction energy is also presented in the table.

The total interaction energy increases (in absolute value) with the increasing size of the graphene model, as expected, and approaches near-converged values for all energy terms with the **C96** model. The differences in interaction energy between the **C96** and **C150** models are approximately 1 kcal/mol. To ensure that the observed differences among various functionals are solely attributable to changes in electron density, all calculations were performed using geometries obtained with the M06-2X functional.

As shown in Table 3, the differences in electrostatic, exchange, and repulsion energies among the various functionals are relatively small, even when compared to HF values. However, significant differences are observed in polarization energies, which are primarily attributed to dispersion, the most substantial contribution to the term labeled as Res-Pol in the table. Induction energies are generally small and exhibit minimal variation across different functionals, with HF values being consistently lower in all complexes. The Res-Pol results clearly highlight that certain functionals almost entirely neglect dispersion effects, as their values closely resemble the HF results. Examples of such functionals include B97 and BLYP.

Table 3 Interaction energy components (in kcal/mol) for the adsorption of pyridine on different graphene surface models obtained at different computational levels and the 6-31G(d,p) basis set

	HF-D	BLYP-D	B97-D	B3LYP-D	ωB97X-D[a]	M06-2X-D
C24						
Electrostatic	−7.75	−7.88	−7.01	−7.35	−6.47	−6.72
Pauli	21.31	22.52	21.43	21.85	20.74	20.78
Exchange	−31.66	−31.51	−29.35	−30.61	−28.95	−29.48
Repulsion	52.97	54.03	50.78	52.45	49.68	50.26
Polarization	−2.11	−3.61	−2.96	−5.70	−10.00	−19.15
Induction	−0.65	−1.13	−1.19	−1.05	−1.09	−1.07
Res-Pol	−1.46	−2.48	−1.77	−4.65	−8.91	−18.08
D3-corr	−16.92	−18.09	−17.49	−14.55	−11.45	−1.33
Total	−5.46	−7.06	−6.04	−5.76	−7.18	−6.43
C54						
Electrostatic	−9.54	−9.27	−8.41	−8.81	−8.02	−8.31
Pauli	22.88	24.11	22.94	23.39	22.19	22.25
Exchange	−32.93	−32.89	−30.59	−31.92	−30.14	−30.77
Repulsion	55.82	57.00	53.53	55.31	52.33	53.02
Polarization	−2.28	−3.62	−2.99	−5.91	−10.52	−20.46
Induction	−0.75	−1.30	−1.37	−1.20	−1.25	−1.21
Res-Pol	−1.53	−2.32	−1.62	−4.71	−9.27	−19.25
D3-corr	−19.95	−21.18	−20.33	−17.34	−13.92	−2.14
Total	−8.90	−9.96	−8.79	−8.66	−10.27	−8.66
C96						
Electrostatic	−9.65	−9.47	−8.59	−8.99	−8.19	−8.50
Pauli	22.98	24.23	23.05	23.51	22.28	22.34
Exchange	−32.83	−32.80	−30.48	−31.81	−30.03	−30.64
Repulsion	55.81	57.03	53.53	55.32	52.31	52.97
Polarization	−2.39	−3.59	−3.00	−5.91	−10.54	−20.42
Induction	−0.81	−1.35	−1.42	−1.26	−1.31	−1.28
Res-Pol	−1.58	−2.24	−1.58	−4.65	−9.23	−19.14
D3-corr	−20.29	−21.50	−20.61	−17.72	−14.33	−2.55
Total	−9.35	−10.32	−9.15	−9.10	−10.78	−9.13
C150						
Electrostatic	−8.81	−8.75	−7.92	−8.27	−7.53	−7.82
Pauli	20.68	21.74	20.70	21.10	20.00	20.04
Exchange	−30.00	−29.91	−27.79	−29.00	−27.33	−27.91
Repulsion	50.68	51.66	48.49	50.10	47.34	47.95
Polarization	−2.08	−2.75	−2.37	−5.05	−9.54	−18.94
Induction	−0.72	−1.19	−1.25	−1.11	−1.16	−1.16
Res-Pol	−1.36	−1.56	−1.12	−3.94	−8.38	−17.78
D3-corr	−19.92	−21.06	−19.99	−17.42	−14.22	−2.70
Total	−10.13	−10.83	−9.57	−9.65	−11.29	−9.41

Reproduced from Ref. [40] with permission from the PCCP Owner Societies

[a] This functional uses an own version of DFT-D2 correction instead of DFT-D3

Conversely, other functionals demonstrate improved performance with respect to dispersion. For instance, the B3LYP functional accounts for some dispersion implicitly, resulting in lower Res-Pol energies and requiring a smaller D3 dispersion correction. The ωB97X functional performs even better, showing significantly lower Res-Pol energies and further reduced D3 corrections. Among all the functionals, M06-2X exhibits the best performance by far. The Res-Pol energies predicted by this functional are much lower than those of the others, and the D3 correction accounts for only about 25% of the total interaction energy. In contrast, the rest of the functionals fail to predict stable complexes without the inclusion of the empirical dispersion correction.

It is insightful to investigate whether the significant differences in polarization energies obtained with the DFT functionals are also reflected in the polarization density plots. The polarization densities, calculated using HF, B3LYP, ωB97X, and M06-2X functionals, are presented in Fig. 2 for the complex formed by pyridine with the largest graphene model. As observed, the characteristic positive electron polarization in the intermolecular region – commonly seen in dispersion-dominated complexes, as discussed in the previous section – is barely noticeable in the HF results, somewhat apparent in B3LYP, notably more pronounced in ωB97X, and clearly evident in M06-2X. These plots suggest that the large dispersion energy predicted by the M06-2X functional arises from its accurate description of the polarization density.

It is also worthwhile to compare the polarization densities obtained with different functionals to those derived from post-SCF methods, such as MP2 or CCSD. As shown in Fig. 3 for the pyridine-benzene complex, the polarization density plots from MP2 and CCSD calculations closely resemble those obtained using the M06-2X functional. In contrast, significant differences are observed in the intermolecular region when comparing these results to those from HF, B97, and B3LYP functionals. For the pyridine-**C24** complex, the comparison was made solely at the MP2 level due to the high computational cost of CCSD. The results for this complex are consistent, with MP2 and M06-2X showing good agreement in the intermolecular region and notable discrepancies when compared to the other functionals.

One of the key strengths of the DD-EDA approach is its low computational cost. This enables the decomposition of the interaction energy in very large complexes with minimal computational resources and relatively short processing times, even when using a single processor. Such efficiency is crucial for studies involving large biomolecules. In this context, Ref. 41 presents an analysis of the interaction energy terms for complexes formed by various biochemically relevant molecules adsorbed on the **C96** graphene model. The selected molecules represent systems of increasing size, including glycine (**Gly**), melamine (**Mel**), pyronine cation (**Pyr**), porphine (**PP**), phthalocyanine (**H_2Pc**), and tetrabenzoporphine (**TBPP**). These molecules can adsorb onto the graphene surface in three possible orientations: a stacking configuration (**A**) and up to two perpendicular configurations (**B** and **C**). Since the stacking configuration is by far the most stable, our analysis here will be focused exclusively on this conformation. Further details about the geometries of the

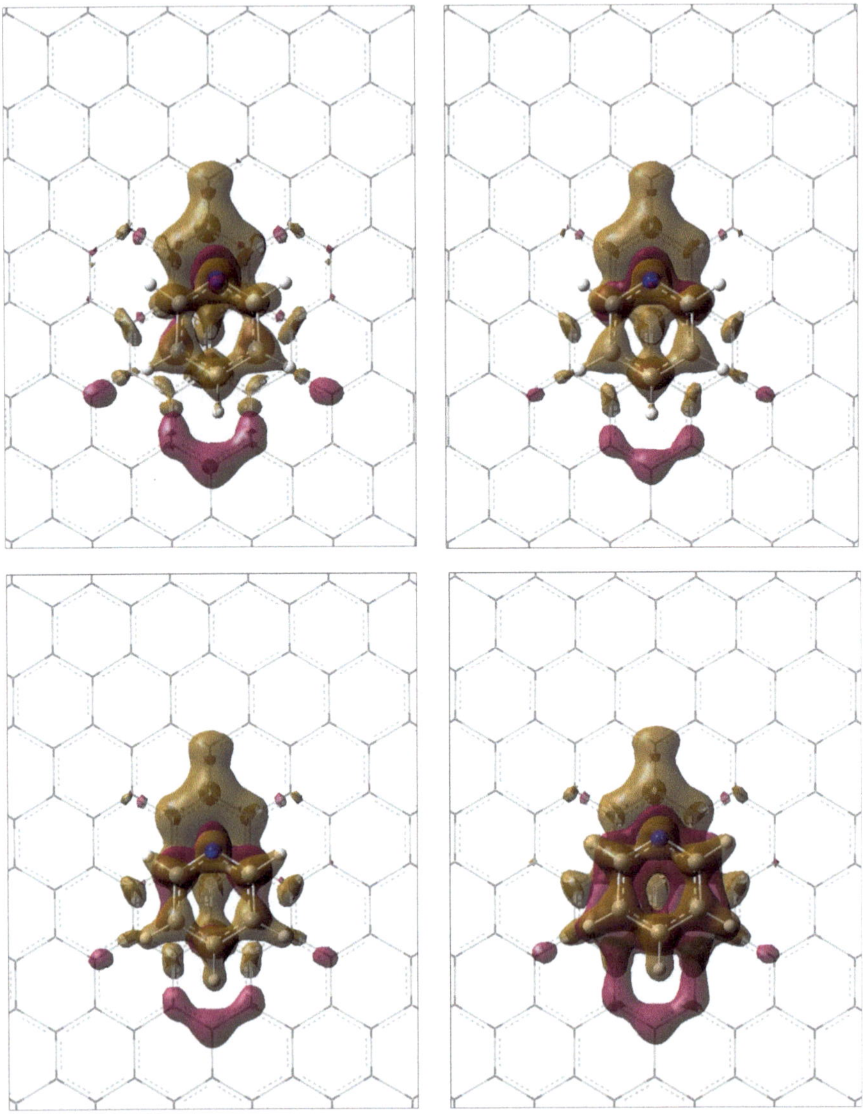

Fig. 2 Electron polarization densities for the adsorption of pyridine on the **C150** graphene model obtained with HF-D (upper left), B3LYP-D (upper right), ωB97X-D (bottom left), and M06-2X-D (bottom right) in combination with the 6-31G(d,p) basis set. The density is shown for an isosurface value of 2×10^{-4} a.u. Magenta and brown colors represent positive and negative densities, respectively. (Reproduced from Ref. [40] with permission from the PCCP Owner Societies)

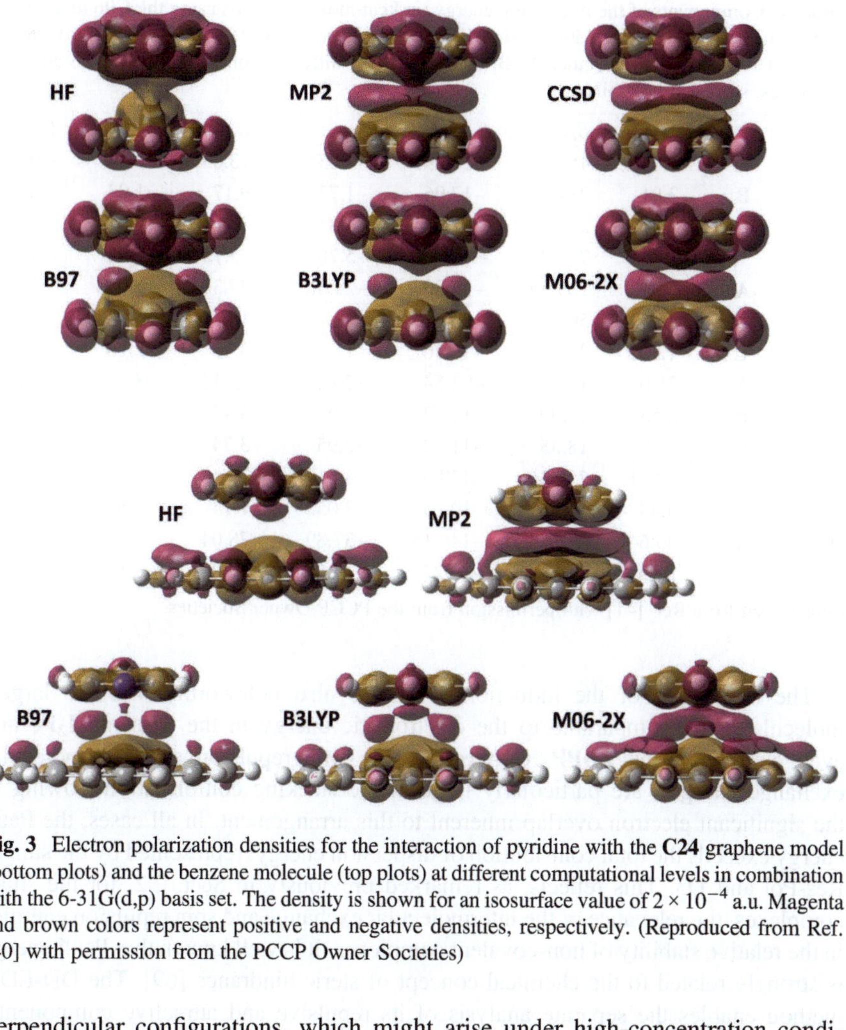

Fig. 3 Electron polarization densities for the interaction of pyridine with the **C24** graphene model (bottom plots) and the benzene molecule (top plots) at different computational levels in combination with the 6-31G(d,p) basis set. The density is shown for an isosurface value of 2×10^{-4} a.u. Magenta and brown colors represent positive and negative densities, respectively. (Reproduced from Ref. [40] with permission from the PCCP Owner Societies)

perpendicular configurations, which might arise under high-concentration conditions, and a comprehensive discussion of their stability, can be found in the original work [41].

The energy terms are summarized in Table 4, calculated using densities obtained with the M06-2X functional. The table also includes the D3 dispersion correction. All the stacking complexes exhibit significant stability, with total interaction energies for the largest molecules (**H_2Pc** and **TBPP**) approaching the magnitude of a covalent bond. As expected, dispersion is the dominant term in all stacking conformations, with Res-Pol values exceeding twice the electrostatic energy in most cases. The sole exception is the pyronine cation, where the electrostatic and Res-Pol terms are comparable. This behavior arises primarily from the positive charge of this molecule, which significantly increases the electrostatic energy compared to larger neutral molecules, such as porphine.

Table 4 Components of the interaction energy (in kcal/mol) calculated using the **C96** finite model for the graphene surface, the M06-2X-D DFT functional and the 6-31G(d,p) basis set for **C96**, and 6-311++G(d,p) for the molecule. **A**, **B**, and **C** refer to different configurations of the adsorbed molecules; see text for details

		E_{Elec}	E_{Rep}	E_x	E_{Ind}	$E_{Res\text{-}Pol}$	$D3$	E_{Int}
Gly	**A**	−6.00	45.65	−28.79	−4.83	−13.63	−2.26	−9.86
	B	−3.91	28.44	−17.96	−1.73	−9.17	−1.93	−6.26
	C	−3.37	27.73	−17.92	−1.81	−8.71	−2.01	−6.09
Mel		−11.60	75.50	−46.80	−5.70	−25.75	−3.98	−18.33
Pyr	**A**	−36.12	133.99	−77.20	−24.46	−37.31	−5.96	−47.06
	B	−16.61	56.31	−33.92	−13.27	−15.77	−4.61	−27.87
	C	−12.18	30.20	−18.36	−11.88	−3.42	−2.76	−18.40
PP	**A**	−25.03	160.08	−93.54	−20.52	−52.12	−8.46	−39.59
	B	−2.55	17.99	−11.17	−2.92	−4.42	−2.94	−6.01
	C	−3.00	18.35	−11.42	−2.95	−3.74	−2.93	−5.69
H_2Pc	**A**	−32.41	236.66	−138.61	−32.17	−78.46	−11.88	−56.88
	C	−4.45	16.68	−10.25	−3.05	−1.74	−2.50	−5.31
TBPP	**A**	−35.64	250.04	−146.18	−37.89	−78.04	−12.28	−59.99
	C	−3.49	16.81	−10.27	−3.06	−2.80	−2.54	−5.36

Reproduced from Ref. [41] with permission from the PCCP Owner Societies

The magnitude of the induction energy is also noteworthy. For the largest molecules, it is comparable to the electrostatic energy in the case of $\mathbf{H_2Pc}$ and even exceeds it for **TBPP**. Conversely, the spin repulsion and intermolecular exchange energies are particularly large in the stacking conformations, owing to the significant electron overlap inherent to this arrangement. In all cases, the Pauli energy exceeds the total contribution of dispersion energy, represented by the sum of Res-Pol and D3. This reflects, as remarked previously in Sect. 2.2, for the small complexes, the relevance of the intermolecular exchange and spin repulsion energies in the relative stability of non-covalent complexes. It is well known that Pauli energy is strongly related to the chemical concept of steric hindrance [69]. The DD-EDA method enables the separate analysis of its repulsive and attractive components, offering deeper insight into its role in steric interactions. Further research should explore this role in more detail.

The polarization densities for the various stacking complexes are shown in Fig. 4. In almost all cases, electron polarization involves a migration of electron density from the molecule and the carbon surface to the intermolecular region. As previously discussed for the set of small complexes and for the pyridine complexes, this behavior is characteristic of interactions primarily driven by dispersion forces. The extent of this positive electron polarization in the intermolecular region correlates with the strength of the interaction. However, the electron polarization in the stacking conformation of **Pyr** exhibits a markedly different behavior. Instead of being directed toward the intermolecular region, there is a migration of electron density from the external hydrogen atoms of the carbon nanodisk to the active

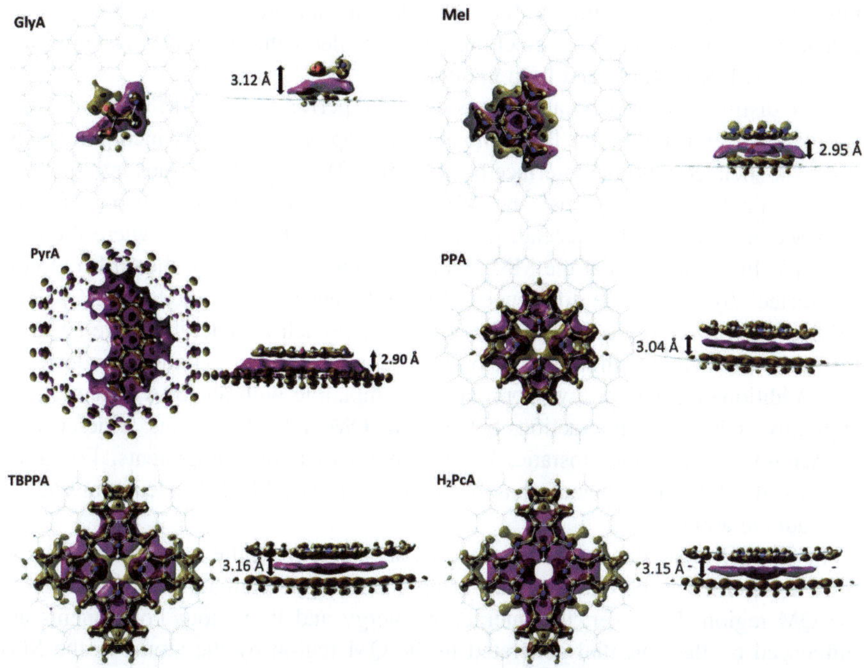

Fig. 4 Adsorption complexes formed by **Gly**, **Pyr**, **PP**, **H_2Pc**, and **TBPP** with **C96** in stacking conformation (**A**) together with the corresponding molecule–surface averaged distance and electron polarization density obtained with the M06-2X-D functional in combination with the 6-31G(d,p) basis set for **C96**, and 6-311++G(d,p) for the molecules. Magenta and brown colors represent positive and negative densities, respectively (isosurface value of 0.004 a.u.). (Reproduced from Ref. [41] with permission from the PCCP Owner Societies)

centers of the surface. The significant role of terminal hydrogen atoms in charge transfer processes is well documented, for example, in studies of protonation and basicity [70, 71], where this behavior is attributed to the high atomic softness of hydrogen. The positive charge of **Pyr** likely induces this charge transfer within the carbon nanodisk, which explains the observed differences in the interaction energy terms of Table 4, compared to the other molecules.

4 QM/MM-EDA Scheme

4.1 Theoretical Framework

In order to apply the DD-EDA to the study of intermolecular interactions in complex systems, the scheme was extended to hybrid QM/MM methods [45–47]. These more intricate scenarios encompass, for example, the adsorption of molecules onto extended materials and the interactions of molecules with biological systems such as proteins, DNA, RNA, and lipid membranes.

Two distinct approaches can be considered for performing QM/MM calculations: additive [72] and subtractive [73]. The additive approach explicitly incorporates the energy associated with the interface between the QM and MM regions into the total energy calculation. In contrast, the subtractive approach does not account for the interface energy directly. Instead, it employs a subtractive scheme where the total energy is first calculated at the MM level, and then the energy of the QM region is "corrected" by adding the difference between the energies obtained at the QM and MM levels for that region. While the additive approach is generally more accurate than the subtractive method, it is also more complex and computationally demanding. Additionally, the additive approach is compatible with an EDA scheme, as it explicitly includes the interactions between the QM and MM regions, which can be further divided into electrostatic, Pauli, and polarization components. For these reasons, the DD-EDA scheme has been extended to QM/MM calculations using the additive approach.

In QM/MM calculations of intermolecular interactions, the interaction center is fully represented at the QM level, ensuring that all significant atoms are included in the QM region. However, the interaction energy and its various components are influenced by the potential generated in the QM region by the atoms in the MM region. Therefore, it is essential to incorporate this potential into the calculations. The specific form of the potential depends on the type of embedding used when calculating the electron densities of the complex *AB* and its fragments, *A* and *B*. For an electrostatic embedding, where the atoms of the MM region are exclusively represented by permanent atomic charges, the potential is given by

$$\widehat{V}_{MM}(\boldsymbol{r}) = \sum_{i=1}^{N_{MM}} \frac{q_i}{|\boldsymbol{r} - \boldsymbol{R}_i|} \tag{31}$$

Here, $\boldsymbol{R}_i$ and q_i represent the position and charge of a given atom *i*, respectively, while N_{MM} denotes the total number of atoms in the MM region. In the case of electrostatic embedding, the atomic charges in the MM region are fixed, meaning that only the polarization of the QM region by the MM region is considered. In contrast, a polarizable embedding takes into account the polarization of classical atoms by quantum atoms as well. Consequently, induced atomic dipoles are incorporated into the MM potential in the following form:

$$\widehat{V}_{MM}(\boldsymbol{r}) = \sum_{i=1}^{N_{MM}} \left(\frac{q_i}{|\boldsymbol{r} - \boldsymbol{R}_i|} + \frac{\mu_i(\boldsymbol{r} - \boldsymbol{R}_i)}{|\boldsymbol{r} - \boldsymbol{R}_i|^3} \right) \tag{32}$$

Here, μ_i denotes the induced dipole moment of atom *i*. For practical purposes, the induced atomic dipole moments in the MM region are implemented in the DD-EDA scheme using a point charge model. In this model, each dipole is represented by a pair of opposite induced charges, δ_i^+ and δ_i^-, which are positioned on the atom, separated by a distance d_i, and aligned with the dipole vector. Using this model, the MM potential is written as

$$\widehat{V}_{MM}(\boldsymbol{r}) = \sum_{i=1}^{N_{MM}} \left(\frac{q_i}{|\boldsymbol{r}-\boldsymbol{R}_i|} + \frac{\delta_i^+}{|\boldsymbol{r}-\boldsymbol{R}_i + d_i/2|} + \frac{\delta_i^-}{|\boldsymbol{r}-\boldsymbol{R}_i - d_i/2|} \right) \tag{33}$$

It is important to note that, to accurately reproduce the potential generated by the induced atomic dipoles, the distance between the pair of induced charges must be significantly shorter than the shortest distance between the QM and MM atoms. As a result, the induced charges need to be large to effectively replicate the values of the induced dipole moments. These short distances and large induced charges lead to a substantial total electrostatic energy within the MM region. This undesirable electrostatic interaction arises from the point charge model of the induced atomic dipole and lacks physical meaning. To prevent unreliable results during the energy decomposition analysis, the electrostatic energy resulting from the interaction between the charges of an induced dipole is calculated and subsequently removed from the total energy of the MM region.

The current strategy followed to incorporate the classical potential in the calculation of the different energy terms in the DD-EDA scheme for the electrostatic and polarizable embedding is the same. The potential is added to the nuclear potential of the fragment that includes the classical region. If the classical atoms belong to fragment B, then the nuclear potential of this fragment also includes the classical potential:

$$\widehat{v}_{N_B}^{QM/MM}(\boldsymbol{r}) = \widehat{v}_{N_B}(\boldsymbol{r}) + \widehat{V}_{MM}(\boldsymbol{r}) \tag{34}$$

It is important to note that the MM potential also interacts electrostatically with the nuclei of fragment A. In addition to replacing $\widehat{v}_{N_B}(\boldsymbol{r})$ with $\widehat{v}_{N_B}^{QM/MM}(\boldsymbol{r})$ in the DD-EDA equations, the energy from this interaction must be included to calculate the electrostatic energy at the QM/MM level:

$$E_{Elec} = \int \widehat{v}_{N_A}\rho_B(\boldsymbol{r})d\boldsymbol{r} + \int \widehat{v}_{N_B}^{QM/MM}\rho_A(\boldsymbol{r})d\boldsymbol{r} + \iint \frac{\rho_A(\boldsymbol{r}_1)\rho_B(\boldsymbol{r}_2)}{|\boldsymbol{r}_2-\boldsymbol{r}_1|}d\boldsymbol{r}_1 d\boldsymbol{r}_2 + \sum_{I=1}^{N_A}\sum_{J=N_A+1}^{N}\frac{Z_I Z_J}{|\boldsymbol{R}_{IJ}|} + \sum_{I=1}^{N_A}\int Z_I \widehat{V}_{MM}(\boldsymbol{r})\delta(\boldsymbol{r}-\boldsymbol{R}_I)d\boldsymbol{r} \tag{35}$$

If both fragments include an MM region, a different MM potential must be defined for each fragment. In the context of DD-EDA, these potentials are integrated into their corresponding nuclear potentials. Furthermore, an additional term is included in the electrostatic energy to account for the interaction between the MM potential of fragment A and the nuclei of fragment B [47].

4.2 Illustrative Examples

The application of DD-EDA within the context of QM/MM calculations has so far been combined with classical molecular dynamic (CMD) simulations. These classical simulations allow the exploration of the configurational space and the determination of the free energy profile through a dynamic process. Singular points of the free energy profile, such as maxima or minima, can be selected for analyzing the magnitude and nature of the intermolecular interactions occurring throughout the process. By choosing a representative set of equispaced geometries within each of these points, statistical analysis of the different interaction energy terms can be carried out with QM/MM DD-EDA single-point calculations. Therefore, the interaction energy terms obtained are thermally averaged using this semiclassical approach.

In this section, the results shown were obtained in the study of the intermolecular interactions occurring along the permeation process of an antitumor compound, cisplatin, and two pollutants, 2,3,6,7-tetrachlorodibenzodioxin (TCDD) and 2,3,6,7-tetrachlorodibenzofuran (TCDF), through a lipid bilayer formed by 128 molecules of 1,2-dioleoyl-sn-glycero-3-phosphocholine (DOPC). The permeation processes were simulated by combining CMD simulations with the umbrella sampling technique. Detailed information about the CMD simulations and a comprehensive analysis of the results can be found in Refs. [45, 46].

Schematic representations of the compounds, along with the orientation of the DOPC chains in the membrane, are shown in Fig. 5. The free energy profiles derived from the CMD simulations are also presented in the figure. The DOPC molecules are oriented such that the polar heads face the exterior of the membrane, while the nonpolar tails are directed toward the interior. This arrangement creates regions with varying electrostatic potentials, resulting in free energy profiles that include both maxima and minima. In the case of cisplatin permeation, the molecule first encounters a subtle local energy minimum in the polar region. As it moves into the nonpolar region, the energy increases rapidly, reaching a pronounced maximum at the center of the membrane. In the case of TCDD and TCDF permeation, the free energy profile exhibits a distinctly different pattern. The energy decreases as the molecules penetrate the membrane, reaching a pronounced energy minimum near the boundary between the polar heads and the nonpolar tails. As the molecules move toward the center of the membrane, the energy increases slightly, forming a gentle "low hill" with a subtle local maximum.

The most relevant points for the analysis of the intermolecular interactions governing the permeation process of cisplatin are the local minimum found in the polar region and the global maximum at the center of the membrane. Before selecting geometries from the window containing these points to perform the QM/MM calculations and apply the DD-EDA scheme, it is essential to evaluate first the optimal size of the QM region. The QM region should be large enough to ensure convergence of all interaction energy terms.

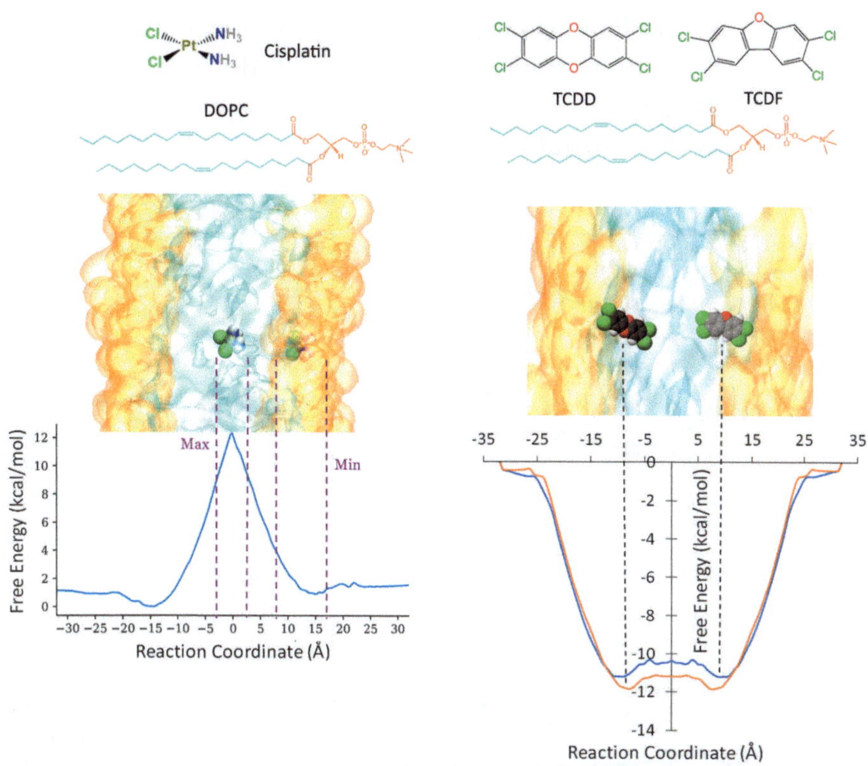

Fig. 5 Schematic representation of the cisplatin (up left) and TCDD/TCDF (up right) molecules embedded in a DOPC lipid bilayer. The polar and nonpolar regions of the lipid bilayer are represented in orange and cyan, respectively. Free energy profiles of the cisplatin (bottom left) and TCDD/TCDF (bottom right, the red and blue lines correspond to TCDD and TCDF, respectively) permeation processes through the lipid bilayer. (Left plots: Reproduced from Ref. [45] with permission from the PCCP Owner Societies. Right plots: Reused from Ref. [46] Open access under a CC BY 4.0 license, https://creativecommons.org/licenses/by/4.0/)

Therefore, the first step is to assess the convergence of different energy terms with respect to the size of the QM region using an arbitrarily chosen geometry. The results of this study are summarized in Fig. 6, where the electrostatic, Pauli, induction, and dispersion energies are plotted against the number of DOPC molecules included in the QM region, for both a geometry in the minimum region and one in the maximum. The DOPC molecules were added to the QM region based on their distance from the cisplatin molecule. The potential of the MM region was represented using electrostatic embedding, with fixed atomic charges derived from the general amber force field (GAFF).

As can be observed in plots c and d of Fig. 6, all energy terms converge with five DOPC molecules, both at the maximum and minimum regions. For comparison, the convergence obtained from pure QM calculations is also displayed in plots a and b of

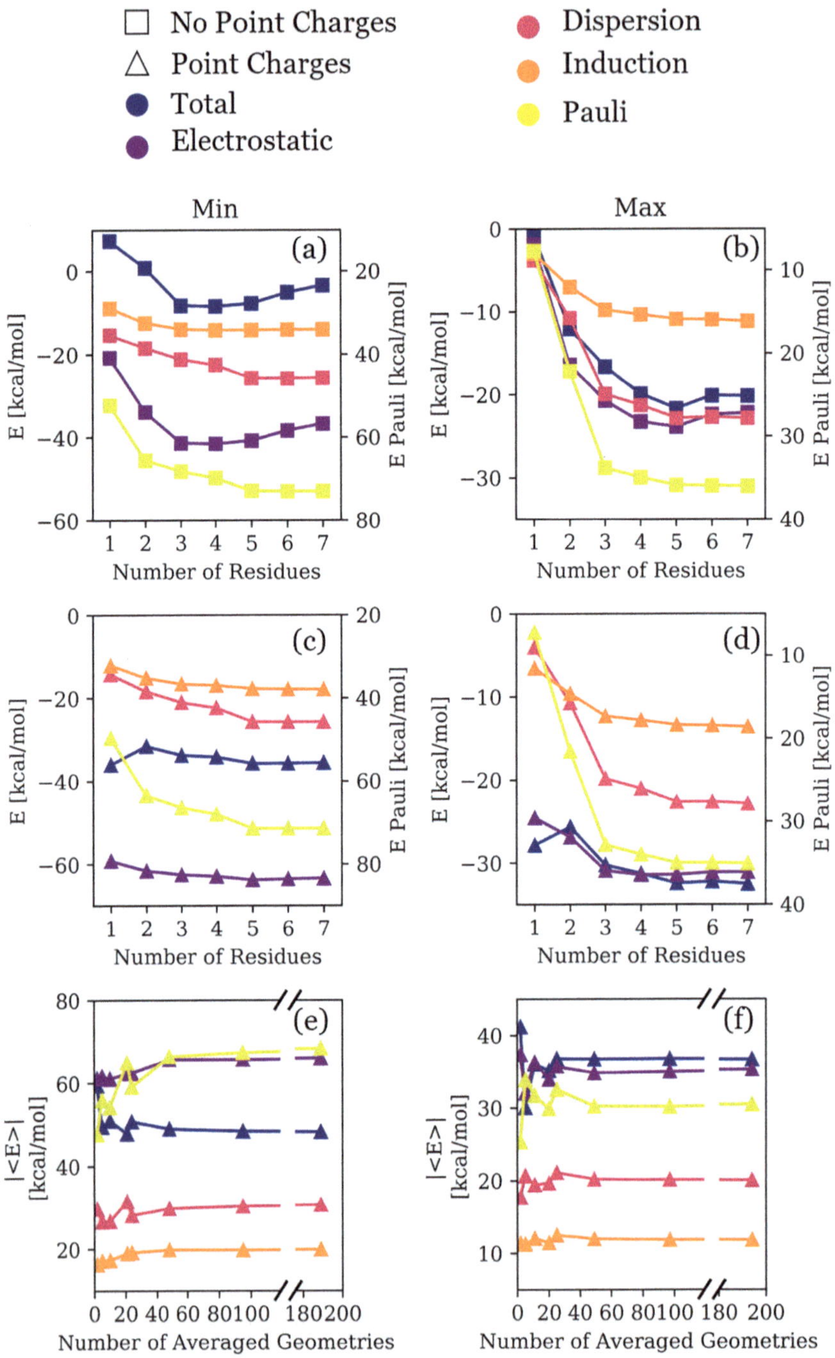

Fig. 6 Analysis of the convergence of the interaction energy and the EDA components as the number of DOPC molecules in the QM region increases without (**a** Min, **b** Max) and with (**c** Min, **d** Max) point charges around the free energy minima of the cisplatin permeation. Convergence of the absolute value of the interaction energy and the EDA components with respect to the number of

Fig. 6. In this case, convergence of the electrostatic and total energies is not achieved even with seven DOPC molecules.

Once the optimal size of the QM region for the QM/MM and DD-EDA calculations is determined, the next step is to select a number of equispaced geometries within the windows of the umbrella sampling that encompass the minimum and maximum. To ensure well-converged statistical results, a preliminary analysis of the required number of equispaced geometries must be conducted. This analysis is presented in Fig. 6, in plots e and f, for the minimum and maximum regions, respectively. The plots showing the average values of the different interaction energy terms as a function of the number of equispaced geometries indicate that the statistics converge with 100 geometries.

The statistical distributions of the different interaction energy terms obtained for 100 geometries are shown in Fig. 7, with plots a and c representing the minimum and maximum regions, respectively. The averaged values are also included in plots b and d, where the electrostatic and non-electrostatic (Pauli and polarization) terms are grouped for further analysis. A common feature across the distribution plots for both the minimum and maximum regions is that the distribution broadens as the averaged energy increases (in absolute value). Notably, the width of distributions for the Pauli energy stands out. This can be attributed to both the large magnitude of this term and its short-range nature, which makes it more sensitive to small changes in geometries.

On the other hand, the differences in the nature of intermolecular interactions established in the minimum and maximum regions are reflected in both the distribution plots and the average energy values. Focusing on the latter, the interaction in the minimum is clearly dominated by the electrostatic energy between cisplatin and polar heads of the DOPC lipids. Although the Pauli repulsion is significant in this region, likely due to the proximity between cisplatin and DOPC molecules induced by the favorable electrostatic interaction, the total interaction energy remains relatively low, around −50 kcal/mol.

In the maximum, where cisplatin interacts with the nonpolar tails of the DOPCs, the total interaction energy decreases by more than 10 kcal/mol compared to the minimum. This reduction is primarily due to a significant drop in electrostatic stabilization, more than 30 kcal/mol, which is not fully compensated by the decrease in Pauli repulsion energy.

For the permeation of TCDD and TCDF, the most relevant points for the analysis of the intermolecular interactions governing the process are the global minimum, located at the boundary between polar and nonpolar regions, and the local maximum at the center of the membrane. Reference [46] reports no significant differences in the interaction energy terms at these two points. Consequently, only the energy distributions obtained at the free energy minimum are presented here.

Fig. 6 (continued) (equispaced) geometries selected from the free energy minimum (**e**) and maximum (**f**) of the cisplatin permeation. The calculations were done with the 6-31G(d) basis set for all the atoms but the Pt where the LANL2DZ basis set + pseudopotential was employed. (Reproduced from Ref. [45] with permission from the PCCP Owner Societies)

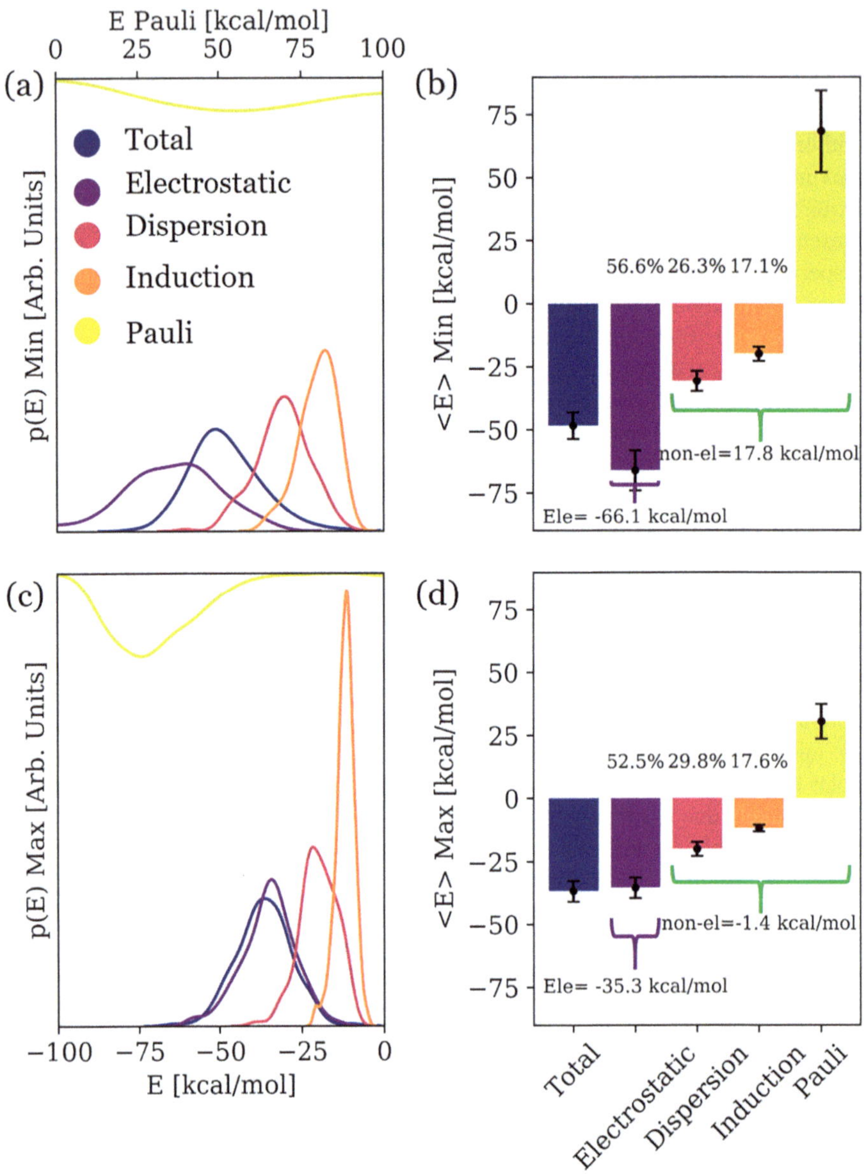

Fig. 7 Energy distributions of contributions to the total interaction energy stemming from the EDA on 190 (194) geometries on (**a**) the minimum (Min) and (**c**) the maximum (Max) of the free energy profile of cisplatin permeation. Averages and standard deviations of the distributions obtained at Min (**b**) and Max (**d**). The percentages are relative to the overall attractive component to the interaction energy. Electrostatic and non-electrostatic energies (sum of Pauli, induction, and dispersion components) are also shown. The calculations were done with the 6-31G(d) basis set for all the atoms but the Pt where the LANL2DZ basis set + pseudopotential was employed. (Reproduced from Ref. [45] with permission from the PCCP Owner Societies)

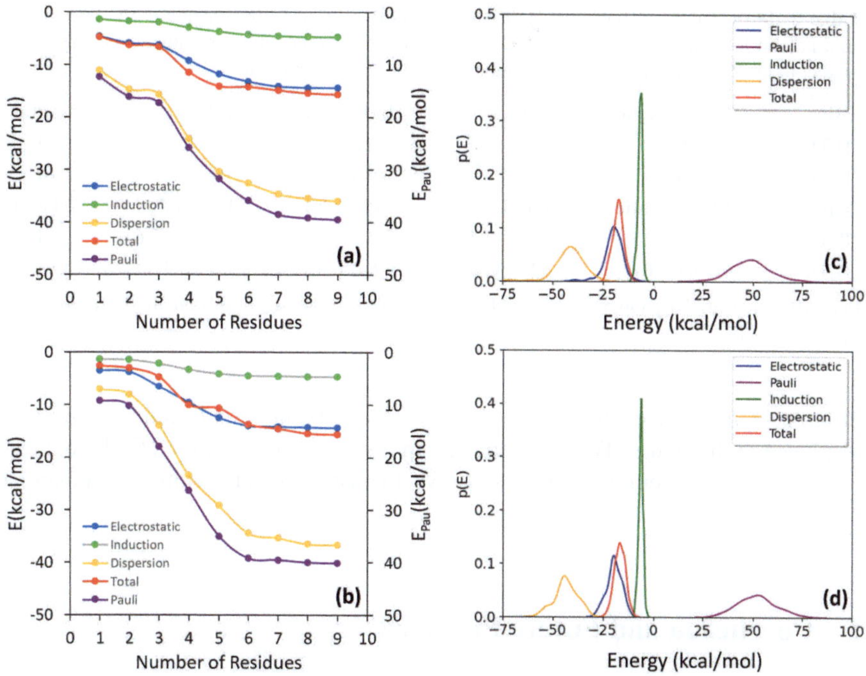

Fig. 8 Analysis of the convergence of the interaction energy and its different components as the number of the DOPC molecules increases in the QM region for an arbitrary geometry around the free energy minima of the TCDD (**a**) and TCDF (**b**) permeation. Distributions of the interaction energy and its different components on 100 geometries selected from the free energy minima of the TCDD (**c**) and TCDF (**d**) permeation. The calculations were done with the 6-31G(d,p) basis set. (Reused from Ref. [46] Open access under a CC BY 4.0 license, https://creativecommons.org/licenses/by/4.0/)

The determination of the optimal QM region size was performed following the same procedure used for cisplatin. As shown in Fig. 8, convergence of all energy terms is achieved when the QM region includes 8 DOPC molecules, for both TCDD (plot a) and TCDF (plot b). However, since the differences between 7 and 8 DOPC molecules are quite small, the calculations were conducted with 7 molecules instead of 8 to reduce the computational cost of the QM/MM calculations. The same electrostatic embedding employed for cisplatin was used here to model the MM potential. Similarly, 100 equispaced geometries were included in the statistical analysis of various energy terms.

The energy distributions at the global minimum are presented in Fig. 8 for TCDD (plot c) and TCDF (plot d). The results reveal almost no differences between the two pollutants, consistent with the similar free energy profiles observed in the CMD simulations. Table 5 summarizes the average values of the various energy terms at the global minimum and local maximum. A more detailed term-by-term analysis is available in Ref. [46]. Here, we simply note that the differences in interaction energy between the minimum and maximum are small for both pollutants. The percentage

Table 5 Mean values of the interaction energy and its different components at the free energy minima and local maxima of the TCDD and TCDF permeation. The percentage of change with respect to the minimum is given in parenthesis for the local maximum

Min	**Electrostatic**	**Pauli**	**Induction**	**Dispersion**	**Total**
TCDD	−19.6	49.7	−6.5	−41.1	−17.5
TCDF	−19.3	52.1	−6.4	−43.0	−16.5
Max	**Electrostatic**	**Pauli**	**Induction**	**Dispersion**	**Total**
TCDD	−15.0(23.5)	41.2(17.1)	−4.7(27.7)	−35.5(13.6)	−14.0
TCDF	−19.3(0.0)	54.1(3.8)	−6.0(6.3)	−43.8(1.9)	−15.0

differences in the average energies between the maximum and minimum are also displayed in the figure. These differences are more pronounced for TCDD, indicating that this compound is more sensitive to displacements within the nonpolar region.

5 Conclusion and Future Prospects

This chapter provides a detailed exploration of the theoretical framework of the DD-EDA scheme, tracing its development from the original QM implementation to its recent extension within the QM/MM framework. The usefulness of the DD-EDA scheme in characterizing the strength and nature of intermolecular interactions, ranging from small molecular complexes to large biological systems, has been demonstrated in this chapter through the review and analysis of selected results previously published in the literature. The method is robust, easy to implement, and computationally inexpensive. Even more importantly, it can be applied at various computational levels, from SCF (HF and DFT) to post-SCF, making it a widely applicable and versatile approach.

However, the DD-EDA method is currently implemented exclusively for use with the Gaussian and AMBER software packages. Further steps can be taken to make the method accessible to a wider range of researchers in the computational chemistry, physics, and biology communities. These steps include developing interfaces with a broader array of electronic structure determination programs and dynamic simulation tools.

A scheme designed for analyzing intermolecular interactions in excited states, along with its implementation, would be highly valuable, as many important chemical and biological processes occur when one or more molecules are excited by electromagnetic radiation. Currently, the DD-EDA method is limited to applications involving ground-state complexes.

Additionally, defining and implementing an atom-pair DD-EDA would be valuable for the parameterization of classical force fields. The contribution of a specific atom pair to electrostatic, exchange, repulsion, and polarization energies can be determined by decomposing the corresponding unperturbed densities, as well as the Pauli and polarization densities, into atom-pair contributions. This can be achieved through an atomic partition of either real space or Hilbert space.

References

1. Leckband D, Israelachvili JN (2001) Intermolecular forces in biology. Q Rev Biophys 34:105
2. Kaplan IG (2006) Intermolecular interactions: physical picture, computational methods and model potentials. John Wiley and Sons Ltd., Chichester
3. Stone A (2013) The theory of intermolecular forces. 2nd edn. Oxford Academic, Oxford
4. Moszynski R (2007) Theory of intermolecular forces: an introductory account. In: Sokalski WA, Leszczynski JJ (eds) Molecular materials with specific interactions – modeling and design. Challenges and advances in computational chemistry and physics. Springer, Dordrecht
5. Hobza P, Müller-Dethlefs K (2010) Non-covalent interactions: theory and experiment. Royal Society of Chemistry, Cambridge
6. Chalasinski G, Szczesniak MM (1994) Chem Rev 94:1723
7. Chałasiński G, Szczęśniak MM (2000) Chem Rev 100:4227
8. Kestner NR (1968) J Chem Phys 48:252
9. Liu B, McLean AD (1973) J Chem Phys 59:4557
10. Boys SF, Bernardi F (1970) Mol Phys 19:553
11. Morokuma K (1971) J Chem Phys 55:1236
12. Kitaura K, Morokuma K (1976) Int J Quant Chem 10:325
13. Ziegler T, Rauk A (1977) Theor Chim Acta 46:1
14. Krapp A, Frenking G (2006) Chem Eur J 12:9196
15. Becke A (2014) J Chem Phys 140:18A301
16. Grimme S, Antony J, Ehrlich S, Krieg H (2010) J Chem Phys 132:154104
17. Grimme S, Ehrlich S, Goerigk L (2011) J Comput Chem 32:1456
18. Cardozo TM, Chaer Nascimento MA (2009) J Chem Phys 130:104102
19. Cardozo TM, Nascimento MA (2009) J Phys Chem A 113:12541
20. Mo YR, Gao JL, Peyerimhoff SD (2000) J Chem Phys 112:5530
21. Mo Y, Song L, Lin Y (2007) J Phys Chem A 111:8291
22. Mo Y (2014) The block-localized wavefunction (BLW) perspective of chemical bonding. In: Frenking G, Shaik S (eds) The chemical bond 1. Fundamental aspects of chemical bonding. Wiley-VCH, Weinheim
23. Blanco MA, Martín-Pendás A, Francisco E (2005) J Chem Theory Comput 1:1096
24. Pendás AM, Blanco MA, Francisco E (2007) J Comput Chem 28:161
25. Khaliullin RZ, Head-Gordon M, Bell AT (2006) J Chem Phys 124:204105
26. Khaliullin RZ, Cobar EA, Lochan RC, Bell AT, Head-Gordon M (2007) J Phys Chem A 111: 8753
27. Glendening ED, Streitwieser Jr A (1994) J Chem Phys 100:2900
28. Glendening ED (2005) J Phys Chem A 109:11936
29. Mitoraj MP, Michalak A, Ziegler T (2009) J Chem Theory Comput 5:962
30. Mitoraj M, Michalak A (2007) J Mol Model 13:347
31. Jeziorski B, Moszynski R, Szalewicz K (1994) Chem Rev 94:1887
32. Engkvist O, Åstrand PO, Karlström G (2000) Chem Rev 100:4087
33. Hirschfelder JO (1967) Chem Phys Lett 1:325
34. Hirschfelder JO (1967) Chem Phys Lett 1:363
35. Patkowski K (2020) WIREs Comput Mol Sci 10:e1452
36. Misquitta AJ, Podeszwa R, Jeziorski B, Szalewicz K (2005) J Chem Phys 123:214103

37. Hesselmann A, Jansen G, Schütz M (2005) J Chem Phys 122:014103
38. Williams HL, Chabalowski CF (2001) J Phys Chem A 105:646
39. Mandado M, Hermida-Ramón JM (2011) J Chem Theory Comput 7:633
40. Ramos-Berdullas N, Pérez-Juste I, Van Alsenoy C, Mandado M (2015) Phys Chem Chem Phys 17:575
41. López-Carballeira D, Ramos-Berdullas N, Pérez-Juste I, Cagide-Fajín JL, Cordeiro MNDS, Mandado M (2016) Phys Chem Chem Phys 18:15312
42. Iribarren I, Mates-Torres E, Trujillo C (2024) Dalton Trans 53:1322
43. Pérez-Barcia A, Montero-Campillo MM, Lamsabhi AM, Salpin J-Y, Yáñez M (2022) Phys Chem Chem Phys 24:20624
44. Martín-Fernández C, Ferrer M, Alkorta I, Montero-Campillo MM, Elguero J, Mandado M (2023) Inorg Chem 62:16523
45. Cárdenas G, Pérez-Barcia Á, Mandado M, Nogueira JJ (2021) Phys Chem Chem Phys 23: 20533
46. Alvarado R, Cárdenas G, Nogueira JJ, Ramos-Berdullas N, Mandado M (2022) Membranes 13: 28
47. Pérez-Barcia Á, Cárdenas G, Nogueira JJ, Mandado M (2023) J Chem Inf Model 63:882
48. Mao Y, Shao Y, Dziedzic J, Skylaris CK, Head-Gordon T, Head-Gordon M (2017) J Chem Theory Comput 13:1963
49. Dziedzic J, Head-Gordon T, Head-Gordon M, Skylaris CK (2019) J Chem Phys 150:074103
50. Fox SJ, Pittock C, Fox T, Tautermann CS, Malcolm N, Skylaris CK (2011) J Chem Phys 135: 224107
51. Xiong X, Friedman R, Wu W, Su P (2024) J Phys Chem A 128:3529
52. Olivares del Valle FJ, Tolosa S, Lopez-Piñeiro A, Requena A (1985) J Comput Chem 6:39
53. Johansson A, Kollman P, Rothenberg S (1973) Theor Chim Acta 29:167
54. Gray M, Bowling PE, Herbert JM (2022) J Chem Theory Comput 18:6742
55. Löwdin P-O (1950) J Chem Phys 18:365
56. Müller AMK (1984) Phys Lett A 105:44
57. Goedecker S, Umrigar CJ (1998) Phys Rev Lett 81:866
58. Reed AE, Weinhold F (1983) J Chem Phys 78:4066
59. Reed AE, Weinstock RB, Weinhold F (1985) J Chem Phys 83:735
60. Bukowski R, Cencek W, Garcia J, Jankowski P, Jeziorski B, Jeziorska M, Korona T, Kucharski SA, Lotrich VF, Metz MP, Misquitta AJ, Moszynski R, Patkowski K, Podeszwa R, Rob F, Rybak S, Szalewicz K, Williams HL, Wheatley RJ, Wormer PES, Żuchowski PS. SAPT2020: an ab initio program for many-body symmetry-adapted perturbation theory calculations of intermolecular interaction energies
61. Garcia J, Podeszwa R, Szalewicz K (2020) J Chem Phys 152:184109
62. Mandado M, Van Alsenoy K. EDA-NCI: energy decomposition analysis of non-covalent interactions. https://github.com/marcos-mandado/EDA-NCI. Last accessed 26 Feb 2025
63. Zhao Y, Truhlar DG (2005) J Chem Theory Comput 1:415
64. Zhao Y, Schultz NE, Truhlar DG (2006) J Chem Theory Comput 2:364
65. Zhao Y, Schultz NE, Truhlar DG (2005) J Chem Phys 123:161103
66. Zhao Y, Truhlar DG (2005) J Phys Chem A 109:5656
67. Ángyán JG (2008) Phys Rev A 78:022510
68. Israelachvili JN (2011) Interactions involving the polarization of molecules. In: Intermolecular and surface forces. 3rd edn. Academic Press, Amsterdam, pp. 91–106
69. Pinter D, Fievez T, Bickelhaupt FM, Geerlings P, De Proft F (2012) Phys Chem Chem Phys 14: 9846
70. Stutchbury NCJ, Cooper DL (1983) J Chem Phys 79:4967
71. González-Moa MJ, Mandado M, Mosquera RA (2006) Chem Phys Lett 428:255
72. Senn HM, Thiel W (2009) Angew Chem Int Ed 48:1198
73. Maseras F, Morokuma K (1995) J Comput Chem 16:1170

Struct Bond (2026) 190: 109–136
https://doi.org/10.1007/430_2025_104

Published online: 31 July 2025

GKS-EDA Method for Intermolecular Interactions in Complex Systems

Zijing Wu and Peifeng Su

Contents

Abstract Energy decomposition analysis (EDA) is a theoretical framework that quantitatively analyzes intermolecular interactions and reveals their physical origins through numerical insights. Since its introduction in 2014, the generalized Kohn-Sham EDA (GKS-EDA) method has emerged as a robust family based on KS-DFT calculations. This chapter systematically presents the GKS-EDA family for various intermolecular interactions with a focus on its applications to complex systems including spin states, excited states, and environment effects.

Keywords Energy decomposition analysis · GKS-EDA · Intermolecular interactions · Magnetic interactions · Solvent effects

Z. Wu and P. Su (✉)
The State Key Laboratory of Physical Chemistry of Solid Surfaces, Fujian Provincial Key Laboratory of Theoretical and Computational Chemistry, and College of Chemistry and Chemical Engineering, Xiamen University, Xiamen, Fujian, China
e-mail: supi@xmu.edu.cn

1 Introduction

Intermolecular interactions play important roles for various processes in physics, chemistry, and life sciences. Theoretical approximations of intermolecular interactions focus on using appropriate theoretical models to "translate" the data of theoretical computations into chemical concepts, identifying the key factors that influence structure-activity relationships, and thereby guiding molecular design [1–3]. One category of these approximations tends to obtain information from electronic density topological analysis. These kinds of qualitative analysis methods, for example, independent gradient model (IGM) [4], atoms in molecules (AIM) [5, 6], and non-covalent interaction (NCI) analysis [7], provide synergistic frameworks for characterizing non-covalent interactions, combining geometric visualization with rigorous quantum mechanical descriptors.

Another category of methods provides numerical interpretations by decomposing the total interaction energy into several components, which is called energy decomposition analysis (EDA) [8–15]. Current popular EDA methods can be divided into variational EDA methods and perturbational EDA methods. Variational EDA methods use the difference between several intermediate states to compute the EDA individual terms in the framework of the supermolecule approach. One of the main advantages of variational EDA methods is that they are flexible and can be readily used with various electronic structure methods, such as post-Hartree-Fock (post-HF) or Kohn-Sham density functional theory (KS-DFT). Since the 1970s, many variational EDA methods, including KM-EDA [16], extended transition state EDA (ETS-EDA) [17], absolutely localized molecular orbital EDA (ALMO-EDA) [18, 19], block-localized wavefunction EDA (BLW-EDA) [20, 21], and localized molecular orbital EDA (LMO-EDA) [22], have been proposed [23].

In LMO-EDA, by using either HF or KS orbitals, the total interaction energy from calculations via mean field approximations can be divided into EDA terms of electrostatic, exchange-repulsion, polarization, and dispersion. LMO-EDA employs restricted (R), restricted-open (RO), or unrestricted (U) type of HF or KS orbitals for closed-shell systems or open-shell systems with high spin states. In 2014, Su et al. proposed the generalized Kohn-Sham EDA (GKS-EDA) [24] based on LMO-EDA and generalized Kohn-Sham theory [25]. This approach can be integrated with various functionals like LDA, GGA, hybrid, double hybrid, range separated, and dispersion-corrected functionals. Based on KS-DFT calculations, GKS-EDA has been widely employed in studies of chemical bonds and non-covalent interactions.

Perturbation EDA is represented by symmetry-adapted perturbation theory (SAPT) [9, 10, 14], which employs double perturbation operators, i.e., intramonomer correlation operators and intermolecular interaction operators, to explore two- or three-body non-covalent interactions. It provides a rigorous expression of the dispersion term via perturbation theory. This method is typically restricted to non-covalent interactions since it is based on the assumption that the magnitude of intermolecular interaction is small.

The EDA terms in the variational EDA methods are computed by the supermolecule approach with several intermediate states, while the terms in perturbation EDA methods are derived from perturbation theory. Typical EDA terms are shown below:

1. **Electrostatic term**: the classical Coulomb interaction between monomers.
2. **Pauli/exchange-repulsion term**: the energy penalty arising from the antisymmetrized and renormalized form of the monomer's wavefunctions.
3. **Orbital/polarization/induction/charge transfer term**: denotes the orbital relaxation due to the presence of neighbor monomers. These terms can be obtained by the self-consistent field procedure in variational methods or linear response techniques in perturbative methods. Among these EDA terms, charge transfer (CT) term is highly basis set dependent [26, 27]. With a complete basis set, charge transfer and polarization are indistinguishable. Thus, some EDA methods explicitly include CT, while the others do not. The latter methods include LMO-EDA, GKS-EDA, ETS-EDA, SAPT, etc.
4. **Dispersion/correlation term**: provides the electronic correlation contribution that cannot be accounted by calculations at the Hartree-Fock level. This term is often used for the description of dispersion interaction. In many variational EDA methods, the correlation term is absorbed in the other EDA terms. In several EDA methods, the dispersion interaction can be considered by post-HF methods with large computational scaling, such as Møller-Plesset perturbation theory or coupled cluster methods, while in several KS-DFT-based EDA methods, a classical force field type of description can be applied for dispersion, which means that dispersion is considered, for example, by the Grimme dispersion correction [28]. The summation of correlation and dispersion terms in GKS-EDA is comparable to the dispersion term in DFT-SAPT.

Despite the different definitions of EDA terms across these various methods, in principle, they should not provide conflicting interpretations for well-characterized interacting systems. Our recent benchmarking study for the performance of ETS-EDA, GKS-EDA, and DFT-SAPT reveals that (1) computational results of the three EDA methods are comparable despite variations in basis sets and (2) all energy decomposition components in them show remarkable consistency, demonstrating robust agreement in quantifying distinct interaction types [29].

Despite their great successes and wide applications, EDA methods for complex systems still encounter some challenges. With the rapid developments of molecular science for materials, energy, and life sciences, studies of intermolecular interactions are no longer confined to simple closed-shell systems or high-spin open-shell systems. The inherent complexity of intermolecular interactions requires the systematic development of new EDA methods.

Since 2014, GKS-EDA has become a family of methods, including GKS-EDA (sol) for implicit solvent effects, GKS-EDA(QM/MM) for explicit solvent effects, and GKS-EDA(BS) and GKS-EDA(CU) for spin effects, while GKS-EDA(TD) addresses excited states. In the following, the basic formulations and applications of GKS-EDA and these variants are introduced.

2 Methodology

2.1 GKS-EDA formulation

By GKS-EDA, the total interaction energy is divided into the electrostatic, exchange, repulsion, polarization, correlation, and dispersion terms:

$$\Delta E^{\mathrm{TOT}} = E_{\mathrm{S}} - \sum_{\mathrm{A}} E_{\mathrm{A}} = \Delta E^{\mathrm{ele}} + \Delta E^{\mathrm{ex}} + \Delta E^{\mathrm{rep}} + \Delta E^{\mathrm{pol}} + \Delta E^{\mathrm{corr}} + \Delta E^{\mathrm{disp}} \quad (1)$$

Here, the subscript S denotes supermolecule, and A denotes monomers.

The electrostatic term can be expressed as

$$\Delta E^{\mathrm{ele}} = E_{\mathrm{S}}^{(1)} - E_{\mathrm{S}}^{(0)} = \left\langle \Phi^{\mathrm{DP}} \middle| \widehat{F}_{\mathrm{S}}^{0} \middle| \Phi^{\mathrm{DP}} \right\rangle - \sum_{\mathrm{A}} \left\langle \Phi_{\mathrm{A}} \middle| \widehat{F}_{\mathrm{A}}^{0} \middle| \Phi_{\mathrm{A}} \right\rangle + \Delta \widehat{V}_{\mathrm{NN}} \quad (2)$$

$$\widehat{F}^{0} = \widehat{T} + \widehat{\nu}_{\mathrm{X}}^{\mathrm{HF}} + \widehat{\nu}_{\mathrm{J}} + \widehat{\nu}_{\mathrm{ne}} \quad (3)$$

Here, Φ^{DP} stands for the direct product of monomer wavefunction, and $\widehat{T}$, $\widehat{\nu}_{\mathrm{X}}^{\mathrm{HF}}$, $\widehat{\nu}_{\mathrm{J}}$, and $\widehat{\nu}_{\mathrm{ne}}$ are the kinetic operator, exact HF exchange operator, electron-electron Coulomb operator, and nuclear-electron potential operator, respectively. The term $\Delta \widehat{V}_{\mathrm{NN}}$ is the nuclear-nuclear repulsive potential difference between the supermolecule and monomers. With KS spin orbitals, the electrostatic term can also be written as

$$\Delta E^{\mathrm{ele}} = \sum_{i \in \Phi^{\mathrm{DP}}}^{\alpha,\beta} \langle i|h|i\rangle + \frac{1}{2} \sum_{i \in \Phi^{\mathrm{DP}}}^{\alpha,\beta} \sum_{j \in \Phi^{\mathrm{DP}}}^{\alpha,\beta} \langle ij|ij\rangle - \sum_{\mathrm{A}} \left(\sum_{i \in \Phi_{\mathrm{A}}}^{\alpha,\beta} \langle i|h|i\rangle + \frac{1}{2} \sum_{i \in \Phi_{\mathrm{A}}}^{\alpha,\beta} \sum_{j \in \Phi_{\mathrm{A}}}^{\alpha,\beta} \langle ij|ij\rangle \right) + \Delta \widehat{V}_{\mathrm{NN}} \quad (4)$$

Here, i and j denote spin orbitals.

The exchange-repulsion term ($\Delta E^{\mathrm{exrep}}$) is defined as

$$\Delta E^{\mathrm{exrep}} = \left\langle \Phi^{\mathrm{ASN}} \middle| \widehat{F}_{\mathrm{S}}^{0} \middle| \Phi^{\mathrm{ASN}} \right\rangle - \left\langle \Phi^{\mathrm{DP}} \middle| \widehat{F}_{\mathrm{S}}^{0} \middle| \Phi^{\mathrm{DP}} \right\rangle \quad (5)$$

$\Delta E^{\mathrm{exrep}}$ can be further separated into exchange and repulsion terms:

$$\Delta E^{\mathrm{ex}} = E_{\mathrm{X}}^{\mathrm{GKS}}(2) - E_{\mathrm{X}}^{\mathrm{GKS}}(1) = \left\langle \Phi^{\mathrm{AS}} \middle| \widehat{F}_{\mathrm{S}}^{0} \middle| \Phi^{\mathrm{AS}} \right\rangle - \left\langle \Phi^{\mathrm{DP}} \middle| \widehat{F}_{\mathrm{S}}^{0} \middle| \Phi^{\mathrm{DP}} \right\rangle \quad (6)$$

$$\Delta E^{\mathrm{rep}} = \Delta E^{\mathrm{exrep}} - \Delta E^{\mathrm{ex}} = \left\langle \Phi^{\mathrm{ASN}} \middle| \widehat{F}_{\mathrm{S}}^{0} \middle| \Phi^{\mathrm{ASN}} \right\rangle - \left\langle \Phi^{\mathrm{AS}} \middle| \widehat{F}_{\mathrm{S}}^{0} \middle| \Phi^{\mathrm{AS}} \right\rangle \quad (7)$$

Here, Φ^{AS} and Φ^{ASN} denote the antisymmetrized, and the antisymmetrized and renormalized forms of Φ^{DP}, respectively. ΔE^{ex} is the exact exchange energy between different monomers, and ΔE^{rep}, in turn, is consequently the result of forced normalization after previous antisymmetrization. ΔE^{ex} should be negative, showing the energetic stabilization due to the fact that electrons with the same spin try to keep away in the whole space; ΔE^{rep} should be positive, showing the energy penalty due to enforcing the simultaneous orthonormalization of all the orbitals from all monomers. With KS spin orbitals, the two terms can be expressed as

$$\Delta E^{ex} = \frac{1}{2}\sum_{A}\left(\sum_{i\in\Phi_A}^{\alpha}\sum_{j\in\Phi_A}^{\alpha}\langle ij|ji\rangle + \sum_{i\in\Phi_A}^{\beta}\sum_{j\in\Phi_A}^{\beta}\langle ij|ji\rangle\right) - \frac{1}{2}\left(\sum_{i\in\Phi_{DP}}^{\alpha}\sum_{j\in\Phi_{DP}}^{\alpha}\langle ij|ji\rangle + \sum_{i\in\Phi_{DP}}^{\beta}\sum_{j\in\Phi_{DP}}^{\beta}\langle ij|ji\rangle\right) \tag{8}$$

$$\begin{aligned}\Delta E^{rep} = {} & \sum_{i\in\Phi^{DP}}^{\alpha,\beta}\sum_{j\in\Phi^{DP}}^{\alpha,\beta}\langle i|h|j\rangle\left(\left(S^{-1}\right)_{ij} - \delta_{ij}\right) \\ & + \frac{1}{2}\sum_{i\in\Phi^{DP}}^{\alpha,\beta}\sum_{j\in\Phi^{DP}}^{\alpha,\beta}\sum_{k\in\Phi^{DP}}^{\alpha,\beta}\sum_{l\in\Phi^{DP}}^{\alpha,\beta}\langle ij|kl\rangle\left(\left(S^{-1}\right)_{ik}\left(S^{-1}\right)_{jl} - \delta_{ik}\delta_{jl}\right) \\ & - \frac{1}{2}\sum_{i\in\Phi^{DP}}^{\alpha}\sum_{j\in\Phi^{DP}}^{\alpha}\sum_{k\in\Phi^{DP}}^{\alpha}\sum_{l\in\Phi^{DP}}^{\alpha}\langle ij|lk\rangle\left(\left(S^{-1}\right)_{il}\left(S^{-1}\right)_{jk} - \delta_{il}\delta_{jk}\right) \\ & - \frac{1}{2}\sum_{i\in\Phi^{DP}}^{\beta}\sum_{j\in\Phi^{DP}}^{\beta}\sum_{k\in\Phi^{DP}}^{\beta}\sum_{l\in\Phi^{DP}}^{\beta}\langle ij|lk\rangle\left(\left(S^{-1}\right)_{il}\left(S^{-1}\right)_{jk} - \delta_{il}\delta_{jk}\right)\end{aligned} \tag{9}$$

Here, S^{-1} denotes the inverse of the overlap matrix for occupied KS orbitals.

The polarization term (ΔE^{pol}) arises from the orbitals' relaxation during the SCF procedure. Φ_S represents the wavefunction of the supermolecule:

$$\Delta E^{pol} = \left\langle \Phi_S \middle| \widehat{F}_S^0 \middle| \Phi_S \right\rangle - \left\langle \Phi^{ASN} \middle| \widehat{F}_S^0 \middle| \Phi^{ASN} \right\rangle \tag{10}$$

With KS spin orbitals, this term can be rewritten as

$$
\begin{aligned}
\Delta E^{\text{pol}} = {} & \sum_{i\in\Phi_S}^{\alpha,\beta} i|h|i + \frac{1}{2}\sum_{i\in\Phi_S}^{\alpha,\beta}\sum_{j\in\Phi_S}^{\alpha,\beta} ij|ij - \frac{1}{2}\left(\sum_{i\in\Phi_S}^{\alpha}\sum_{j\in\Phi_S}^{\alpha} ij|ji + \sum_{i\in\Phi_S}^{\beta}\sum_{j\in\Phi_S}^{\beta} ij|ji\right) \\
& - \left[\sum_{i\in\Phi^{\text{DP}}}^{\alpha,\beta}\sum_{j\in\Phi^{\text{DP}}}^{\alpha,\beta} i|h|j\left(S^{-1}\right)_{ij} + \frac{1}{2}\sum_{i\in\Phi^{\text{DP}}}^{\alpha,\beta}\sum_{j\in\Phi^{\text{DP}}}^{\alpha,\beta}\sum_{k\in\Phi^{\text{DP}}}^{\alpha,\beta}\sum_{l\in\Phi^{\text{DP}}}^{\alpha,\beta} ij|kl\left(S^{-1}\right)_{ik}\left(S^{-1}\right)_{jl}\right] \\
& - \frac{1}{2}\left(\begin{array}{l} \displaystyle\sum_{i\in\Phi^{\text{DP}}}^{\alpha}\sum_{j\in\Phi^{\text{DP}}}^{\alpha}\sum_{k\in\Phi^{\text{DP}}}^{\alpha}\sum_{l\in\Phi^{\text{DP}}}^{\alpha} ij|lk\left(S^{-1}\right)_{il}\left(S^{-1}\right)_{jk} \\ \displaystyle + \sum_{i\in\Phi^{\text{DP}}}^{\beta}\sum_{j\in\Phi^{\text{DP}}}^{\beta}\sum_{k\in\Phi^{\text{DP}}}^{\beta}\sum_{l\in\Phi^{\text{DP}}}^{\beta} ij|lk\left(S^{-1}\right)_{il}\left(S^{-1}\right)_{jk} \end{array}\right)
\end{aligned}
\tag{11}
$$

The correlation term (ΔE^{corr}) shows the correlation contributions from the KS-DFT calculation. It is defined as

$$
\Delta E^{\text{corr}} = E_{\text{S}}^{\text{corr}} - \sum_{\text{A}} E_{\text{A}}^{\text{corr}} \tag{12}
$$

where $E_{\text{S}}^{\text{corr}}$ and $E_{\text{A}}^{\text{corr}}$ represent the correlation energies of supermolecule S and monomer A, respectively. With KS orbitals, $E_{\text{S}}^{\text{corr}}$ and $E_{\text{A}}^{\text{corr}}$ can be rewritten as

$$
E_{\text{S}}^{\text{corr}} = (1-a)\left[E_{\text{X}}\left(\rho_{\text{S}}^{\alpha},\rho_{\text{S}}^{\beta}\right) + \frac{1}{2}\left(\sum_{i,j\in\phi_{\text{S}}}^{\alpha}\langle ij|ji\rangle + \sum_{i,j\in\phi_{\text{S}}}^{\beta}\langle ij|ji\rangle\right)\right] + E_{\text{C}}\left(\rho_{\text{S}}^{\alpha},\rho_{\text{S}}^{\beta}\right) \tag{13}
$$

$$
E_{\text{A}}^{\text{corr}} = (1-a)\left[E_{\text{X}}\left(\rho_{\text{A}}^{\alpha},\rho_{\text{A}}^{\beta}\right) + \frac{1}{2}\left(\sum_{i,j\in\phi_{\text{A}}}^{\alpha}\langle ij|ji\rangle + \sum_{i,j\in\phi_{\text{A}}}^{\beta}\langle ij|ji\rangle\right)\right] + E_{\text{C}}\left(\rho_{\text{A}}^{\alpha},\rho_{\text{A}}^{\beta}\right) \tag{14}
$$

where a is the portion of exact HF exchange in a hybrid functional, within the range of 0.0 to 1.0; $E_{\text{X}}\left(\rho_{\text{S}}^{\alpha},\rho_{\text{S}}^{\beta}\right)$ and $E_{\text{C}}\left(\rho_{\text{S}}^{\alpha},\rho_{\text{S}}^{\beta}\right)$ are the exchange and correlation energy of the hybrid functional, respectively.

The dispersion term is added if a dispersion correction is applied for DFT functionals:

$$
\Delta E^{\text{disp}} = E_{\text{S}}^{\text{disp}} - \sum_{\text{A}} E_{\text{A}}^{\text{disp}} \tag{15}
$$

If Grimme's DFT-D dispersion correction [30, 31] is used, an additional term ΔE^{disp} can be written as

$$\Delta E^{\mathrm{disp}} = -\frac{1}{2}\sum_{p\neq q\in \mathrm{S}}\sum_{n=6,8,10} s_n \frac{C_n^{pq}}{R_{pq}^n} f_{\mathrm{damp},n}(R_{pq}) + \frac{1}{2}\sum_{p\neq q\in \mathrm{A}}\sum_{n=6,8,10} s_n \frac{C_n^{pq}}{R_{pq}^n} f_{\mathrm{damp},n}(R_{pq}) + \frac{1}{2}\sum_{p\neq q\in \mathrm{B}}\sum_{n=6,8,10} s_n \frac{C_n^{pq}}{R_{pq}^n} f_{\mathrm{damp},n}(R_{pq}) \quad (16)$$

where s_n is the global (functional dependent) scaling factor, and C_n^{pq} and R_{pq} are the averaged nth dispersion coefficient and distance between atoms p and q, respectively; $f_{\mathrm{damp},\ n}$ is the damping function determining the range of the dispersion correction. Conventionally, for evaluating the vdW character in intermolecular interactions, the dispersion and correlation terms are often combined as a single term of $\Delta E^{\mathrm{corr/disp}}$. Finally, to overcome basis set superposition error (BSSE), the monomer orbitals in these terms are expanded within the supermolecule's basis set.

2.2 *GKS-EDA for Intermolecular Interactions in Various Environments*

Most of the EDA methods are proposed for gas-phase interactions. Due to the complexity of intermolecular interactions influenced by solvent effects, the analysis methods for intermolecular interactions with solvated environments remain limited [32–34]. With the development of materials science and life science, understanding solute-solvent or solute-solute interactions has become increasingly important. In quantum chemistry, solvent effects can be modeled using either implicit or explicit solvation approaches [35, 36]. Implicit solvation models, including the polarizable continuum model (PCM) [37–39], universal solvation model (abbreviated as SMD, here D denotes density) [40], or conductor-like screening model (COSMO) [41, 42], approximate a large number of solvent molecules as a continuous dielectric medium. They have become widely used in studies in the solution phase. Implicit solvation models are simple and inexpensive but lack the description for specific short-range effects. In explicit solvation models, solvent molecules can be represented explicitly by quantum mechanics or molecular mechanics (MM). Considering the balance between accuracy and efficiency, quantum mechanics/molecular mechanics (QM/MM) offers a versatile framework for investigating chemical reactions [43–45], enzyme catalysis [46, 47], and behavior of large biomolecules [48, 49]. This approach partitions a molecular system into two distinct regions: an active region and an environment. The active region is treated with quantum mechanical methods, such as KS-DFT or wavefunction-based methods, while the environment is modeled by molecular mechanics (MM). In general, QM/MM is able to provide more detailed information on solvent-solute interactions but is expensive compared to implicit solvation models.

We choose different strategies to consider environment effects, leading to the two extensions of GKS-EDA below:

2.2.1 GKS-EDA(sol)

GKS-EDA(sol) is the extended version of GKS-EDA with implicit solvent effects [24]. When a supermolecule S consisting of several monomers A in a polarizable medium is considered, the total free energy change (ΔG^{TOT}) becomes

$$\Delta G^{\mathrm{TOT}} = G_{\mathrm{S}}^{\mathrm{TOT}} - \sum_{\mathrm{A}} G_{\mathrm{A}}^{\mathrm{TOT}} = \left(\left\langle \Phi_{\mathrm{S}}^{\mathrm{SOL}} \middle| F_{\mathrm{S}}^{\mathrm{GKS}} \middle| \Phi_{\mathrm{S}}^{\mathrm{SOL}} \right\rangle - \sum_{\mathrm{A}} \left\langle \Phi_{\mathrm{A}}^{\mathrm{SOL}} \middle| F_{\mathrm{A}}^{\mathrm{GKS}} \middle| \Phi_{\mathrm{A}}^{\mathrm{SOL}} \right\rangle \right) + \left(\Delta G_{\mathrm{S}}^{\mathrm{SOL}} - \sum_{\mathrm{A}} \Delta G_{\mathrm{A}}^{\mathrm{SOL}} \right) \tag{17}$$

The total interaction free energy can be further decomposed into electrostatic (ΔG^{ele}), exchange (ΔG^{ex}), repulsion (ΔG^{rep}), polarization (ΔG^{pol}), and correlation (ΔG^{corr}) terms. The definitions of these terms are the same as the corresponding terms in GKS-EDA, except that the orbitals are optimized by SCRF in implicit solvation model.

The last parenthesis on the right-hand side of Eq. (17) defines the desolvation term ($\Delta G^{\mathrm{desol}}$), the free energy penalty by the environment due to the interaction between monomers. It should be noted that $\Delta G^{\mathrm{desol}}$ is not equal to the difference of solvation free energies between the sum of monomers and the supermolecule, because their computation with the corresponding monomer's or supermolecule's own cavities neglects the loss of solvent-solute interactions due to short distances. Thus, the monomer solvation free energies would be too negative, leading to the underestimation of total binding energy. As illustrated in Fig. 1, GKS-EDA(sol) overcomes this error by constructing the supermolecule's cavity according to interacting distances. When interacting distances are short, monomers are located in the same cavity, properly describing the solvent environments of monomers.

GKS-EDA(sol) can be used in combination with various implicit solvation models, including SMD [40], CPCM [50, 51], IEF-PCM [52–55], and Het-CPCM [56], to consider the influence of heterogeneous or homogeneous solvent effects on bonding and nonbonding intermolecular interactions.

2.2.2 GKS-EDA(QM/MM)

In GKS-EDA(QM/MM) [57], according to the QM/MM scheme, without loss of generality, the interacting molecular system can be divided into two monomers, where each monomer can be further divided into MM regions and QM regions, as shown in Fig. 2. It means that one monomer contains regions A and C, while the other contains regions B and D.

If the system energy satisfies pairwise additivity, the total interaction energy can be expressed as the summation of four parts:

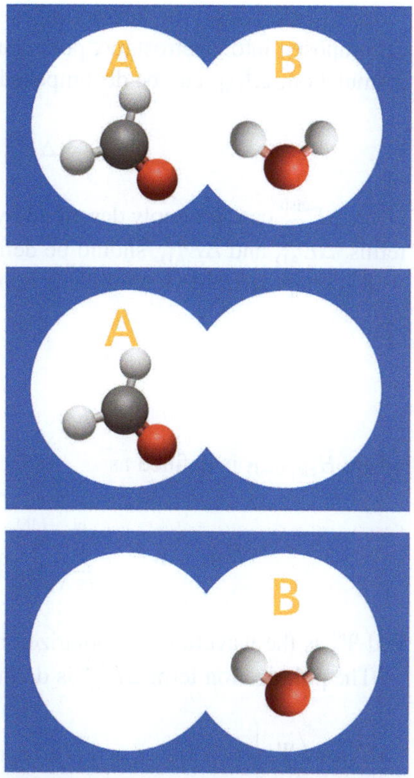

Fig. 1 Cavity definitions in the GKS-EDA(sol). The blue boxes represent the implicit solvation environment

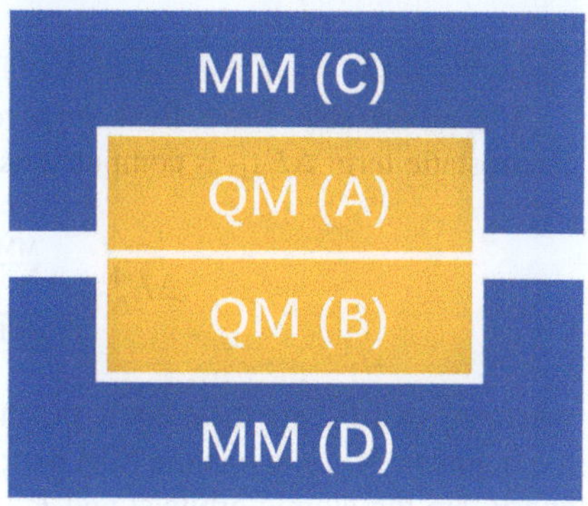

Fig. 2 Fragmentation scheme in GKS-EDA (QM/MM)

$$\Delta E^{\mathrm{TOT}} = \Delta E_{\mathrm{AB}} + \Delta E_{\mathrm{CD}} + \Delta E_{\mathrm{AD}} + \Delta E_{\mathrm{BC}} \tag{18}$$

ΔE_{AB}, the interaction energy between the QM regions A and B, can be decomposed into electrostatic, exchange-repulsion, polarization, correlation, and dispersion terms by the standard GKS-EDA.

ΔE_{AD} and ΔE_{BC}, the interaction energies between QM and MM regions, can be decomposed into electrostatic, polarization, and dispersion terms, sharing the same formulation. ΔE_{AD} can be decomposed as

$$\Delta E_{\mathrm{AD}} = \Delta E_{\mathrm{AD}}^{\mathrm{ele}} + \Delta E_{\mathrm{AD}}^{\mathrm{pol}} + \Delta E_{\mathrm{AD}}^{\mathrm{disp}} \tag{19}$$

Here, $\Delta E_{\mathrm{AD}}^{\mathrm{disp}}$ can be simply described by a dispersion correction, while the other two terms, $\Delta E_{\mathrm{AD}}^{\mathrm{ele}}$ and $\Delta E_{\mathrm{AD}}^{\mathrm{pol}}$, should be defined with the formulations of force fields.

If a nonpolarizable force field is applied, the electrostatic term $\Delta E_{\mathrm{AD}}^{\mathrm{ele}}$ would be

$$\Delta E_{\mathrm{AD}}^{\mathrm{ele}} = \left\langle \Psi_{\mathrm{A}}^{0} \middle| \widehat{H}_{\mathrm{ele,A}\leftrightarrow\mathrm{D}} \middle| \Psi_{\mathrm{A}}^{0} \right\rangle \tag{20}$$

where $\widehat{H}_{\mathrm{ele,A}\leftrightarrow\mathrm{D}}$ is defined as

$$\widehat{H}_{\mathrm{ele,A}\leftrightarrow\mathrm{D}} = -\sum_{i\in\mathrm{A}}^{\mathrm{el.}} \sum_{a\in\mathrm{D}}^{\mathrm{MM}} \frac{q_a}{|\mathbf{r}_i - \mathbf{R}_a|} + \sum_{I\in\mathrm{A}}^{\mathrm{nuc.}} \sum_{a\in\mathrm{D}}^{\mathrm{MM}} \frac{Z_I q_a}{|\mathbf{R}_I - \mathbf{R}_a|} \tag{21}$$

and Ψ_{A}^{0} is the wavefunction polarized by the MM part of D.

The polarization term $\Delta E_{\mathrm{AD}}^{\mathrm{pol}}$ is defined as

$$\Delta E_{\mathrm{AD}}^{\mathrm{pol}} = \left\langle \Psi_{\mathrm{A}} \middle| \widehat{H}_{\mathrm{A}} + \widehat{H}_{\mathrm{A}\leftrightarrow\mathrm{C,ele}} + \widehat{H}_{\mathrm{A}\leftrightarrow\mathrm{D,ele}} \middle| \Psi_{A} \right\rangle - \left\langle \Psi_{\mathrm{A}}^{0} \middle| \widehat{H}_{\mathrm{A}} + \widehat{H}_{\mathrm{A}\leftrightarrow\mathrm{C,ele}} + \widehat{H}_{\mathrm{A}\leftrightarrow\mathrm{D,ele}} \middle| \Psi_{\mathrm{A}}^{0} \right\rangle \tag{22}$$

where Ψ_{A} is the wavefunction of A polarized by C and D.

If a polarizable force field (induced point dipole approach) is applied, the electrostatic term ΔE_{AD} is computed as

$$\Delta E_{\mathrm{AD}}^{\mathrm{ele}} = \sum_{a\in\mathrm{D}}^{\mathrm{MM}} \phi_a^{\mathrm{A}} q_a - \sum_{a\in\mathrm{D}}^{\mathrm{MM}} \boldsymbol{F}_{\boldsymbol{a}}^{\mathrm{A}} \boldsymbol{\mu}_{\boldsymbol{a}} \tag{23}$$

where ϕ_a^{A} and $\boldsymbol{F}_{\boldsymbol{a}}^{\mathrm{A}}$ are the electric potential and electric field of the wavefunction in the fragment A at site a in the MM part; $\boldsymbol{\mu}_{\boldsymbol{a}}$ is the equilibrated induced dipole at a; χ_μ and χ_ν are the atomic orbitals; and $P_{\mu\nu}$ is the density matrix in the basis of atomic orbitals:

$$\phi_a^{\mathrm{A}} = \sum_{I\in\mathrm{A}}^{\mathrm{nuc.}} \frac{Z_I}{|\mathbf{R}_I - \mathbf{R}_a|} - \sum_{\mu,\nu\in\mathrm{A}}^{\mathrm{el.}} P_{\mu\nu} \left\langle \chi_\mu \middle| \frac{1}{|\boldsymbol{r} - \mathbf{R}_a|} \middle| \chi_\nu \right\rangle_r \tag{24}$$

$$\boldsymbol{F}_a^{\mathrm{A}} = -\nabla_a \phi_a^{\mathrm{A}} = -\sum_{I \in \mathrm{A}}^{\text{nuc.}} \nabla_a \left(\frac{Z_I}{|\mathbf{R}_I - \mathbf{R}_a|} \right) + \sum_{\mu,\nu \in \mathrm{A}}^{\text{el.}} P_{\mu\nu} \nabla_a \left\langle \chi_\mu \middle| \frac{1}{|\boldsymbol{r} - \mathbf{R}_a|} \middle| \chi_\nu \right\rangle_r \tag{25}$$

The mutual polarization between A and D leads to the relaxation of wavefunction at A and update of induced dipole at D. The electric potential ϕ_a^{A} and electric field $\boldsymbol{F}_a^{\mathrm{A}}$ of the wavefunction at A will in turn become $\tilde{\phi}_a^{\mathrm{A}}$ and $\tilde{\boldsymbol{F}}_a^{\mathrm{A}}$. Therefore, the polarization term can be expressed as

$$\begin{aligned}
\Delta E_{\mathrm{AD}}^{\mathrm{pol}} = & \left\langle \Psi_{\mathrm{A}} \middle| \widehat{H}_{\mathrm{A}} \middle| \Psi_{\mathrm{A}} \right\rangle - \left\langle \Psi_{\mathrm{A}}^0 \middle| \widehat{H}_{\mathrm{A}} \middle| \Psi_{\mathrm{A}}^0 \right\rangle + \sum_{a \in \mathrm{D}}^{\mathrm{MM}} \tilde{\phi}_a^{\mathrm{A}} q_a - \sum_{a \in \mathrm{D}}^{\mathrm{MM}} \tilde{\mathbf{F}}_a^{\mathrm{A}} \tilde{\boldsymbol{\mu}}_a \\
& + \frac{1}{2} \sum_{a \in \mathrm{D}}^{\mathrm{MM}} \sum_{b \neq a \in \mathrm{D}}^{\mathrm{MM}} \tilde{\boldsymbol{\mu}}_a \nabla_b \left(\frac{1}{|\mathbf{R}_a - \mathbf{R}_b|} \right) q_b \\
& + \frac{1}{2} \sum_{a \in \mathrm{D}}^{\mathrm{MM}} \sum_{b \neq a \in \mathrm{D}}^{\mathrm{MM}} \tilde{\boldsymbol{\mu}}_a \nabla_a \nabla_b \left(\frac{1}{|\mathbf{R}_a - \mathbf{R}_\mathbf{b}|} \right) \tilde{\boldsymbol{\mu}}_\mathbf{b} \\
& + \sum_{a \in \mathrm{D}}^{\mathrm{MM}} \frac{\tilde{\boldsymbol{\mu}}_a \cdot \tilde{\boldsymbol{\mu}}_a}{2\alpha_a} - \sum_{a \in \mathrm{D}}^{\mathrm{MM}} \phi_a^{\mathrm{A}} q_a + \sum_{a \in \mathrm{D}}^{\mathrm{MM}} \mathbf{F}_a^{\mathrm{A}} \boldsymbol{\mu}_a
\end{aligned} \tag{26}$$

where α_a is the polarizability of the site a.

ΔE_{CD}, the interaction energy between the MM regions, includes electrostatics, polarization, and dispersion terms if a polarizable force field is applied or only electrostatics and dispersion terms when a classical force field is employed. The electrostatic term is generally expressed using a classical point charge model, and the dispersion term using a Lennard-Jones potential.

Finally, the electrostatic, exchange-repulsion, polarization, and dispersion terms are defined as

$$\Delta E^{\mathrm{ele}} = \Delta E_{\mathrm{AB}}^{\mathrm{ele}} + \Delta E_{\mathrm{AD}}^{\mathrm{ele}} + \Delta E_{\mathrm{BC}}^{\mathrm{ele}} + \Delta E_{\mathrm{CD}}^{\mathrm{ele}} \tag{27}$$

$$\Delta E^{\mathrm{exrep}} = \Delta E_{\mathrm{AB}}^{\mathrm{exrep}} \tag{28}$$

$$\Delta E^{\mathrm{pol}} = \Delta E_{\mathrm{AB}}^{\mathrm{pol}} + \Delta E_{\mathrm{AD}}^{\mathrm{pol}} + \Delta E_{\mathrm{BC}}^{\mathrm{pol}} + \Delta E_{\mathrm{CD}}^{\mathrm{pol}} + \Delta O \tag{29}$$

$$\Delta E^{\mathrm{corr}} = \Delta E_{\mathrm{AB}}^{\mathrm{corr}} \tag{30}$$

$$\Delta E^{\mathrm{disp}} = \Delta E_{\mathrm{AB}}^{\mathrm{disp}} + \Delta E_{\mathrm{AD}}^{\mathrm{disp}} + \Delta E_{\mathrm{BC}}^{\mathrm{disp}} + \Delta E_{\mathrm{CD}}^{\mathrm{disp}} \tag{31}$$

Because it has been illustrated that hybrid density functionals include some part of the dispersion which creates an overlap with the classical treatment [58], both correlation and dispersion terms are summed in our final form of GKS-EDA (QM/MM). Therefore, the total energy from QM/MM calculations would be separated as

$$\Delta E^{\mathrm{TOT}} = \Delta E^{\mathrm{ele}} + \Delta E^{\mathrm{exrep}} + \Delta E^{\mathrm{pol}} + \Delta E^{\mathrm{corr/disp}} \tag{32}$$

It is noted that in contrast to GKS-EDA(sol), the influence of solvent effects by GKS-EDA(QM/MM) is incorporated in electrostatic, polarization, correlation, and dispersion terms defined above, and no additional term is provided.

2.3 GKS-EDA for Intermolecular Interactions with Various Spin States

Intermolecular interactions involving radicals, especially for cases where there are two unpaired electrons in a molecular system, are prevalent in biochemistry, photophysics, materials science, etc. Given the multireference character arising from the degenerate frontier molecular orbitals, this kind of intermolecular interaction is challenging to single configuration-based EDA methods. Besides multireference methods, an alternative and widely used approach for this type of interactions is broken symmetry-unrestricted density functional theory (BS-UDFT) [59, 60].

In BS-UDFT theory, spin decontamination schemes are required because the unrestricted KS orbital is not an eigenfunction of the spin operator ($\widehat{S}^2$). Several schemes have been proposed for spin decontamination, including spin projection approximations such as the Yamaguchi scheme [61, 62], the Noodleman scheme [63, 64], or the constrained unrestricted mean field method (CUHF) [65–67].

GKS-EDA(BS) is the variant of GKS-EDA with the Yamaguchi approximation. The expression of the ground-state (GS) energy is

$$E_{\mathrm{GS}} = (1 + c)E_{\mathrm{BS}} - cE_{\mathrm{HS}} \tag{33}$$

$$c = \frac{\left\langle \widehat{S}^2 \right\rangle_{\mathrm{BS}}}{\left\langle \widehat{S}^2 \right\rangle_{\mathrm{HS}} - \left\langle \widehat{S}^2 \right\rangle_{\mathrm{BS}}} \tag{34}$$

E_{BS} and E_{HS} represent total energies in the broken symmetry (BS) singlet state and high-spin (HS) state [59, 68]. Here, HS state means that the spins of valence electrons are parallel. As such, the total interaction energy at the ground state of the interacting system is expressed as

$$\Delta E_{\mathrm{GS}}^{\mathrm{TOT}} = (1 + c)\Delta E_{\mathrm{BS}}^{\mathrm{TOT}} - c\Delta E_{\mathrm{HS}}^{\mathrm{TOT}} \tag{35}$$

where $\Delta E_{\mathrm{BS}}^{\mathrm{TOT}}$ and $\Delta E_{\mathrm{HS}}^{\mathrm{TOT}}$ are the total interaction energies in the BS state and HS state, respectively:

$$\Delta E_{\mathrm{BS}}^{\mathrm{TOT}} = E_{\mathrm{BS}} - E_{\mathrm{A}} - E_{\mathrm{B}} \tag{36}$$

$$\Delta E_{\mathrm{HS}}^{\mathrm{TOT}} = E_{\mathrm{HS}} - E_{\mathrm{A}} - E_{\mathrm{B}} \tag{37}$$

The total interaction energies in the BS state and HS state are both spin contaminated. In GKS-EDA(BS), the spin contamination in the HS state is neglected.

The interaction energy at a high-spin state $\Delta E_{\mathrm{HS}}^{\mathrm{TOT}}$ can be obtained by a GKS-EDA calculation with parallel spins of monomers, which shows that $\Delta E_{\mathrm{HS}}^{\mathrm{TOT}}$ can be decomposed as

$$\Delta E_{\mathrm{HS}}^{\mathrm{TOT}} = \Delta E_{\mathrm{HS}}^{\mathrm{ele}} + \Delta E_{\mathrm{HS}}^{\mathrm{exrep}} + \Delta E_{\mathrm{HS}}^{\mathrm{pol}} + \Delta E_{\mathrm{HS}}^{\mathrm{corr}} + \Delta E_{\mathrm{HS}}^{\mathrm{disp}} \tag{38}$$

The BS wavefunction ϕ_0^{BS} is constructed from the direct product of the monomers' wavefunctions. Unpaired electrons with different spins are located at different monomers; say, the α electron is located at monomer A, while the β electron is located at monomer B. With the BS wavefunction, $E_{\mathrm{BS}}^{\mathrm{TOT}}$ can be decomposed as

$$\Delta E_{\mathrm{BS}}^{\mathrm{TOT}} = \Delta E_{\mathrm{BS}}^{\mathrm{ele}} + \Delta E_{\mathrm{BS}}^{\mathrm{ex}} + \Delta E_{\mathrm{BS}}^{\mathrm{rep}} + \Delta E_{\mathrm{BS}}^{\mathrm{pol}} + \Delta E_{\mathrm{BS}}^{\mathrm{corr}} + \Delta E_{\mathrm{BS}}^{\mathrm{disp}} \tag{39}$$

Therefore, the total interaction energy is decomposed as

$$\Delta E_{\mathrm{GS}}^{\mathrm{TOT}} = \Delta E_{\mathrm{GS}}^{\mathrm{ele}} + \Delta E_{\mathrm{GS}}^{\mathrm{exrep}} + \Delta E_{\mathrm{GS}}^{\mathrm{pol}} + \Delta E_{\mathrm{GS}}^{\mathrm{corr}} + \Delta E_{\mathrm{GS}}^{\mathrm{disp}} \tag{40}$$

$$\Delta E_{\mathrm{GS}}^{\mathrm{X}} = (1+c)\Delta E_{\mathrm{BS}}^{\mathrm{X}} - c\Delta E_{\mathrm{HS}}^{\mathrm{X}}, \quad \mathrm{X} = \mathrm{ele}, \mathrm{exrep}, \mathrm{pol}, \mathrm{corr}, \mathrm{disp} \tag{41}$$

GKS-EDA(BS) has been successfully applied in various open-shell singlet (OSS)-type intermolecular interactions, such as the pancake bond in the phenalenyl dimer [69] and donor-acceptor interactions in molecular switches [70]. The shortcoming of GKS-EDA(BS) is that when the spin contamination of high-spin state cannot be neglected, GKS-EDA(BS) would meet difficulty. As pointed out by literature [71–73], the spin contamination of high-spin state sometimes cannot be neglected.

GKS-EDA(CU) is the variant of GKS-EDA combined with the CUHF/CUDFT method [74]. In CUHF/CUDFT, the CUHF/CUDFT Fock matrices, which are constructed from UHF/UKS orbitals, can be divided into core, active, and virtual spaces according to the distribution of occupation numbers in the natural orbital basis. For open-shell singlet state and high-spin state, the active space contains two natural orbitals for two electrons with antiparallel spins and parallel spins, respectively. In order to eliminate spin contamination, the $\boldsymbol{\lambda}$ matrix is constructed by the core-virtual (c-v) block of $\mathbf{F}^{\alpha}$ and $\mathbf{F}^{\beta}$ matrices of the UHF method and then used to transform the conventional UHF Fock matrices $\mathbf{F}^{\alpha}$ and $\mathbf{F}^{\beta}$ into the corresponding CUHF/CUDFT Fock matrices $\mathbf{F}_{\mathrm{CUHF}}^{\alpha}$ and $\mathbf{F}_{\mathrm{CUHF}}^{\beta}$. Thereafter, the CUHF/CUDFT energy and wavefunction could be obtained with the SCF iteration by using the newly formed Fock matrices.

In GKS-EDA(CU), based on CUHF/CUDFT calculations, the total interaction energy, $\Delta E_{\text{CU}}^{\text{TOT}}$, can be divided into electrostatic $\Delta E_{\text{CU}}^{\text{ele}}$, exchange-repulsion $\Delta E_{\text{CU}}^{\text{exrep}}$, polarization $\Delta E_{\text{CU}}^{\text{pol}}$, correlation $\Delta E_{\text{CU}}^{\text{corr}}$, and dispersion $\Delta E_{\text{CU}}^{\text{disp}}$:

$$\Delta E_{\text{CU}}^{\text{TOT}} = \Delta E_{\text{CU}}^{\text{ele}} + \Delta E_{\text{CU}}^{\text{ex}} + \Delta E_{\text{CU}}^{\text{rep}} + \Delta E_{\text{CU}}^{\text{pol}} + \Delta E_{\text{CU}}^{\text{corr}} + \Delta E_{\text{CU}}^{\text{disp}} \tag{42}$$

The definitions of electrostatic, exchange, repulsion, and polarization terms in GKS-EDA(CU) are the same as the corresponding terms in the GKS-EDA, except that the KS orbitals in GKS-EDA(CU) are optimized by the SCF procedure of CUDFT. The $\Delta E_{\text{CU}}^{\text{disp}}$ term is the same as that in GKS-EDA because the Grimme dispersion correction (D2 and D3 version) is not related to the wavefunction.

GKS-EDA(CU) can handle spin contamination for both open-shell singlet states and high-spin states without a spin projection correction. All the energy terms in Eq. (42) share the same definitions as in other GKS-EDA variants, except that the KS orbitals are optimized with the CUDFT method. Different from GKS-EDA(BS), GKS-EDA(CU) only computes the EDA terms once.

Understanding magnetic interactions in molecules and highly correlated materials is an important task in theoretical chemistry [75]. In the study of magnetic interactions with two unpaired electrons, by the scheme of Noodleman [64], the exchange coupling constant (J) is defined as the energy difference between the singlet spin state and triplet spin state. Computation of the exchange coupling requires the correct description for both these two spin states [74]. According to the advantage of GKS-EDA(CU), it can be employed in the decomposition of the variation of exchange coupling constant ΔJ:

$$\Delta J = \Delta E_{\text{CU,S}}^{\text{int}} - \Delta E_{\text{CU,T}}^{\text{int}} = \Delta\Delta E_{\text{CU}}^{\text{ele}} + \Delta\Delta E_{\text{CU}}^{\text{exrep}} + \Delta\Delta E_{\text{CU}}^{\text{pol}} + \Delta\Delta E_{\text{CU}}^{\text{corr}} \tag{43}$$

Here, $\Delta\Delta E_{\text{CU}}^{\text{ele}}$, $\Delta\Delta E_{\text{CU}}^{\text{exrep}}$, $\Delta\Delta E_{\text{CU}}^{\text{pol}}$, and $\Delta\Delta E_{\text{CU}}^{\text{corr}}$ are defined as

$$\Delta\Delta E_{\text{CU}}^{\text{X}} = \Delta E_{\text{CU,S}}^{\text{X}} - \Delta E_{\text{CU,T}}^{\text{X}}, \quad \text{X} = \text{ele, exrep, pol, corr} \tag{44}$$

Thus, the variation of magnetic coupling by external effects can be explored from the GKS-EDA(CU) results.

2.4 *GKS-EDA for Excited States*

Molecular systems in electronic excited states are typically short-lived and highly reactive, presenting significant challenges for theoretical chemistry. Given a satisfactory balance between computational efficiency and accuracy, time-dependent density functional theory (TD-DFT) [76, 77], which is derived by considering the linear response of the time-dependent approximate ground-state wavefunction, is widely used for medium-sized molecules in excited states.

Since the 2010s, several EDA methods for excited states have been introduced. In 2018, based on the configuration interaction singles (CIS) method and TD-DFT, ALMO-EDA was extended to explore intermolecular interactions in excimers and exciplexes [78]. Lately, the SAPT method was extended to consider molecular complexes with localized excitons by means of multiconfigurational SAPT: SAPT (MC). It describes electrostatic, exchange, induction, and dispersion interaction energy components based on a multiconfiguration wavefunction [79]. In 2023, a multistate energy decomposition analysis (MS-EDA) method for excimers using density functional theory was introduced by Gao et al. [80] MS-EDA provides a clear separation of energy terms that can be variationally optimized for excited-state complexes based on the Lu-Gao theorem, which is able to straightforwardly classify exciton coupling, super-exchange resonance, and orbital configuration expansion.

Based on TD-DFT calculations, we recently presented the GKS-EDA(TD) method, which analyzes intermolecular interactions in the case that one of the monomers A is in its singly excited state while the other monomers are in their ground states [81]. The excitation energies of monomer A and supermolecule S are ω_A and ω_S, respectively. The wavefunction of monomer B is denoted as Φ_B, while the corresponding wavefunctions of monomer A and supermolecule S can be expressed as

$$\Psi_A^* = \sum_{i,a \in A} c_i^a \Phi_i^a \tag{45}$$

$$\Psi_S^* = \sum_{i,a \in S} d_i^a \Theta_i^a \tag{46}$$

Here, the asterisk in the superscript means excited state. Index i represents occupied orbitals, while a represents unoccupied orbitals; c_i^a and d_i^a are the coefficients of excited configurations for monomer A (Φ_i^a) and supermolecule S (Θ_i^a), respectively. A pseudo-direct-product wavefunction Ψ_{DP}^* is defined as

$$\Psi_{DP}^* = \sum_{i,a \in A} c_i^a \Phi_i^a \Phi_B \tag{47}$$

The total interaction energy in an excited state ΔE^{TOT*} is decomposed into these energy components (electrostatic, exchange-repulsion, polarization, correlation, and dispersion):

$$\Delta E^{TOT*} = \Delta E^{ele*} + \Delta E^{exrep*} + \Delta E^{pol*} + \Delta E^{corr*} + \Delta E^{disp} \tag{48}$$

The asterisk denotes the interaction term for the excited state. The definition of dispersion in GKS-EDA(TD) is the same as in GKS-EDA when the Grimme dispersion correction is applied.

For ΔE^{ele*},

$$\Delta E^{\text{ele}*} = \sum_{i,a\in \text{A}} \left(c_i^a\right)^2 \left\langle \Phi_i^a \Phi_\text{B} \middle| \widehat{F}^0 \middle| \Phi_i^a \Phi_\text{B} \right\rangle - \sum_{i,a\in \text{A}} \left(c_i^a\right)^2 \left\langle \Phi_i^a \middle| \widehat{F}^0 \middle| \Phi_i^a \right\rangle - \left\langle \Phi_\text{B} \middle| \widehat{F}^0 \middle| \Phi_\text{B} \right\rangle + \Delta V_\text{NN} \tag{49}$$

$\Delta E^{\text{exrep}*}$ is the sum of $\Delta E^{\text{ex}*}$ and $\Delta E^{\text{rep}*}$:

$$\Delta E^{\text{ex}*} = \sum_{i,a\in \text{A}} \left(c_i^a\right)^2 \left(\left\langle \widehat{A}\Phi_i^a \Phi_\text{B} \middle| \widehat{F}^0 \middle| \widehat{A}\Phi_i^a \Phi_\text{B} \right\rangle - \left\langle \Phi_i^a \Phi_\text{B} \middle| \widehat{F}^0 \middle| \Phi_i^a \Phi_\text{B} \right\rangle \right) \tag{50}$$

$$\Delta E^{\text{rep}*} = \sum_{i,a\in \text{A}} \left(c_i^a\right)^2 \left(\left\langle \widehat{N}\widehat{A}\Phi_i^a \Phi_\text{B} \middle| \widehat{F}^0 \middle| \widehat{N}\widehat{A}\Phi_i^a \Phi_\text{B} \right\rangle - \left\langle \widehat{A}\Phi_i^a \Phi_\text{B} \middle| \widehat{F}^0 \middle| \widehat{A}\Phi_i^a \Phi_\text{B} \right\rangle \right) \tag{51}$$

The exchange term ($\Delta E^{\text{ex}*}$) arises from enforcing the direct product of the determinant of monomer B (Φ_B) and each determinant of monomer A (Φ_i^a) to obey antisymmetry requirements, while the repulsion term ($\Delta E^{\text{rep}*}$) arises from the energy penalty from the orthonormalization of monomer's wavefunction in each pair of determinants.

For $\Delta E^{\text{pol}*}$,

$$\Delta E^{\text{pol}^*} = \sum_{i,a\in \text{S}} \left(d_i^a\right)^2 \langle \Theta_i^a | \hat{F}^0 | \Theta_i^a \rangle - \sum_{i,a\in \text{A}} \left(c_i^a\right)^2 \langle \Phi_i^a | \hat{F}^0 | \Phi_i^a \rangle - \langle \Phi_\text{B} | \hat{F}^0 | \Phi_\text{B} \rangle + \Delta E_\text{OV} \tag{52}$$

The additional ΔE_OV term denotes the energy contribution arising from the interplay between occupied orbitals and virtual (unoccupied) orbitals from monomer A to the supermolecule S:

$$\begin{aligned} \Delta E_\text{OV} = & \sum_{i,j,a,b\in \text{S}} d_i^a d_j^b \left\langle ja \middle\| ib \right\rangle \left(1 - \delta_{ij}\delta_{ab}\right) - \sum_{i,j,a,b,\in \text{A}} c_i^a c_j^b \left\langle ja \middle\| ib \right\rangle \left(1 - \delta_{ij}\delta_{ab}\right) \\ & + \sum_{i,j,a,b\in \text{S}} d_i^a d_j^b \langle ja | f_\text{C}^\text{GKS} | ib \rangle - \sum_{i,j,a,b,\in \text{A}} c_i^a c_j^b \langle ja | f_\text{C}^\text{GKS} | ib \rangle \end{aligned} \tag{53}$$

$$f_\text{C}^\text{GKS} = \frac{\delta E^\text{corr}}{\delta \rho} \tag{54}$$

where indices i and j denote the occupied KS orbitals, while a and b represent the unoccupied KS orbitals and ρ is the electron density, achieved from the occupied KS orbitals.

Figure 3 summarizes the relationship of GKS-EDA and its extensions. With the exception of GKS-EDA(sol), the EDA terms of these extensions are the same as GKS-EDA. As such, by examination of the difference of EDA values in GKS-EDA and the extension, the complexity of intermolecular interactions can be readily

Fig. 3 The GKS-EDA family

interpreted. It is noted that with a given condition, these extensions will come back to GKS-EDA. For example, when the dielectric constant ε is zero, or the MM part is not considered, the results of GKS-EDA(sol) and GKS-EDA(QM/MM) are equal to the GKS-EDA ones.

3 Applications

Applications of the GKS-EDA for bonding and nonbonding interactions have been introduced in the recently published reviews [23] and chapter [82]. In this section, we present several representative cases to show the capability of GKS-EDA for complex systems.

3.1 Intermolecular Interactions in Solvated Environments

3.1.1 Ion–Probe Interactions

Chemical sensing in aqueous solution is a very active field for both experimental and theoretical research. The interactions of a urea-based probe with a wide variety of various ions (both cations and anions) in aqueous solution were explored by GKS-EDA(sol) [83].

For the cations (e.g., Na^+, Mg^{2+}, Ca^{2+}), the total interactions are dominated by the electrostatic term because of the high charge density of cations (e.g., Mg^{2+}, as shown in Fig. 4) interacting with the electron-rich phenyl group and urea oxygen. Polarization plays a secondary role, showing significant induction effects as cations polarize the probe's electron density (e.g., Mg^{2+} induces large polarization in the aromatic rings). The desolvation term is significantly positive due to disruption of the cation's tightly bound hydration shell in water. For example, Mg^{2+} experiences a large desolvation penalty when interacting with the probe via urea-enhanced cation-π interactions, as the ion's proximity to the phenyl group disrupts its hydration.

For the anions (e.g., ClO^-, SO_4^{2-}), electrostatic terms are also important, showing the primary driver for hydrogen bonding between urea's N–H groups and anion's electronegative atoms (e.g., O in ClO^-, as shown in Fig. 5). Polarization terms are

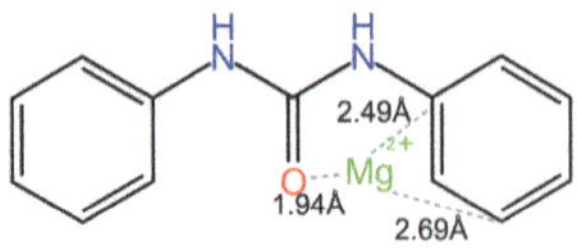

Fig. 4 The diagram of the probe···Mg^{2+} complex [83]

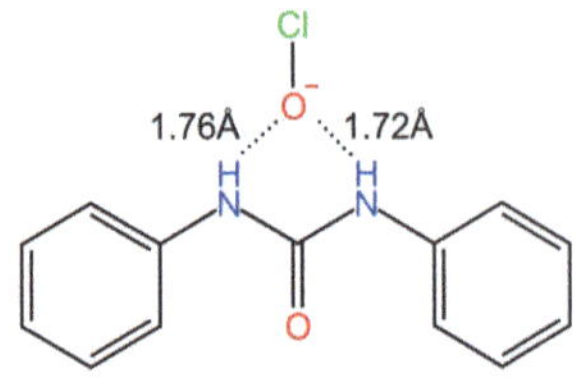

Fig. 5 The diagram of the probe···ClO^- complex [83]

moderate, arising from anion's charge redistribution during H-bond formation. Desolvation contributions are smaller compared to cations, as anions retain more of their solvation shell during hydrogen bonding with the urea group.

For neutral molecules (e.g., H_2O, HClO), the larger contributions come from the electrostatic term because of the weak H-bonding interactions with urea's N–H groups and the dispersion term because of the aromatic ring in the urea-based probe. According to the small desolvation terms, solvent effects are negligible due to their weaker solvation in water compared to ions.

As such, GKS-EDA(sol) was employed to understand the competing interaction mechanisms and establish a computational methodology for designing selective chemical sensors in aqueous environments.

3.1.2 Bonding Nature of Fe···CO in Protein Environment

The nature of Fe···CO bonding in heme has been extensively studied by various EDA methods [84, 85]. Liu et al. perform the GKS-EDA(sol) calculations at the level of B3LYP-D3BJ/def2-TZVP ($\varepsilon = 5.0$ for CPCM calculation) for the ligand interactions between CO and the Fe-active sites [84]. Figure 6 demonstrates the GKS-EDA(sol) results of the Fe(II)···CO and Fe(III)···CO bonding with two kinds of axis ligands opposite to CO. First, it was found that the electrostatic term dominates the total interaction while the contributions of polarization and correlation/dispersion terms are more sensitive to the ligand and the valence state of the Fe ion. Although the polarization term cannot be neglected, it does not have good correlation with the variation of Fe ions and the ligands. Different from the ion–probe interactions mentioned above, the desolvation terms are very small, showing that the protein environment has no significant effect on the iron-ligand interactions.

Second, the EDA results indicate that the ferrous···CO (Fe(II)···CO) ligand interactions are weaker than those of ferric···CO (Fe(III)···CO), primarily due to smaller electrostatic and correlation/dispersion contributions. The imidazole ligand exhibits larger correlation/dispersion terms compared to the thiolate ligand, attributable to the presence of its aromatic ring. This significant correlation/dispersion

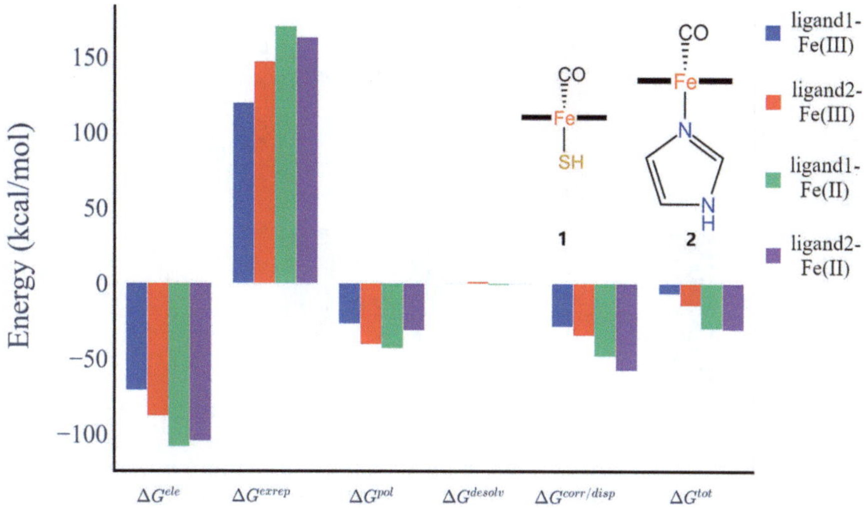

Fig. 6 GKS-EDA(sol) results of Fe(II)···CO and Fe(III)···CO bonding. "II" (singlet state) and "III" (doublet state) denote the valence of the iron atom

component explains why the iron center with the imidazole ligand demonstrates a stronger stabilization effect than that observed in the Fe···CO bond.

3.1.3 Protein-Ligand Interaction Between an MMP-13 Enzyme and Its Inhibitor Pyrimidinetrione

By using GKS-EDA(QM/MM), the protein-ligand interaction between an MMP-13 enzyme and its inhibitor pyrimidinetrione (PDB ID: 1YOU [86]) was explored [57]. As shown by Fig. 7, the QM part includes the pyrimidinetrione and the active site of MMP-13, which contains residues A186, H222, H226, H232, and ion Zn^{2+}. The remainder of MMP-13 is chosen as the MM part. For the GKS-EDA(QM/MM) calculation, the QM part is considered at the level of ωB97X-D/6-311G∗, while the MM part is treated by Amberff14SB force field.

According to the GKS-EDA(QM/MM) result, it can be shown that for the protein-ligand interaction, the electrostatic, polarization, and correlation/dispersion terms are all important although the electrostatic term is the largest contribution. The large electrostatic and polarization terms can be attributed to the $Zn^{2+}\cdots N$ interaction because Zn^{2+} is highly polarized by N-containing ligands, while the attractive correlation/dispersion term arises from the multiple van der Waals interactions between the hydrophobic rings and carbon chain in both the inhibitor and nearby protein residues.

As such, with the introduction of the QM/MM scheme, GKS-EDA is able to interpret the protein-ligand interaction within the explicit biomolecular environment. Compared to GKS-EDA(sol), there is no additional term for the surrounding

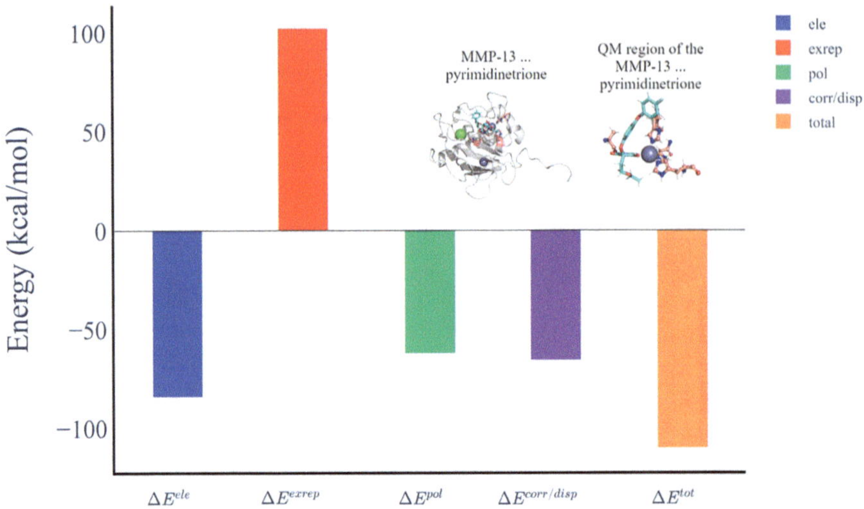

Fig. 7 GKS-EDA(QM/MM) result for MMP-13 complexed with a potent pyrimidinetrione inhibitor (QM: ωB97X-D/def2-TZVP; MM: ff14SB)

environment effects, which is absorbed in the other terms. As for the example above, it can be found that charge-charge interactions are not the only contributing factor for the protein-ligand interaction. The development of MMP-13 inhibitors should make use of this information and aim for ligands that include hydrophobic and polarizable groups.

3.2 Intermolecular Interactions Involving Radicals

3.2.1 Radical Pairing Interactions in Cyclobis(Paraquat-*p*-Phenylene) Inclusion Complexes

To explore the nature of intermolecular interactions in mechanically interlocked molecules (MIMs), Wang et al. applied the GKS-EDA(BS) to study the nature of intermolecular interactions in cyclobis(paraquat-*p*-phenylene) ($CBPQT^{n+}$, $n = 0$–4) inclusion complexes with various recognition units (RUs) [70]. The authors systematically analyzed how the charge states and multiplicities of $CBPQT^{n+}$ and RUs influence the intermolecular interactions in $RU\cdots CBPQT^{n+}$ complexes ($RU = BIPY^{2+}$, NDI, TTF, $BIPY^{\cdot+}$, $NDI^{\cdot-}$, $BTA^{\cdot}$) [70]. With the variation of spin states and charge state, the $RU\cdots CBPQT^{n+}$ interactions change from donor-acceptor interactions to radical pairing interactions.

Figure 8 shows the GKS-EDA(BS) result for the radical pairing interaction between $BIPY^{\cdot+}$ and $CBPQT^{\cdot 3+}$ at the level of ωB97X-D/6-31+G*. It can be found that the radical pairing interactions (e.g., between open-shell RUs and $CBPQT^{n+}$) are

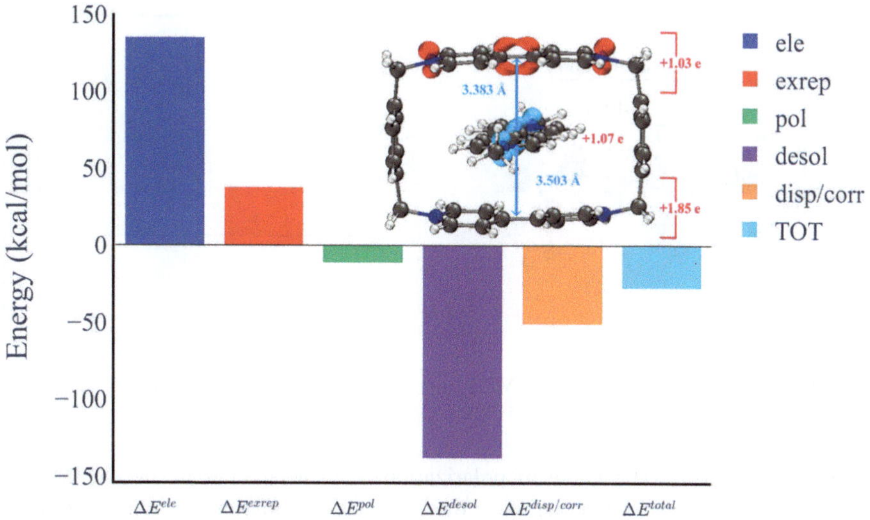

Fig. 8 GKS-EDA(BS) result of the $BIPY^{\cdot+}\cdots CBPQT^{\cdot3+}$ interaction at the level of ωB97X-D/6-31+G* in MeCN solution (kcal/mol); different spin densities are demonstrated in the diagram of the complex

dominated by correlation/dispersion terms, with small contributions of polarization, while for the donor-acceptor interactions (e.g., between closed-shell RUs and $CBPQT^{n+}$), there are significant polarization terms arising from electron transfer. As such, the physical origin of the radical pairing interactions is totally different from the donor-acceptor interactions. On the other hand, the electrostatic interaction plays a critical role when RUs are negatively charged, while the desolvation term counteracts repulsive electrostatic forces in cationic host-guest systems. Therefore, by GKS-EDA(BS), the subtle difference in interaction mechanisms (e.g., radical pairing vs. donor-acceptor) is revealed.

3.2.2 Magnetic Interaction

As shown by the literature [74], the value of the exchange coupling constant (J) of m-phenylene-based bis(aminoxyl) diradicals (denoted as MPB) by CUDFT is closer to high-level MRCI calculation and experimental value, while the data by BS-UDFT is seriously overestimated. It is shown that CUDFT is able to describe the triplet ground state while BS-UDFT cannot due to the spin contamination of high-spin state [60, 71, 87]. As such, compared to GKS-EDA(BS), GKS-EDA(CU) is more appropriate for the interactions involving the radical in the triplet state.

In the ground triplet state of the MPB···ion (ion = Na^+, or Cl^-) complexes, according to the GKS-EDA(CU) results at the M06-2X/cc-pVTZ level, the physical origin of the cation···π is quite different from the anion···π; for the cation···π in the MPB···Na^+ complex, the total interaction is controlled by the polarization and

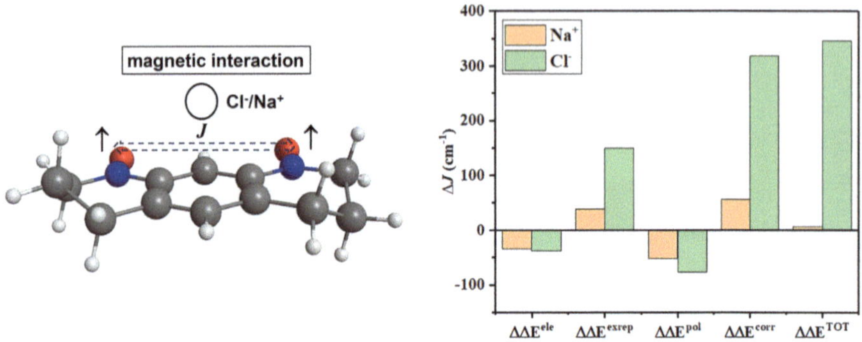

Fig. 9 Decomposition of ΔJ for the MPB···ion complexes by GKS-EDA(CU) at the level of M06-2X/cc-pVTZ

electrostatic terms. The value of polarization is even larger than the electrostatic term. For the anion···π interaction in the MPB···Cl^- complex, the anion···π interaction is governed by the electrostatic term, and the correlation term provides the secondary contribution. The polarization term in MPB···Cl^- is larger than the MPB···Na^+ one, showing the strong orbital relaxation effect.

Furthermore, GKS-EDA (CU) can be used to study the effects of external influences (such as ions) on the exchange coupling constant J. Figure 9 shows the variation of exchange coupling constant J in the MPB···ion complex when the MPB···ion distance is 2.0 Å. The results show that the anion significantly enhances the magnetic coupling constant, primarily attributed to the correlation term, while the exchange-repulsion term plays a secondary role. On the other hand, the cation has a minor impact on the variation of the J value. Whether cation or anion, the contributions of electrostatic and polarization are negative, showing that these two terms favor the antiparallel singlet state, while the correlation term always favors the parallel triplet state.

GKS-EDA(CU) shows the capability for quantitatively analyzing the differences in the contribution of each term to the magnetic coupling constants of singlet and triplet states, ΔJ, which helps understand the nature of the magnetic interaction in organic complexes.

3.3 *Intermolecular Interactions in Excited States*

The interaction between benzene and tetracyanoethylene (TCNE) with the lowest π–π* excitation belongs to a typical charge transfer excitation, which is totally different from that in the ground state. Arising from the excitation, electron transfer takes place between benzene and TCNE. The GKS-EDA(TD) analysis result for the benzene···TCNE complex with the 6-311++G** basis set is displayed in Table 1 [81]. For the total interaction in the ground state S_0, we found that all density functionals provide similar results, which indicate that the correlation/dispersion term dominates the π–π stacking interaction [81].

However, for the interaction in the benzene···TCNE complex with the S_1 excited state, the polarization term dominates because with the excitation, there are strong orbital interactions between benzene and TCNE, leading to the formation of $benzene^{+}$···$TCNE^{-}$ validated by the natural population analysis. For example, by CAM-B3LYP, the total interaction energy with the $\pi \rightarrow \pi^*$ excited state is −64.69 kcal/mol, which is mainly contributed to by the large value of polarization term, −77.0 kcal/mol. From the ground state to the excited state, all the functionals predict the increase of polarization term and the decrease of correlation term. However, the electrostatic interaction is almost unchanged as this reflects the Coulomb interaction between the excited benzene and TCNE. No electron transfer takes place. The variation from the neutral configuration to the ionic configuration arises from the π* (the excited benzene)–π* (TCNE) orbital mixture, which can be accounted for by the large polarization term.

4 Conclusions

In this chapter, the methodologies and applications of GKS-EDA family are introduced. It is shown that based on KS-DFT calculations, the GKS-EDA family can be successfully used for various interactions, from gas phase to solution phase, from closed-shell systems to open-shell systems, and from small systems to large biomolecular systems.

GKS-EDA family methods have been implemented in Xiamen Energy Decomposition Analysis (XEDA) program developed by our group [88]. Besides GKS-EDA methods, XEDA is able to perform LMO-EDA calculation at the post-Hartree-Fock level. Algorithms ensure that computational costs and memory usage are comparable to single-point energy calculations, enabling analysis of large systems containing thousands of atoms. XEDA combines speed, accuracy, and

Table 1 GKS-EDA(TD) results of the benzene···TCNE interaction in the $\pi \rightarrow \pi^*$ excited state with the 6-311++G** basis set (kcal/mol)

		ΔE^{ele}	ΔE^{exrep}	ΔE^{pol}	ΔE^{disp}	ΔE^{corr}	ΔE^{TOT}
LC-PBE	S_0	−7.85	14.99	−6.86	−6.89	−3.45	−10.06
	S_1: $\pi \rightarrow \pi^*$	−7.33	16.01	−62.12	−6.89	5.11	−55.21
M06-2X	S_0	−10.43	18.12	−5.76	−7.68	−12.57	−18.33
	S_1: $\pi \rightarrow \pi^*$	−11.64	23.19	−77.10	−7.68	0.11	−73.12
CAM-B3LYP	S_0	−10.79	19.00	−6.24	−7.01	−4.38	−9.42
	S_1: $\pi \rightarrow \pi^*$	−11.39	23.10	−77.00	−7.01	7.62	−64.69
LC-BLYP	S_0	−11.13	19.24	−6.14	−12.96	−8.92	−19.93
	S_1: $\pi \rightarrow \pi^*$	−11.81	28.52	−61.97	−12.96	−5.26	−63.48

flexibility, making it a powerful tool for quantitative analysis of intermolecular interactions, ranging from weak non-covalent bonds to strong covalent bonds.

Acknowledgments This project is supported by the National Natural Science Foundation of China (Nos. 22473092 and 22173076).

References

1. Stone A (2013) The theory of intermolecular forces. Oxford University Press
2. Kryachko ES (2006). In: Kaplan IG (ed) Intermolecular interactions: physical picture, computational methods and model potentials. John Wiley & Sons Ltd., Chichester. ISBN-13: 978-0-470-86332-9. Int J Quantum Chem 2008;108(1):198–201
3. Otero de la Roza A, DiLabio GA (eds) (2017) Non-covalent interactions in quantum chemistry and physics. Elsevier, p iv
4. Lefebvre C, Khartabil H, Boisson J-C, Contreras-García J, Piquemal J-P, Hénon E (2018) The independent gradient model: a new approach for probing strong and weak interactions in molecules from wave function calculations. ChemPhysChem 19(6):724–735
5. Bader RFW, Carroll MT, Cheeseman JR, Chang C (1987) Properties of atoms in molecules: atomic volumes. J Am Chem Soc 109(26):7968–7979
6. Bader RFW (1991) A quantum theory of molecular structure and its applications. Chem Rev 91(5):893–928
7. Johnson ER, Keinan S, Mori-Sánchez P, Contreras-García J, Cohen AJ, Yang W (2010) Revealing Noncovalent Interactions. J Am Chem Soc 132(18):6498–6506
8. Hohenstein EG, Sherrill CD (2012) Wavefunction methods for noncovalent interactions. WIREs Comput Mol Sci 2(2):304–326
9. Jansen G (2014) Symmetry-adapted perturbation theory based on density functional theory for noncovalent interactions. WIREs Comput Mol Sci 4(2):127–144
10. Jeziorski B, Moszynski R, Szalewicz K (1994) Perturbation theory approach to intermolecular potential energy surfaces of van Der Waals complexes. Chem Rev 94(7):1887–1930
11. Andrés J, Ayers PW, Boto RA, Carbó-Dorca R, Chermette H, Cioslowski J, Contreras-García J, Cooper DL, Frenking G, Gatti C, Heidar-Zadeh F, Joubert L, Martín Pendás Á, Matito E, Mayer I, Misquitta AJ, Mo Y, Pilmé J, Popelier PLA, Rahm M, Ramos-Cordoba E, Salvador P, Schwarz WHE, Shahbazian S, Silvi B, Solà M, Szalewicz K, Tognetti V, Weinhold F, Zins É-L (2019) Nine questions on energy decomposition analysis. J Comput Chem 40(26):2248–2283
12. Pastorczak E, Corminboeuf C (2017) Perspective: found in translation: quantum chemical tools for grasping non-covalent interactions. J Chem Phys 146(12):120901
13. Phipps M, Fox T, Tautermann CS, Skylaris C-K (2015) Energy decomposition analysis approaches and their evaluation on prototypical protein–drug interaction patterns. Chem Soc Rev 44(10):3177–3211
14. Szalewicz K (2012) Symmetry-adapted perturbation theory of intermolecular forces. WIREs Comput Mol Sci 2(2):254–272
15. Zhao L, von Hopffgarten M, Andrada DM, Frenking G (2018) Energy decomposition analysis. WIREs Comput Mol Sci 8(3):e1345
16. Kitaura K, Morokuma K (1976) A new energy decomposition scheme for molecular interactions within the Hartree-Fock approximation. Int J Quantum Chem 10(2):325–340
17. Ziegler T, Rauk A (1977) On the calculation of bonding energies by the Hartree Fock Slater method. Theor Chim Acta 46(1):1–10
18. Mao Y, Horn PR, Head-Gordon M (2017) Energy decomposition analysis in an adiabatic picture. Phys Chem Chem Phys 19(8):5944–5958

19. Khaliullin RZ, Cobar EA, Lochan RC, Bell AT, Head-Gordon M (2007) Unravelling the origin of intermolecular interactions using absolutely localized molecular orbitals. J Phys Chem A 111(36):8753–8765
20. Mo Y, Bao P, Gao J (2011) Energy decomposition analysis based on a block-localized wavefunction and multistate density functional theory. Phys Chem Chem Phys 13(15): 6760–6775
21. Mo Y, Gao J, Peyerimhoff SD (2000) Energy decomposition analysis of intermolecular interactions using a block-localized wave function approach. J Chem Phys 112(13):5530–5538
22. Su P, Li H (2009) Energy decomposition analysis of covalent bonds and intermolecular interactions. J Chem Phys 131(1):014102
23. Su P, Tang Z, Wu W (2020) Generalized Kohn-Sham energy decomposition analysis and its applications. WIREs Comput Mol Sci 10(5):e1460
24. Su P, Jiang Z, Chen Z, Wu W (2014) Energy decomposition scheme based on the generalized Kohn-Sham scheme. J Phys Chem A 118(13):2531–2542
25. Seidl A, Görling A, Vogl P, Majewski JA, Levy M (1996) Generalized Kohn-Sham schemes and the band-gap problem. Phys Rev B Condens Matter 53(7):3764–3774
26. Lao KU, Herbert JM (2016) Energy decomposition analysis with a stable charge-transfer term for interpreting intermolecular interactions. J Chem Theory Comput 12(6):2569–2582
27. Deng S, Wang Q, Ren P (2017) Estimating and modeling charge transfer from the SAPT induction energy. J Comput Chem 38(26):2222–2231
28. Grimme S, Hansen A, Brandenburg JG, Bannwarth C (2016) Dispersion-corrected mean-field electronic structure methods. Chem Rev 116(9):5105–5154
29. Xu Y, Zhang S, Wu W, Su P (2024) Assessments of DFT-based energy decomposition analysis methods for intermolecular interactions. J Chem Phys 158:124116
30. Grimme S, Antony J, Ehrlich S, Krieg H (2010) A consistent and accurate ab initio parametrization of density functional dispersion correction (DFT-D) for the 94 elements H-Pu. J Chem Phys 132(15):154104
31. Grimme S, Ehrlich S, Goerigk L (2011) Effect of the damping function in dispersion corrected density functional theory. J Comput Chem 32(7):1456–1465
32. Mao Y, Loipersberger M, Kron KJ, Derrick JS, Chang CJ, Sharada SM, Head-Gordon M (2021) Consistent inclusion of continuum solvation in energy decomposition analysis: theory and application to molecular CO2 reduction catalysts. Chem Sci 12(4):1398–1414
33. Giovannini T (2024) Kohn–Sham fragment energy decomposition analysis. J Chem Phys 161(10):104110
34. Gora RW, Bartkowiak W, Roszak S, Leszczynski J (2004) Intermolecular interactions in solution: elucidating the influence of the solvent. J Chem Phys 120(6):2802–2813
35. Manian A, Shaw RA, Lyskov I, Russo SP (2022) The quantum chemical solvation of indole: accounting for strong solute–solvent interactions using implicit/explicit models. Phys Chem Chem Phys 24(5):3357–3369
36. Monard, G.; Rivail, J.-L. Solvent effects in quantum chemistry. In Leszczynski, J., Kaczmarek-Kedziera, A., Puzyn, T., Papadopoulos, M.G., Reis, H., Shukla, M.K., Eds. Handbook of computational chemistry. Springer International Publishing: Cham, 2017; pp. 727–739
37. Miertus S, Tomasi J (1982) Approximate evaluations of the electrostatic free energy and internal energy changes in solution processes. Chem Phys 65(2):239–245
38. Cammi R, Tomasi J (1995) Remarks on the use of the apparent surface charges (ASC) methods in solvation problems: iterative versus matrix-inversion procedures and the renormalization of the apparent charges. J Comput Chem 16(12):1449–1458
39. Miertuš S, Scrocco E, Tomasi J (1981) Electrostatic interaction of a solute with a continuum. A direct utilization of AB initio molecular potentials for the prevision of solvent effects. Chem Phys 55(1):117–129
40. Marenich AV, Cramer CJ, Truhlar DG (2009) Universal solvation model based on solute electron density and on a continuum model of the solvent defined by the bulk dielectric constant and atomic surface tensions. J Phys Chem B 113(18):6378–6396

41. Klamt A, Schüürmann G (1993) COSMO: a new approach to dielectric screening in solvents with explicit expressions for the screening energy and its gradient. J Chem Soc Perkin Trans 2 5: 799–805
42. Klamt A (1995) Conductor-like screening model for real solvents: a new approach to the quantitative calculation of solvation phenomena. J Phys Chem 99(7):2224–2235
43. Yang Z, Doubleday C, Houk KN (2015) QM/MM protocol for direct molecular dynamics of chemical reactions in solution: the water-accelerated Diels–Alder reaction. J Chem Theory Comput 11(12):5606–5612
44. Kubicskó K, Farkas Ö (2020) Quantum chemical (QM:MM) investigation of the mechanism of enzymatic reaction of tryptamine and N,N-dimethyltryptamine with monoamine oxidase A. Org Biomol Chem 18(47):9660–9674
45. Berraud-Pache R, Garcia-Iriepa C, Navizet I (2018) Modeling chemical reactions by QM/MM calculations: the case of the tautomerization in fireflies bioluminescent systems. Front Chem 6: 116
46. Lawan N, Chasing P, Santatiwongchai J, Muangpil S (2019) QM/MM molecular modelling on mutation effect of chorismate synthase enzyme catalysis. J Mol Graph Model 87:250–256
47. Neves RPP, Fernandes PA, Ramos MJ (2022) Role of enzyme and active site conformational dynamics in the catalysis by α-amylase explored with QM/MM molecular dynamics. J Chem Inf Model 62(15):3638–3650
48. Yagi K, Yamada K, Kobayashi C, Sugita Y (2019) Anharmonic vibrational analysis of biomolecules and solvated molecules using hybrid QM/MM computations. J Chem Theory Comput 15(3):1924–1938
49. Fan J, Liu H, Wang Y, Xie Z, Lin Z, Pang K (2024) Hydrostatic pressure effect on excited state properties of room temperature phosphorescence molecules: a QM/MM study. Spectrochim Acta A Mol Biomol Spectrosc 320:124626
50. Barone V, Cossi M (1998) Quantum calculation of molecular energies and energy gradients in solution by a conductor solvent model. J Phys Chem A 102(11):1995–2001
51. Cossi M, Rega N, Scalmani G, Barone V (2003) Energies, structures, and electronic properties of molecules in solution with the C-PCM solvation model. J Comput Chem 24(6):669–681
52. Mennucci B, Cammi R, Tomasi J (1999) Analytical free energy second derivatives with respect to nuclear coordinates: complete formulation for electrostatic continuum solvation models. J Chem Phys 110(14):6858–6870
53. Cossi M, Barone V (1998) Analytical second derivatives of the free energy in solution by polarizable continuum models. J Chem Phys 109(15):6246–6254
54. Cancès E, Mennucci B (1998) Analytical derivatives for geometry optimization in solvation continuum models. I. Theory. J Chem Phys 109(1):249–259
55. Cancès E, Mennucci B, Tomasi J (1998) Analytical derivatives for geometry optimization in solvation continuum models. II. Numerical Applications. J Chem Phys 109(1):260–266
56. Si D, Li H (2009) Heterogeneous conductorlike solvation model. J Chem Phys 131(4):044123
57. Xiong X, Friedman R, Wu W, Su P (2024) QM/MM-based energy decomposition analysis method for large systems. J Phys Chem A 128(18):3529–3538
58. Grimme S (2011) Density functional theory with London dispersion corrections. WIREs Comput Mol Sci 1(2):211–228
59. Nair NN, Schreiner E, Pollet R, Staemmler V, Marx D (2008) Magnetostructural dynamics with the extended broken symmetry formalism: antiferromagnetic [2Fe-2S] complexes. J Chem Theory Comput 4(8):1174–1188
60. Soda T, Kitagawa Y, Onishi T, Takano Y, Shigeta Y, Nagao H, Yoshioka Y, Yamaguchi K (2000) Ab initio computations of effective exchange integrals for H–H, H–He–H and Mn2O2 complex: comparison of broken-symmetry approaches. Chem Phys Lett 319(3):223–230
61. Yamaguchi K, Takahara Y, Fueno T (1986) Ab-initio molecular orbital studies of structure and reactivity of transition metal-OXO compounds. In: Smith VH, Schaefer HF, Morokuma K (eds) Applied quantum chemistry. Springer, Dordrecht, pp 155–184

62. Yamanaka S, Kawakami T, Nagao H, Yamaguchi K (1994) Effective exchange integrals for open-shell species by density functional methods. Chem Phys Lett 231(1):25–33
63. Noodleman L, Davidson ER (1986) Ligand spin polarization and antiferromagnetic coupling in transition metal dimers. Chem Phys 109(1):131–143
64. Noodleman L (1981) Valence bond description of antiferromagnetic coupling in transition metal dimers. J Chem Phys 74(10):5737–5743
65. Tsuchimochi T, Scuseria GE (2011) Constrained active space unrestricted mean-field methods for controlling spin-contamination. J Chem Phys 134(6):064101
66. Tsuchimochi T, Scuseria GE (2010) Communication: ROHF theory made simple. J Chem Phys 133(14):141102
67. Jiménez-Hoyos CA, Henderson TM, Tsuchimochi T, Scuseria GE (2012) Projected Hartree–Fock theory. J Chem Phys 136(16):164109
68. Bovi D, Guidoni L (2012) Magnetic coupling constants and vibrational frequencies by extended broken symmetry approach with hybrid functionals. J Chem Phys 137(11):114107
69. Tang Z, Jiang Z, Chen H, Su P, Wu W (2019) Energy decomposition analysis based on broken symmetry unrestricted density functional theory. J Chem Phys 151(24):244106
70. Wang W, Wu W, Su P (2023) Radical pairing interactions and donor–acceptor interactions in cyclobis(paraquat-p-phenylene) inclusion complexes. Molecules 28(5):2057
71. Biktagirov T, Schmidt WG, Gerstmann U (2020) Spin decontamination for magnetic dipolar coupling calculations: application to high-spin molecules and solid-state spin qubits. Phys Rev Res 2(2):022024
72. Tada K, Koga H, Okumura M, Tanaka S (2018) Estimation of spin contamination error in dissociative adsorption of Au2 onto MgO(0 0 1) surface: first application of approximate spin projection (AP) method to plane wave basis. Chem Phys Lett 701:103–108
73. Goel S, Masunov AE (2009) Pairwise spin-contamination correction method and DFT study of MnH and H2 dissociation curves. In: Allen G, Nabrzyski J, Seidel E, Van Albada GD, Dongarra J, Sloot PMA (eds) Computational science—ICCS 2009. Lecture notes in computer science, vol 5545. Springer, Berlin, pp 141–150
74. Chen S, Ying F, Wu W, Su P (2024) Energy decomposition analysis method based on the constrained active space unrestricted mean-field method. J Phys Chem A 128(49):10680–10688
75. Malrieu JP, Caballol R, Calzado CJ, de Graaf C, Guihéry N (2014) Magnetic interactions in molecules and highly correlated materials: physical content, analytical derivation, and rigorous extraction of magnetic hamiltonians. Chem Rev 114(1):429–492
76. Dreuw A, Head-Gordon M (2005) Single-reference Ab initio methods for the calculation of excited states of large molecules. Chem Rev 105(11):4009–4037
77. Furche F, Rappoport D (2005) III—Density functional methods for excited states: equilibrium structure and electronic spectra. In: Olivucci M (ed) Theoretical and computational chemistry, vol 16. Elsevier, pp 93–128
78. Ge Q, Head-Gordon M (2018) Energy decomposition analysis for excimers using absolutely localized molecular orbitals within time-dependent density functional theory and configuration interaction with single excitations. J Chem Theory Comput 14(10):5156–5168
79. Hapka M, Przybytek M, Pernal K (2021) Symmetry-adapted perturbation theory based on multiconfigurational wave function description of monomers. J Chem Theory Comput 17(9): 5538–5555
80. Hettich CP, Zhang X, Kemper D, Zhao R, Zhou S, Lu Y, Gao J, Zhang J, Liu M (2023) Multistate energy decomposition analysis of molecular excited states. JACS Au 3(7): 1800–1819
81. Tang Z, Shao B, Wu W, Su P (2023) Energy decomposition analysis methods for intermolecular interactions with excited states. Phys Chem Chem Phys 25(27):18139–18148
82. Su P (2024) Energy decomposition analysis and its applications. In: Exploring chemical concepts through theory and computation. Wiley, pp 433–454

83. Benda R, Vezin T, Lebental B (2022) Prediction of the interaction strength of an urea-based probe towards ions in water by means of DFT/PCM calculations. Int J Quantum Chem 122: e26901
84. Liu S, Xia S, Yue D, Sun H, Hirao H (2022) The bonding nature of Fe–CO complexes in heme proteins. Inorg Chem 61(44):17494–17504
85. Altun A, Saitow M, Neese F, Bistoni G (2019) Local energy decomposition of open-shell molecular systems in the domain-based local pair natural orbital coupled cluster framework. J Chem Theory Comput 15(3):1616–1632
86. Blagg JA, Noe MC, Wolf-Gouveia LA, Reiter LA, Laird ER, Chang S-PP, Danley DE, Downs JT, Elliott NC, Eskra JD, Griffiths RJ, Hardink JR, Haugeto AI, Jones CS, Liras JL, Lopresti-Morrow LL, Mitchell PG, Pandit J, Robinson RP, Subramanyam C, Vaughn-Bowser ML, Yocum SA (2005) Potent pyrimidinetrione-based inhibitors of MMP-13 with enhanced selectivity over MMP-14. Bioorg Med Chem Lett 15(7):1807–1810
87. Sousa C, Bagus PS, Illas F (2019) Approaching multiplet splitting in X-ray photoelectron spectra by density functional theory methods: NO and O2 molecules as examples. Chem Phys Lett 731:136617
88. Tang Z, Song Y, Zhang S, Wang W, Xu Y, Wu D, Wu W, Su P (2021) XEDA, a fast and multipurpose energy decomposition analysis program. J Comput Chem 42(32):2341–2351

Struct Bond (2026) 190: 137–222
https://doi.org/10.1007/430_2025_98

Published online: 20 September 2025

Extended Symmetry-Adapted Perturbation Theory (XSAPT): A Cubic-Scaling Platform for Computing Accurate Intermolecular Interaction Energies and *Ab Initio* Energy Decomposition Analysis

John M. Herbert, Montgomery Gray, Kuan-Yu Liu, and Kevin Carter-Fenk

Contents

Abstract This chapter provides an overview of some hybrid forms of symmetry-adapted perturbation theory (SAPT), developed over the past decade and known collectively as "extended" (X)SAPT. Two primary innovations are a self-consistent

J. M. Herbert (✉) and M. Gray
Department of Chemistry & Biochemistry, The Ohio State University, Columbus, OH, USA
e-mail: herbert@chemistry.ohio-state.edu

K.-Y. Liu
Q-Chem Inc., Pleasanton, CA, USA

K. Carter-Fenk
Department of Chemistry, University of Pittsburgh, Pittsburgh, PA, USA

charge embedding scheme to capture many-body polarization (the "XPol" procedure) and the use of low-cost dispersion models as replacements for SAPT's perturbative description of dispersion. The latter modification reduces the formal complexity to $\mathcal{O}(N^3)$ with system size. In conjunction with a many-body dispersion model (XSAPT+MBD) or empirical dispersion potentials fitted to *ab initio* data (XSAPT+*ai*D), the hybrid procedures achieve sub-kcal/mol accuracy with respect to high-level benchmarks. XSAPT is equipped with an energy decomposition that partitions the intermolecular interaction energy into components that include electrostatics, Pauli repulsion, dispersion, and induction, the latter of which can be further separated into polarization and charge transfer. As compared to energy decomposition analyses used in density functional theory, separation of the dispersion energy in XSAPT is less ambiguous, and the energy partition agrees well with accurate third-order SAPT benchmarks. Theoretical foundations of XSAPT are reviewed, and we provide a thorough discussion of its performance in terms of both accuracy and cost. Exemplary applications are presented that illustrate how XSAPT can be used to uncover the fundamental molecular physics of intermolecular interactions.

Keywords Dispersion · Energy decomposition analysis · Intermolecular forces · Noncovalent interactions · Symmetry-adapted perturbation theory

Abbreviations

ALMO	Absolutely localized molecular orbital
BSSE	Basis-set superposition error
CBS	Complete basis set
CCSD(T)	Coupled-cluster theory with single, double, and (perturbative) triple excitations
CP	Counterpoise
CT	Charge transfer
DFT	Density functional theory
EDA	Energy decomposition analysis
HF	Hartree–Fock
HOMO	Highest occupied molecular orbital
IE	Ionization energy
KS	Kohn–Sham
LRC	Long-range corrected
LUMO	Lowest unoccupied molecular orbital
MBD	Many-body dispersion
MO	Molecular orbital
MP2	Second-order Møller–Plesset perturbation theory
SAPT	Symmetry-adapted perturbation theory
SCF	Self-consistent field
vdW	van der Waals

XC Exchange-correlation
XSAPT Extended symmetry-adapted perturbation theory

1 Intermolecular Interactions the SAPT Way

With the nowadays widespread deployment of density functional theory (DFT) by nonexperts, it can easily seem as if *ab initio* description of the chemical bond is a solved problem. The treatment of nonbonded interactions, however, was certainly not a solved problem in the early days of DFT [1]. Since then, DFT's ability to describe noncovalent interactions has been improved significantly [1–4], thanks to the development of both dispersion-corrected and dispersion-inclusive exchange-correlation (XC) functionals, which we have recently reviewed [1]. The availability of low-cost methods for noncovalent interactions has spurred more detailed scrutiny of their role in chemistry and in particular the important role of dispersion interactions in chemical bonding [5–7].

When it comes to *analyzing* intermolecular interactions, however, it is less straightforward to decide how DFT interaction energies can be decomposed into meaningful components, including dispersion (E_{disp}) but also electrostatics (E_{elst}), steric repulsion (E_{steric}), polarization (E_{pol}), and charge transfer (E_{CT}):

$$\Delta E_{\text{int}} = E_{\text{elst}} + E_{\text{steric}} + E_{\text{disp}} + \underbrace{E_{\text{pol}} + E_{\text{CT}}}_{E_{\text{ind}}} \, . \tag{1}$$

Here, ΔE_{int} is the intermolecular interaction energy computed via the supramolecular approach, meaning

$$\Delta E_{\text{int,AB}} = E_{\text{AB}} - E_{\text{A}} - E_{\text{B}} \tag{2}$$

for a dimer A···B. In Eq. (1), we have defined "induction" (E_{ind}) to be the sum of polarization and charge transfer (CT),

$$E_{\text{ind}} = E_{\text{pol}} + E_{\text{CT}}. \tag{3}$$

This will aid in making contact between the present work (*vide infra*) and the partition in Eq. (1) that is often used in DFT-based energy decomposition analysis (EDA). A method to separate E_{pol} from E_{CT} is described in Sect. 1.3.2.

The decomposition of ΔE_{int} in Eq. (1) is not unique, insofar as there is no quantum-mechanical operator that represents dispersion (or polarization, etc.). As a result of this ambiguity, a variety of EDA schemes have been suggested for use with DFT [8–13], and there are a few EDAs for correlated wavefunction methods as well [14–16]. However, there is also no shortage of criticism over the fundamentally ill-defined nature of such analyses [17–20], with some going so far as to suggest that for "weak" interactions ($\lesssim$ 5 kcal/mol), one should not even bother [20]. We agree with the general sentiment that EDAs are often overinterpreted ("how many kJ/mol can dance on the head of a pin?") [1], but a 5 kcal/mol threshold for abandoning any attempt at interpretation feels too conservative to us. After all, the energy of the single hydrogen bond in $(H_2O)_2$ is 3.2 kcal/mol [21], and we *should* demand that a useful EDA can discern a hydrogen bond from some other type of intermolecular interaction.

The present chapter describes an alternative approach to EDA for intermolecular interactions that is not based on DFT but rather on many-body theory and aims to compute the interaction energy directly from first principles (not by energy difference), with quantitative accuracy. This approach, which is known as symmetry-adapted perturbation theory (SAPT) [10, 22–28], has a long history in quantum chemistry that is not reviewed here. Rather, the present work examines a relatively new family of hybrid methods that are built upon the SAPT framework but substitute alternative models for dispersion in place of perturbation theory. This is done in the interest of achieving quantitative accuracy at a cost that is cubic-scaling [$\mathcal{O}(N^3)$] with system size, comparable to the cost of DFT. These methods are known collectively as "extended" (X)SAPT, originally built upon a combination of SAPT with the "XPol" self-consistent charge embedding procedure [29–31]. XSAPT has been developed in the Herbert group since 2010 [32–43] with the goal of providing *qualitative* insight, via a well-defined EDA that is built atop a method that delivers *quantitative* accuracy. Our philosophy is that physical insight should be independent of the ever-evolving milieu of density-functional approximations.

The remainder of this section outlines the SAPT formalism (Sect. 1.1), including definitions for its energy components. Modifications that form the basis of XSAPT are discussed in Sect. 1.2. The presentation is intended to be only moderately technical, an overview with enough detail so that a potential user can make informed choices with regard to the particular setup of an XSAPT calculation. Shorter and even less technical discussions of XSAPT can be found elsewhere [37, 42].

The accuracy and performance of XSAPT are discussed in Sect. 2. Accuracy is assessed (in Sect. 2.1) by comparing to benchmark *ab initio* results for standard noncovalent datasets [44], where interaction energies are available at the CCSD(T) level extrapolated to the complete basis-set (CBS) limit. Performance benchmarks for XSAPT (Sect. 2.2) are based on a newly rewritten version of the code that exists within the Q-Chem software package [45, 46], featuring improved single-node parallel efficiency that is documented here for the first time. Finally, illustrative applications are presented in Sect. 3. After making some elementary remarks about the distinction between electrostatics and Pauli repulsion, and the failure of multipole ideas at nonbonded close-contact distances (Sect. 3.1), we demonstrate that (X)SAPT provides a much cleaner definition of dispersion as compared to DFT-based EDA schemes (Sect. 3.2). We then describe how XSAPT calculations have been used to upend the conventional understanding of π–π interactions (Sect. 3.3) and to reinforce the conventional idea that CT plays an important role in ion–molecule hydrogen bonding (Sect. 3.4).

1.1 Symmetry-Adapted Perturbation Theory

Quite apart from DFT-based EDAs, the SAPT framework is older and has its basis in many-body perturbation theory [22]. It uses isolated-monomer wavefunctions as a zeroth-order starting point and then computes the intermolecular interaction using a form of perturbation theory. For a noncovalent dimer A···B, this amounts to a partition of the dimer's Hamiltonian according to

$$\hat{H}_{\mathrm{AB}} = \underbrace{\hat{F}_{\mathrm{A}} + \hat{F}_{\mathrm{B}}}_{\hat{H}_0} + \xi \underbrace{(\hat{W}_{\mathrm{A}} + \hat{W}_{\mathrm{B}})}_{\hat{W}} + \lambda \hat{V}_{\mathrm{AB}} = \hat{H}_0 + \xi \hat{W} + \lambda \hat{V}_{\mathrm{AB}} \,. \tag{4}$$

Here, $\hat{F}_{\mathrm{X}}$ is a Fock operator for the monomer (X = A or B) and $\hat{W}_{\mathrm{X}}$ is the corresponding Møller–Plesset fluctuation potential. The quantity $\hat{V}_{\mathrm{AB}}$ contains the intermolecular Coulomb operators for both nuclei and electrons. In atomic units where $4\pi\epsilon_0 = 1$, the latter is

$$\begin{aligned} \hat{V}_{\mathrm{AB}} &= \sum_{A\in \mathrm{A}}\sum_{B\in \mathrm{B}} \frac{Z_A Z_A}{R_{AB}} - \sum_{a\in \mathrm{A}}\sum_{B\in \mathrm{B}} \frac{Z_B}{R_{aB}} - \sum_{A\in \mathrm{A}}\sum_{b\in \mathrm{B}} \frac{Z_A}{R_{Ab}} \\ &+ \sum_{a\in \mathrm{A}}\sum_{b\in \mathrm{B}} \frac{1}{R_{ab}} . \end{aligned} \tag{5}$$

Parameters ξ and λ in Eq. (4) serve to count orders in a double perturbation expansion, with the direct product $|\psi_0^{\mathrm{A}}\rangle|\psi_0^{\mathrm{B}}\rangle$ as a reference state.

A caveat is that $|\psi_0^{\mathrm{A}}\rangle|\psi_0^{\mathrm{B}}\rangle$ is not antisymmetric with respect to exchange of electrons between A and B, since $|\psi_0^{\mathrm{A}}\rangle$ and $|\psi_0^{\mathrm{B}}\rangle$ were computed in isolation. The "symmetry adaptation" in SAPT is really an *antisymmetry* adaptation in which A ↔ B exchange is incorporated perturbatively, via an antisymmetrizer $\hat{\mathcal{A}}_{\mathrm{AB}}$ [47]. In practice, this means that the perturbation is $\hat{V}_{\mathrm{AB}}\hat{\mathcal{A}}_{\mathrm{AB}}$, albeit with some additional simplifications for evaluating matrix elements of $\hat{\mathcal{A}}_{\mathrm{AB}}$, most notably the "single-exchange" (or "S^2") approximation [48–50]. The presence of $\hat{\mathcal{A}}_{\mathrm{AB}}$ in the perturbation gives rise to certain exchange terms in this antisymmetrized form of Rayleigh–Schrödinger perturbation theory, in addition to the usual terms that arise in the so-called polarization expansion [51]. The latter is the perturbative approximation to the A···B interaction energy that is based on $\hat{V}_{\mathrm{AB}}$ alone, disregarding $\hat{\mathcal{A}}_{\mathrm{AB}}$. That expansion is valid for the long-range forces when the isolated monomer densities $\rho_{\mathrm{A}}(\mathbf{r})$ and $\rho_{\mathrm{B}}(\mathbf{r})$ do not overlap and there is no need to enforce the Pauli principle for A ↔ B exchange.

SAPT is a direct perturbative expansion for the A···B interaction energy, which we call $E_{\mathrm{int,AB}}$ (or simply E_{int}) in order to distinguish it from the supramolecular energy-difference approach. Whereas ΔE_{int} in Eq. (2) is computed by difference and is subject to basis-set superposition error (BSSE), the SAPT intermolecular interaction energy is not. In any EDA scheme that is based on a supramolecular calculation in a finite basis set, the BSSE is a spurious contribution that must be included somewhere. Finally, the accuracy of SAPT is not beholden to a particular choice of density functional and can be systematically improved using higher orders in perturbation theory.

Application of symmetrized Rayleigh–Schrödinger perturbation theory to the perturbations $\hat{W}$ and $\hat{V}_{AB}\hat{\mathscr{A}}_{AB}$ affords an expression for E_{int} of the form

$$E_{int} = \sum_{n=1}^{\infty} \sum_{m=0}^{\infty} \left(E_{pol}^{(nm)} + E_{exch}^{(nm)} \right), \tag{6}$$

whose terms are explicated elsewhere [22, 47, 52–54]. Superscripts n and m in Eq. (6) indicate orders in perturbation theory, for either the intermolecular perturbation ($\hat{V}_{AB}\hat{\mathscr{A}}_{AB}$) or for monomer electron correlation ($\hat{W}$), respectively. The terms $E_{pol}^{(nm)}$ constitute the polarization expansion (valid at long range) but each is accompanied by an exchange contribution $E_{exch}^{(nm)}$, arising from the antisymmetrizer.

The simplest form of SAPT sets $m = 0$ and thus does not include monomer electron correlation. Further limiting the expansion to $n \leq 2$ affords the simplest form of SAPT that includes all of the energy components identified in Eq. (1), albeit in different guises. Somewhat confusingly, this second-order (in $\hat{V}_{AB}\hat{\mathscr{A}}_{AB}$) approach is often called "SAPT0" [25], because it is zeroth order in $\hat{W}$. It is akin to a Hartree–Fock (HF) description of the monomers combined with dispersion resembling that of second-order Møller–Plesset perturbation theory (MP2). Regrouping the relevant terms in the polarization expansion, this approximation for E_{int} is

$$E_{int}^{SAPT0} = E_{elst}^{(10)} + E_{exch}^{(10)} + \underbrace{E_{disp}^{(20)} + E_{exch\text{-}disp}^{(20)}}_{E_{disp}^{(2)}} + \underbrace{E_{ind}^{(20)} + E_{exch\text{-}ind}^{(20)}}_{E_{ind}^{(2)}}. \tag{7}$$

Electrostatics and exchange repulsion appear at first order in $\hat{V}_{AB}\hat{\mathscr{A}}_{AB}$ [47], as $E_{elst}^{(10)}$ and $E_{exch}^{(10)}$, respectively. The latter is identified with Pauli or steric repulsion [43], for reasons that are discussed in Sect. 3.1. Induction and dispersion do not appear until second order, which is why the second-order approximation is the simplest reasonable form of Eq. (6). As indicated in Eq. (7), we combine the second-order dispersion and induction terms in the polarization expansion with their exchange analogues to define the total dispersion and induction energies at second order,

$$E_{disp}^{(2)} = E_{disp}^{(20)} + E_{exch\text{-}disp}^{(20)} \tag{8a}$$

$$E_{ind}^{(2)} = E_{ind}^{(20)} + E_{exch\text{-}ind}^{(20)}. \tag{8b}$$

The exchange components arise from $\hat{\mathscr{A}}_{AB}$ and reduce the attractiveness of the induction or dispersion terms from the polarization expansion, by eliminating

Pauli-forbidden contributions to that expansion. To make contact with the generic energy decomposition in Eq. (1), recall that the induction energy contains both polarization and CT (Eq. 3), for reasons that are discussed in Sect. 1.3.2 where we introduce a technique to separate these two contributions.

Second-order induction often lacks sufficient accuracy for hydrogen-bonded systems [55], so some higher-order effects are often packaged with SAPT0. Specifically, we can define "response" (or "relaxed") versions of second-order induction,

$$E^{(2)}_{\text{ind,resp}} = E^{(20)}_{\text{ind,resp}} + E^{(20)}_{\text{exch-ind,resp}}, \tag{9}$$

obtained by solving coupled-perturbed HF equations [56]. Substituting $E^{(2)}_{\text{ind,resp}}$ in place of $E^{(2)}_{\text{ind}}$ in Eq. (7) affords the best possible treatment of induction at second order in perturbation theory, but even this may be inadequate to describe the significant induction effects associated with hydrogen bonding [57, 58]. A supramolecular HF calculation contains much of that effect as it formally sums induction to infinite order in perturbation theory, albeit at zeroth order in $\hat{W}$. As such, one may consider a correction for induction of the form

$$\delta E_{\text{HF}} = \Delta E^{\text{CP-HF}}_{\text{int}} - \left(E^{(10)}_{\text{elst}} + E^{(10)}_{\text{exch}} + E^{(20)}_{\text{ind,resp}} + E^{(20)}_{\text{exch-ind,resp}}\right) \tag{10}$$

where $\Delta E^{\text{CP-HF}}_{\text{int}}$ is the counterpoise (CP) corrected HF interaction energy, computed using the supramolecular approach (Eq. 2). The quantity defined in Eq. (10) is known as the "δHF" correction, and it is often added to SAPT0:

$$E^{\text{SAPT0}+\delta\text{HF}}_{\text{int}} = E^{(10)}_{\text{elst}} + E^{(10)}_{\text{exch}} + E^{(2)}_{\text{disp}} + \underbrace{E^{(20)}_{\text{ind,resp}} + E^{(20)}_{\text{exch-ind,resp}} + \delta E_{\text{HF}}}_{E_{\text{ind}}} \,. \tag{11}$$

Note that some authors consider δE_{HF} to be part of the definition of SAPT0 [28], but others do not [25]; in the present work, we take δE_{HF} to be an additional correction. Strictly speaking, the model defined in Eq. (11) does not simply add δE_{HF} to the model in Eq. (7), due to the replacement of the second-order induction amplitudes with their response analogues. However, it is common to refer to Eq. (11) as "SAPT0+δHF", and we shall do so.

For hydrogen-bonded systems, the hybrid SAPT0+δHF approach is sometimes necessary to obtain high accuracy, even when third-order induction terms (beyond SAPT0) are included [25, 57, 58]. However, mixing SAPT with an energy-difference approach has the potential to introduce BSSE into what is otherwise a BSSE-free formalism. As such, $\Delta E^{\text{CP-HF}}_{\text{int}}$ should always be evaluated using CP correction, as indicated by the notation. Even for self-consistent field (SCF) calculations, the CP correction is not negligible unless basis sets of at least quadruple-ζ quality are employed [59].

Substituting Eq. (10) into Eq. (11) affords

$$E_{\text{int}}^{\text{SAPT0}+\delta\text{HF}} = \Delta E_{\text{int}}^{\text{HF}} + E_{\text{disp}}^{(2)} . \tag{12}$$

This demonstrates that the SAPT0+δHF combination is nothing more than second-order dispersion added to the supramolecular HF interaction energy, with its infinite order but uncorrelated treatment of induction. (This is also a simple way to understand that dispersion is entirely an electron correlation effect, absent at the mean-field HF level.) For this reason, SAPT0+δHF and similar methods have sometimes been called "HF plus dispersion" [28, 60–62]. We avoid that terminology for SAPT0+δHF, however, because it is easily confused with very old approaches that augment a HF calculation with an empirical dispersion correction (e.g., of the form $-C_6/R^6$) [63–69], analogous to the manner in which contemporary DFT is typically augmented to obtain dispersion-corrected DFT+D methods [1, 2].

Along those lines, let us pause to discuss terminology. We will sometimes refer to the conventional, HF-based SAPT0 model as "SAPT0(HF)," in order to distinguish it from an analogous method that uses Kohn–Sham (KS) DFT for the monomers. We define SAPT0(KS) as second-order intermolecular perturbation theory based on a generalized description of the monomer wavefunctions [42, 55], meaning Eq. (7) with all SAPT terms evaluated using KS orbitals and one-particle energy levels. (There are important caveats concerning which XC functionals should be used, which are discussed in Sect. 1.2.2.) Ultimately, we want to eliminate $E_{\text{disp}}^{(2)}$ in favor of dispersion models that are both more accurate and less expensive (Sect. 1.2.4). Since SAPT0 implies second-order dispersion, we will refer to these hybrid models as "SAPT(KS)," without the zero. Explicitly,

$$E_{\text{int}}^{\text{SAPT(KS)}} = E_{\text{elst}}^{(10)} + E_{\text{exch}}^{(10)} + E_{\text{ind}}^{(20)} + E_{\text{exch-ind}}^{(20)}. \tag{13}$$

This is simply Eq. (7) without the dispersion or exchange-dispersion terms, but where the name suggests that KS-DFT is used to compute the SCF monomer wavefunctions. The SAPT(KS) model in Eq. (13) needs to be augmented with some form of dispersion, e.g., to obtain SAPT(KS)+D. It might or might not be augmented with a δHF correction, which will be indicated explicitly if so. Even for SAPT(KS), we will always use HF orbitals to evaluate δE_{HF} because a supramolecular KS-DFT calculation mixes different energy components, whereas δHF is intended as a correction for induction [55]. Thus,

$$\begin{aligned} E_{\text{int}}^{\text{SAPT(KS)}+\delta\text{HF}} &= E_{\text{elst}}^{(10),\text{KS}} + E_{\text{exch}}^{(10),\text{KS}} + E_{\text{ind,resp}}^{(20),\text{KS}} + E_{\text{exch-ind,resp}}^{(20),\text{KS}} \\ &\quad + \delta E_{\text{HF}}, \end{aligned} \tag{14}$$

where the "KS" superscripts indicate that these terms are evaluated using DFT orbitals and energy levels. This presents a potential mismatch between the KS response terms in Eq. (14) and their HF analogues that are subtracted in Eq. (10). We leave it to benchmark calculations to determine the extent to which this is a problem.

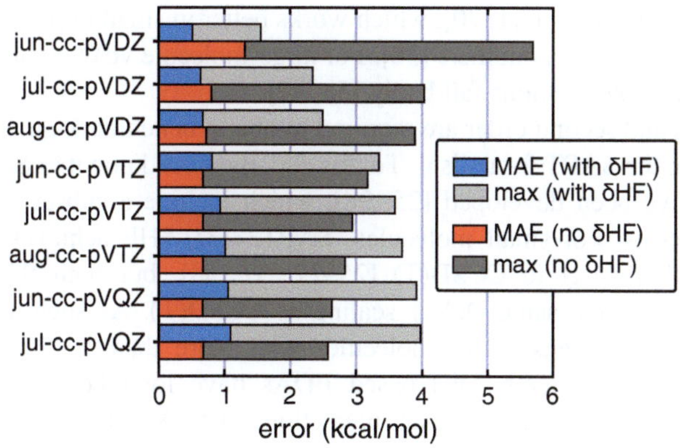

Fig. 1 Error statistics for SAPT0 (with and without the δHF correction) in various basis sets, as applied to the S66 dataset and compared to CCSD(T)/CBS benchmarks. The "calendar" basis sets (jun- and jul-cc-pVXZ) include subsets of the diffuse functions from aug-cc-pVXZ. (Data are from Ref. [77])

Whatever the form of the monomer SCF wavefunctions, SAPT0 or SAPT0 (KS) is MP2-like in its treatment of dispersion. This lends some drawbacks because MP2 itself is not a benchmark-quality method for noncovalent interactions [26, 27, 38] and dispersion energies can be significantly overestimated [70], especially in large molecules [38, 71]. MP2 describes dispersion using uncoupled HF polarizabilities [72], but it is questionable whether any finite-order MPn method is appropriate to describe dispersion in systems whose monomers are large and polarizable [71], e.g., due to extended conjugation. Exaggeration of dispersion interactions is exacerbated as the CBS limit is approached, as demonstrated in Fig. 1 using HF-based SAPT0 applied to the S66 dataset [73]. This is a standard database of 66 small dimers, containing ≤ 10 second-row atoms, for which CCSD(T)/CBS benchmarks are available for comparison.

Mean errors $\lesssim 1$ kcal/mol can be achieved for the S66 complexes using SAPT0+δHF in conjunction with basis sets up to jul-cc-pVQZ, but the outliers grow larger as the basis-set quality improves. These outliers correspond to dispersion-dominated systems such as the π-stacked uracil dimer and other π–π systems including uracil–benzene and uracil–pyridine, whose total interaction energies are relatively large by the standards of small, dispersion-dominated complexes. Interestingly, the importance of the δHF correction diminishes in larger basis sets, and starting with jul-cc-pVTZ, both the MAE and the maximum error are smaller for SAPT0 (sans δE_{HF}) than for SAPT0+δHF. This behavior has not been recognized in previous comprehensive tests of wavefunction-based SAPT [26, 27], because in those studies the δHF correction is always included with SAPT0.

The best accuracy is obtained for SAPT0+δHF/jun-cc-pVDZ, using a basis set that removes certain diffuse functions from aug-cc-pVDZ [74]. That basis, which is called aug-cc-pVDZ$'$ in some older literature [75, 76], has been previously recommended for SAPT0 [26, 77]. However, the limited basis set amounts to a form of error cancellation [1, 70], which works better in small dimers than it does in complexes with larger monomers [38]. For large-molecule vdW complexes, SAPT0 and other MP2-like methods fail badly [38, 42].

Terms beyond second order are required to obtain benchmark-quality interaction energies from the SAPT formalism. To this end, methods known as SAPT2+(3) and SAPT2+3 have been developed [25, 28], which include a subset of the possible third-order terms. For small molecules, SAPT2+(3)+δHF achieves sub-kcal/mol accuracy with respect to CCSD(T) [25–27, 57, 58], but both SAPT2+(3) and SAPT2+3 exhibit the same $\mathcal{O}(N^7)$ scaling as CCSD(T). As such, these methods are prohibitively expensive for molecules with more than 10–20 non-hydrogen atoms. For larger systems, DFT-based EDAs have long been the only means available to analyze intermolecular interactions. The XSAPT family of methods was developed as a way around that bottleneck.

1.2 XSAPT Models

The most expensive terms in the SAPT0 energy expression are $E^{(20)}_{\text{disp}}$ and $E^{(20)}_{\text{exch-disp}}$, which scale as $\mathcal{O}(N^4)$ and $\mathcal{O}(N^5)$, respectively. Other terms including δE_{HF} scale no worse than $\mathcal{O}(N^3)$, which is the same asymptotic scaling as DFT. Thus, what makes SAPT0 expensive is the same second-order dispersion that precludes it from being a benchmark-quality method [70–72]. Recognizing this, we have developed hybrid SAPT-based models in which $E^{(2)}_{\text{disp}}$ is replaced by a more accurate and less costly dispersion model [37, 42]. Several such models have been explored, leading to a family of methods that we call "extended" (X)SAPT [32–42]. In early work, the same approach was called "XPol + SAPT" [32–35], for reasons that will become clear in Sect. 1.2.3 when we discuss XPol-based charge embedding.

The formal scaling of XSAPT is $\mathcal{O}(N^3)$ when evaluated in the atomic orbital basis [38, 78] and combined with density fitting [78]. Because it requires only monomer-based SCF calculations, XSAPT can sometimes be faster than supramolecular DFT [34, 37], although the δHF correction and the many-body EDA do require calculations in line with the cost of DFT. Relative to SAPT0, however, accuracy is improved while cost is reduced [42].

1.2.1 Overview

XSAPT is based on a two-body energy decomposition of the form

$$E^{\text{XSAPT,2B}}_{\text{int}} = E_{\text{elst}} + E_{\text{exch}} + E_{\text{ind}} + \underbrace{E_{\text{disp}}}_{\text{model}} \,. \tag{15}$$

The superscript "2B" indicates that this is a two-body (dimer) EDA, which will be generalized to a many-body (cluster) EDA in Sect. 1.3.3. We equate the first three energy components in Eq. (15) with their SAPT0 analogues:

$$E_{\text{elst}} = E_{\text{elst}}^{(10)} \tag{16a}$$

$$E_{\text{exch}} = E_{\text{exch}}^{(10)} \tag{16b}$$

$$E_{\text{ind}} = \begin{cases} E_{\text{ind}}^{(20)} + E_{\text{exch-ind}}^{(20)} & \text{(simplest)} \\ E_{\text{ind,resp}}^{(20)} + E_{\text{exch-ind,resp}}^{(20)} & \text{(response version)} \\ E_{\text{ind,resp}}^{(20)} + E_{\text{exch-ind,resp}}^{(20)} + \delta E_{\text{HF}} & \text{(infinite-order induction)} \end{cases} . \tag{16c}$$

As suggested in Eq. (16c), induction can be described using various levels of sophistication; these are listed in order of increasing accuracy and cost. By using SAPT0 energy components, electron correlation is excluded in the monomer calculations. This may be undesirable, especially for strong hydrogen bonds where monomer correlation effects can be significant [76]. Within conventional SAPT methodology, this requires terms that are first and second order in $\hat{W}$ (Eq. 4), an approach that has been called "SAPT2" and engenders $O(N^6)$ scaling, similar to CCSD but noniterative [28]. XSAPT instead uses KS-DFT for the monomers, as a low-cost means to incorporate monomer electron correlation effects. Importantly, the XC functional must have correct asymptotic behavior. This is accomplished using long-range corrected (LRC) density functionals [55, 79], as discussed in Sect. 1.2.2.

An additional consideration is how to represent the monomer wavefunctions in systems that contain more than two monomers: $A, B, C, \ldots$. Our choice is to represent the SCF wavefunction on A using strictly those atomic orbital basis functions that are centered on A. Elsewhere, this has been called the "absolutely localized" molecular orbital (ALMO) basis [80], which is a starting point for a DFT-based EDA of the same name [13]. This choice is not entirely obvious from the point of view of conventional (dimer) SAPT calculations, where it is known as a monomer-centered basis, which is atypical because more accurate results are obtained using a dimer-centered basis [81]. The latter approach employs the combined basis functions of A and B to compute both $|\psi_0^A\rangle$ and $|\psi_0^B\rangle$, which allows for some intermolecular CT. For systems with more than two monomers, however, the appropriate definition for a dimer basis becomes ambiguous. For a large collection of monomers, this approach cannot be generalized without prohibitive cost.

As an alternative, XSAPT uses a monomer basis to compute the SCF wavefunctions, ensuring that the total SCF cost is $O(N)$ with respect to the number of monomers. As a means to capture some of the intermolecular CT that is lost in the monomer-centered approach, XSAPT employs a "projected" (pseudocanonicalized) monomer basis to evaluate the dimer SAPT corrections [32, 33]. For the dimer AB, this means first obtaining density matrices $\mathbf{P}^A$ and $\mathbf{P}^B$ from monomer-centered SCF calculations and then building monomer Fock matrices $\mathbf{F}^A$ and $\mathbf{F}^B$ using the block-diagonal dimer density matrix $\mathbf{P}^A \bigoplus \mathbf{P}^B$. Each of these Fock matrices is then transformed into a pseudocanonical basis that diagonalizes the occupied–occupied and virtual–virtual blocks independently, as this does not alter the monomer densities that are invariant to such transformations. We omit the first-order (or "non-Brillouin singles") correction, which is formally introduced into the perturbation expansion by this transformation [82]. As compared to a proper dimer basis, the use of this pseudocanonicalized basis results in acceptable errors in conventional pairwise SAPT0 calculations, e.g., a 10% error in the total interaction energy for $(H_2O)_2$ [32]. For a system of more than two monomers, the aforementioned procedure is repeated for all pairs AB, and SAPT calculations are evaluated in a pairwise-additive manner as described in Sect. 1.3.3.

As indicated already in Eq. (16), XSAPT is not a single model but rather a family of methods [37, 42], or a platform for accurate calculation of intermolecular interactions that is also equipped with an energy decomposition scheme. By its nature, this requires some decisions on the user's part and makes XSAPT less straightforward than a typical DFT-based EDA that simply requires the user to select an XC functional and basis set, push a button, and accept the results at face value. That final facet cannot be overemphasized, although it is overlooked by most users. XSAPT allows a user to test the impact of different approximations, with the goal of obtaining high-quality interaction energies and well-defined energy components, i.e., answers that are physically understandable. (In principle, the DFT-based approach could be tested with a variety of functionals, but few users do this, in our experience.) In our view, the added complexity of XSAPT is a virtue and not a drawback, and it is one that is shared by the more conventional and systematically improvable SAPT formalism [26, 27]. With conventional SAPT, however, the useful levels of approximation subsequent to SAPT0 come with prohibitive $\mathcal{O}(N^7)$ cost that limits their application to very small dimers. XSAPT can be understood as an attempt to achieve the accuracy of higher-order SAPT while retaining the $\mathcal{O}(N^3)$ complexity that is necessary to tackle large systems.

Our cynical view, honed from experience, is that any quantum chemistry method that requires physics-based decisions is too sophisticated for the average user. Nevertheless, we feel there is a need for alternatives to DFT-based EDA schemes. In our view, the latter are often used to reverse-engineer "explanations" that confirm preconceived expectations, with users shopping for an XC functional to achieve that result. This is a lousy approach in any case, but it is supremely dangerous in the context of EDAs, whose energy components do not correspond to observables and thus cannot be interrogated directly against experiment. Nevertheless, the well-defined nature of the XSAPT energy components facilitates comparison with SAPT2+(3) calculations, whose energy components may not be directly verifiable but whose total interaction energies are in good agreement with CCSD(T) results

[25–27, 57, 58], for reasons that are grounded in many-body theory [57, 83]. Such comparisons are discussed in Sect. 1.3.1.

1.2.2 Monomer SCF Calculations

We next consider how to compute individual monomer wavefunctions. Although SAPT's double perturbation formalism is flexible enough to include monomer correlation from many-body theory, the simplest method that does so is SAPT2 [84], which exhibits $\mathcal{O}(N^6)$ cost but does not afford benchmark-quality results for dispersion [25, 28]. As such, we restrict XSAPT to SCF wavefunctions for the monomers. These could be HF wavefunctions, or we could attempt to incorporate monomer electron correlation effects at low cost using KS-DFT, in which case the asymptotic behavior of the XC potential (υ_{xc}) must be considered carefully. That discussion will occupy most of this section.

A SAPT0(KS) formalism, using wavefunction-based perturbation theory expressions for the energy components but orbitals from KS-DFT, was first attempted in the early 2000s with disappointing results [85–87]. This was ultimately attributed to incorrect asymptotic behavior in standard XC functionals [87–89]. For an electron that is well separated from its parent molecule, the asymptotic behavior of the XC potential ought to be

$$\upsilon_{\text{xc}}(r) \sim -\frac{1}{r} + \Delta_\infty \tag{17}$$

for large r [90, 91], where

$$\Delta_\infty = \text{IE} + \varepsilon_{\text{HOMO}}. \tag{18}$$

Here, "IE" is the molecule's first ionization energy, computed using a ΔSCF approach, and $\varepsilon_{\text{HOMO}} < 0$ is the energy level of the highest occupied molecular orbital (HOMO). Proper asymptotic behavior (for exact KS theory) is $\upsilon_{\text{xc}}(r) \sim -1/r$, which is recovered from Eq. (17) provided that the IE theorem of DFT is satisfied [92–95], meaning that

$$\varepsilon_{\text{HOMO}} = -\text{IE}. \tag{19}$$

Improper asymptotic behavior has numerous undesirable consequences for KS-DFT and in the early days of its application to molecular problems, *ad hoc* potentials were sometimes grafted onto the potential obtained from the XC energy functional, to achieve correct asymptotic behavior by construction [91, 96, 97]. For an energy functional $E_{\text{xc}}[\rho(\mathbf{r})]$, the XC potential is

$$v_{\mathrm{xc}}(\mathbf{r}) = \frac{\delta E_{\mathrm{xc}}[\rho(\mathbf{r})]}{\delta\rho(\mathbf{r})}, \tag{20}$$

but the grafting procedure spoils this relationship between energy and potential, leading to "stray" potentials that cannot be derived from any energy functional [98]. Nevertheless, the grafting procedure is still used in the SAPT(DFT) approach [23, 24, 28], in which frequency-dependent density susceptibilities for the monomers, computed from KS-DFT calculations, are used to replace second-order dispersion in the SAPT0 energy formula (Eq. 7).

A simple way to satisfy Eq. (17) is to use LRC functionals [99–101]. These start from a semilocal exchange functional $E_{\mathrm{x}}^{\mathrm{GGA}}$, obtained within the generalized gradient approximation (GGA), then replace the Coulomb potential with a shorter-range attenuated potential. This short-range exchange functional ($E_{\mathrm{x}}^{\mathrm{GGA,SR}}$) is then combined with long-range HF exchange ($E_{\mathrm{x}}^{\mathrm{HF,LR}}$) to afford an XC functional

$$E_{\mathrm{xc}}^{\mathrm{LRC}} = E_{\mathrm{c}} + (1-\alpha)E_{\mathrm{x}}^{\mathrm{GGA,SR}} + \alpha E_{\mathrm{x}}^{\mathrm{HF,SR}} + E_{\mathrm{x}}^{\mathrm{HF,LR}}. \tag{21}$$

Here, E_{c} is a correlation functional and α is any fraction of exact exchange that is included at short range ($E_{\mathrm{x}}^{\mathrm{HF,SR}}$). Functionals that employ range separation of this sort are generally known as range-separated hybrids, but only those that use 100% exact exchange at long range are designated as LRC functionals [102–104]. Of the standard GGAs, meta-GGAs, and hybrid functionals in common use, only LRC functionals afford $v_{\mathrm{xc}}(r) \sim -1/r$. For XSAPT calculations, we typically use the LRC-ωPBE functional [103].

That choice alone does not guarantee that the IE condition in Eq. (19) is satisfied, and adjusting the range-separation parameter (ω) in order to satisfy that condition has been suggested as a nonempirical means to obtain a functional with correct asymptotics [99]. Alternatively, a global density-dependent (GDD) tuning procedure can be used [55], in which the length scale for range separation ($1/\omega$) is set by considering the distance between the outermost electron in a molecule and the region of localized orbitals [105]. The value of ω that is determined in this way is designed to be a good approximation to that obtained via "IE tuning," to satisfy Eq. (19) [105, 106], yet the GDD procedure is automatable and sidesteps certain issues with IE tuning in extended systems [55, 106].

The GDD procedure is especially convenient for SAPT(KS) calculations because ω must be tuned individually for each monomer. The monomer-centric nature of SAPT means that tuned LRC-DFT maintains size consistency because dimer DFT calculations are not required. For any DFT-based EDA that requires both monomer and supramolecular calculations, the "optimal" or IE-tuning procedure generally violates size consistency [107].

Unfortunately, the admonition to maintain correct asymptotic behavior of $v_{\mathrm{xc}}(r)$ is not always heeded. When orbitals from a semilocal or a global hybrid functional are inserted into wavefunction-based SAPT0 formulas, the too-small energy denominators (resulting from HOMO/LUMO gaps that are much smaller than those in HF theory) cause $E_{\text{disp}}^{(2)}$ to increase unrealistically [55, 79]. This is the reason why SAPT0 (KS) was abandoned soon after the first numerical experiments in the early 2000s; it was not resuscitated until the use of optimally tuned LRC functionals became commonplace [79]. Meanwhile the SAPT0(B3LYP) method, meaning Eq. (7) with B3LYP orbitals, was the lone example of the SAPT formalism to be considered in a survey of contemporary EDA methods for protein–drug interactions [108]. From that unfair comparison, the authors concluded that DFT-based EDAs offered a better balance of properties [108], without considering the fact that SAPT0(B3LYP) exacerbates the already too-large dispersion energies predicted by SAPT0 (HF) calculations.

Indeed, direct comparison of SAPT0(KS) methods with different XC functionals demonstrates that SAPT0(B3LYP) is considerably *less* accurate than SAPT0(HF), despite the incorporation of monomer electron correlation [55]. This is illustrated in Table 1 using the S66 dataset. For the dispersion-bound subset of S66, SAPT0 (B3LYP) exhibits a maximum error of 4.5 kcal/mol versus only 1.5 kcal/mol for conventional SAPT0(HF), and the B3LYP-based approach is also less accurate for hydrogen-bonded systems, where monomer correlation is known to be important [76]. Errors for the SAPT0(BLYP) approach are larger still, as the lack of any exact exchange whatsoever leads to worse asymptotic behavior as compared to B3LYP and further exaggeration of dispersion energies as a result. In contrast, the SAPT0 (LRC-ω_{GDD}PBE) approach reduces the maximum error for the dispersion-bound complexes to 2.0 kcal/mol, although this method remains less accurate than SAPT0 (HF) for each of the standard subsets of S66 [55].

Two sets of XSAPT calculations in Table 1 use electrostatic embedding charges that are discussed in Sect. 1.2.3 and dispersion models that are discussed in Sect. 1.2.4. Once those aspects of the formalism have been described, we will present a more complete compendium of accuracy assessments, in Sect. 2.1.

Table 1 Error statistics for SAPT0(KS) methods and XSAPT methods (the latter of which include charge embedding), as applied to the S66 dataset and two subsets[a]

Method	Error (kcal/mol)[b]					
	H-bonded		Disp.-bound		All S66	
	MAE[c]	Max[d]	MAE[c]	Max[d]	MAE[c]	Max[d]
SAPT0(HF)+δHF	0.4	0.8	0.6	1.5	0.5	1.5
SAPT0(BLYP)+δHF	0.8	1.5	2.6	6.1	1.5	6.1
SAPT0(B3LYP)+δHF	0.6	2.1	1.7	4.5	1.1	4.5
SAPT0(LRC-ω_{GDD}PBE)+δHF	1.0	2.3	1.0	2.0	0.8	2.7
XSAPT[e] + *ai*D3+δHF	0.2	0.4	0.4	1.0	0.4	1.0
XSAPT[e]+MBD+δHF	0.2	0.8	0.5	1.1	0.4	1.1

[a] Data are from Ref. [55]. Basis set is jun-cc-pVDZ for SAPT0(KS) and def2-TZVPPD for XSAPT
[b] With respect to CCSD(T)/CBS benchmarks from Ref. [73]
[c] Mean absolute error
[d] Maximum absolute error
[e] LRC-ω_{GDD}PBE and CM5 embedding charges

1.2.3 XPol: Self-Consistent Charge Embedding

XSAPT was originally intended as a method for molecular clusters and liquids, in which many-body polarization effects would be treated self-consistently (via the XPol procedure) while resorting to a pairwise approximation for other energy components, hence, XPol + SAPT [32, 33]. That design reflects the reality that nonadditive effects are generally small, except for polarization [109]. For example, three-body (trimer) dispersion effects contribute no more than 5–6% (and typically $<2\%$) of the total interaction energy in small clusters composed of small molecules [110, 111]. In crystalline benzene, three-body effects contribute 0.8–0.9 kcal/mol to the lattice energy (about 7%) [112–114], with three-body dispersion as the largest contribution [114]. In contrast, three-body polarization ranges from -9 to -13 kcal/mol in the stable structures of $(H_2O)_6$ [115]. In water trimers extracted from a liquid simulation, the total three-body energies range from -6 kcal/mol to $+1$ kcal/mol [116].

These polarization effects are not negligible, and to describe them at a cost that remains tractable for sizable clusters, we use the XPol method of Gao and coworkers [29–31]. XPol starts from single-monomer SCF calculations, following which a set of atomic point charges is computed from those SCF wavefunctions and subsequent monomer SCF calculations are performed in the presence of these embedding charges. This entire procedure is then iterated to self-consistency, with an outer loop over monomers and an inner loop consisting of SCF iterations on a given monomer, in the presence of point charges representing the other $N-1$ monomers.

In more detail, and considering closed-shell fragments for convenience, the XPol energy is [33]

$$E_{\text{XPol}} = \sum_{A=1}^{N} \left[2\sum_{n}^{\text{occ}} \left(\mathbf{c}_n^A\right)^\dagger \left(\mathbf{h}^A + \mathbf{J}^A - \frac{1}{2}\mathbf{K}^A\right)\mathbf{c}_n^A + E_{\text{nuc}}^A \right] + E_{\text{embed}}, \qquad (22)$$

where the term in square brackets is the Hartree–Fock energy for monomer A expressed in terms of ALMOs $\{\mathbf{c}_n^A\}$, meaning MOs that have support only from Gaussian basis functions centered on monomer A. The quantity E_{embed} in Eq. (22) is the sum of electrostatic embedding energies from the wavefunction-derived point charges.

Self-consistent variation of E_{XPol} is accomplished via monomer Fock matrices $\mathbf{F}^A$ whose matrix elements in the atomic orbital basis $(\mu, \nu, \ldots)$ are

$$F_{\mu\nu}^A = f_{\mu\nu}^A - \frac{1}{2}\sum_{B\neq A}\sum_{b\in B} q_b\left(\Phi_b\right)_{\mu\nu} + \sum_{a\in A}\frac{\partial E_{\text{embed}}}{\partial q_a}\frac{\partial q_a}{\partial P_{\mu\nu}^A}. \qquad (23)$$

Here, $\mathbf{f}^A$ is the Fock matrix for isolated monomer A and

$$(\Phi_b)_{\mu\nu} = \left\langle \mu \left| \frac{1}{\| \mathbf{r} - \mathbf{R}_b \|} \right| \nu \right\rangle \tag{24}$$

is the electrostatic potential generated by the function pair $\mu\nu$ at the point $\mathbf{R}_b$, the location of nucleus $b \in B$. The integral that is implicit in Eq. (24) is over the electron coordinate $\mathbf{r}$. The second term in Eq. (23), involving $(\Phi_b)_{\mu\nu}$, comes from the electrostatic interaction between $\rho_A(\mathbf{r})$ and the embedding charges $\{q_b\}$ on monomers $B \neq A$ [117]. The final term in Eq. (23) includes the response of the embedding charges to changes in the fragment wavefunctions $(\partial q_a / \partial P^A_{\mu\nu})$ and ensures that the converged XPol energy is variational [30].

The embedding potential E_{embed} is a straightforward sum of interactions between point charges $q_a \in A$ and monomer densities $\rho_B(\mathbf{r})$, along with interactions between the charges and the nuclei, and $\partial E_{\text{embed}} / \partial q_a$ is easy to evaluate [33]. In contrast, the derivative $\partial q_a / P^A_{\mu\nu}$ depends on how the embedding charges $\{q_a\}$ are determined from the SCF wavefunction for monomer A. Mulliken charges were used in the original XPol method [29–31], but we find them to be unstable when extended basis sets are employed, as are Löwdin charges [32].

A physically appealing choice is to use "ChElPG" charges derived from the molecular electrostatic potential [118]; these are the atomic charges that best reproduce the SCF electrostatic potential near the vdW surface of the molecule. The requisite derivatives $\partial q_a / P^A_{\mu\nu}$ are complicated but they have been implemented [33, 119]. In recent work, however, we have taken the q_a from "Charge Model 5" (CM5) [120], which is an empirical modification of the Hirshfeld scheme [121] that is designed to reproduce molecular dipole moments. This avoids some significant computational overhead associated with evaluating $\partial q_a / P^A_{\mu\nu}$ for ChElPG charges, as illustrated by timing data in Fig. 2 for a C_{60}–buckycatcher complex where the time to compute the derivatives $\partial q_a / P_{\mu\nu}$ is reduced from 16.7 h (ChElPG) to 2.0 h (CM5) [41]. The former value represents about 35% of the total cost of the XSAPT calculation, whereas the cost to differentiate the CM5 charges is almost negligible, resembling the cost to evaluate E_{xc} on the DFT quadrature grid.

Regardless of which charges are used, the XPol SCF time greatly exceeds the time required to evaluate the SAPT corrections. This is predicated on replacing $O(N^5)$ dispersion in SAPT0 with a low-cost dispersion model, and those models are discussed next.

1.2.4 Dispersion Models

Perhaps the most important innovation in XSAPT is its hybrid dispersion model, replacing $E^{(2)}_{\text{disp}}$ in Eq. (7). This, too, has several variants that can be selected, including empirical dispersion potentials [36, 37, 42] and a version of the many-body dispersion (MBD) model [40–42] that was originally developed for DFT by

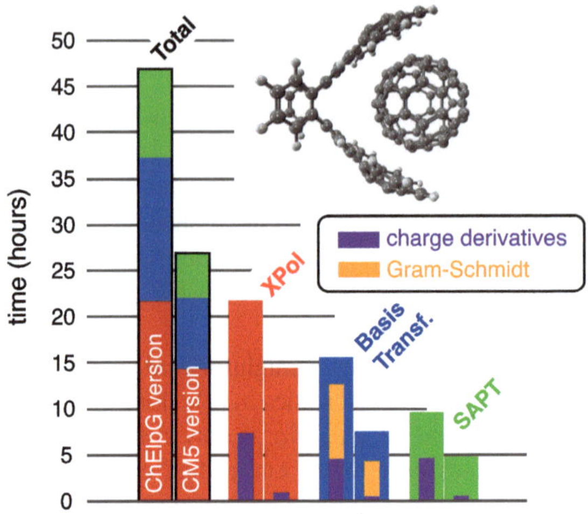

Fig. 2 Timing data for XSAPT+*ai*D3/hp-TZVPP calculations on a C_{60}@$C_{60}H_{28}$ complex that is shown (4592 total basis functions), running on 28 processors (Dell Intel Xeon E5-2680 v4). The total time is broken down into three color-coded steps: the XPol dual SCF procedure (in red), pseudocanonicalization to prepare the dimer basis (in blue), and the SAPT steps (in green). Charge derivatives $\partial q_a/P^A_{\mu\nu}$ are required in all three steps, and the time to evaluate them is indicated in purple. Improvements in Gram–Schmidt orthogonalization (timings in orange) reflect better parallelization beginning with Q-Chem v. 6.0. (Reprinted with permission from Ref. [41]; copyright 2019 American Institute of Physics)

Tkatchenko and coworkers [4, 122–124]. Both approaches are described in this section.

Empirical dispersion potentials take the usual atomic pairwise form [42],

$$E^{aiD}_{\text{disp}} = -\sum_{a\in \text{A}}\sum_{b\in \text{B}}\left[f_6(R_{ab})\frac{C_{6,ab}}{R^6_{ab}} + f_8(R_{ab})\frac{C_{8,ab}}{R^8_{ab}}\right], \tag{25}$$

written here for a dimer A···B. The quantities f_6 and f_8 are damping functions that are needed to prevent divergence as $R\rightarrow 0$ [125]. Pairwise dispersion coefficients $C_{6,ab}$ and $C_{8,ab}$ are obtained from the combination rule

$$C_{n,ab} = (C_{n,a}C_{n,b})^{1/2} \tag{26}$$

for $n=6$ or 8. In the context of XSAPT, we refer to Eq. (25) as an *ab initio* dispersion potential (*ai*D), because the atomic parameters $\{C_{6,a}\}$ and $\{C_{8,a}\}$ were obtained by fitting to *ab initio* dispersion energies [34–37], primarily from SAPT2+(3) calculations.

Notably, Eq. (25) has the same form as the dispersion corrections used in DFT+D [1, 2]. Several other groups have explored the idea of replacing $E^{(2)}_{\text{disp}}$ in SAPT0 with an empirical model [126–128], which has sometimes been called “SAPT0+D” [127, 128], but we prefer to reserve the term SAPT0 to mean second-order dispersion and will instead call these models “SAPT(KS)+D”; see the discussion that accompanies Eq. (13). In terms of the physics, Grimme’s DFT+D3 and DFT+D4 dispersion corrections [129, 130] have been adapted for use with SAPT(HF) by refitting the damping parameters, leaving the C_6 and C_8 parameters unmodified [127, 128]. Unlike earlier dispersion corrections by Grimme [131], beginning with D3 the atomic dispersion coefficients are computed from DFT polarizabilities and should resemble genuine dispersion when the monomers are well separated [129]. However, these corrections *do not* resemble genuine dispersion at vdW contact distances, because the damping is significant in that range. As a result, the energy components in these methods are not reliable even if the total interaction energies may be accurate [128].

The same is true of our first-generation *ai*D1 model [34], whose parameters were fit to obtain total interaction energies in agreement with CCSD(T)/CBS benchmarks. It was subsequently discovered that SAPT(KS) + *ai*D1 benefits significantly from error cancellation, such that the energy *components* are not in good agreement with SAPT2+(3) benchmarks [36], and the same is true for the SAPT(HF)+D3 and SAPT (HF)+D4 models [127, 128]. These methods yield good interaction energies at the expense of significant error cancellation between the dispersion and exchange-repulsion terms [128]. Because energy decomposition is one of the primary motivations to use the SAPT framework, we believe it is crucial to achieve high accuracy not only for E_{int} but also for its components. For this reason, we do not recommend the first-generation *ai*D1 model

In subsequent generations of the model, parameters were fit so that the model dispersion energy $E_{\text{disp}}^{ai\text{D}}$ reproduces *ab initio* dispersion energies obtained from third-order calculations such as SAPT2+(3) [36, 37]. The most recent *ai*D3 version [37] uses an improved training set and is the recommend version. Unlike *ai*D1, these models offer consistent accuracy for E_{int} and each of its individual components, indicating that the total does not benefit from error cancellation so the energy components ought to be reliable [36, 37]. As shown in Fig. 3 using the S66 dataset, the accuracy of XSAPT+*ai*D3 (which includes charge embedding) is competitive with a variety of post-SCF methods including SAPT2+(3), with a maximum error of 1.1 kcal/mol and a mean absolute error (MAE) less than 0.3 kcal/mol.

Nevertheless, the performance of the atomic pairwise dispersion approximation in Eq. (25) breaks down for large molecules [38], degrading the performance of XSAPT+*ai*D and any similar SAPT+D model when the monomers are large. This can be seen in benchmark tests on the L7 dataset [132], which consists of larger complexes with up to 73 second-row atoms (Fig. 4). Errors obtained using XSAPT +*ai*D3 are much larger for L7 (up to 5 kcal/mol) as compared to those for the small S66 dimers, and a variety of dispersion-inclusive and dispersion-corrected DFT approaches are significantly more accurate. At the other end of the spectrum, MP2/CBS significantly overestimates dispersion energies for dimers involving highly conjugated molecules, such as coronene and circumcoronene in L7. The

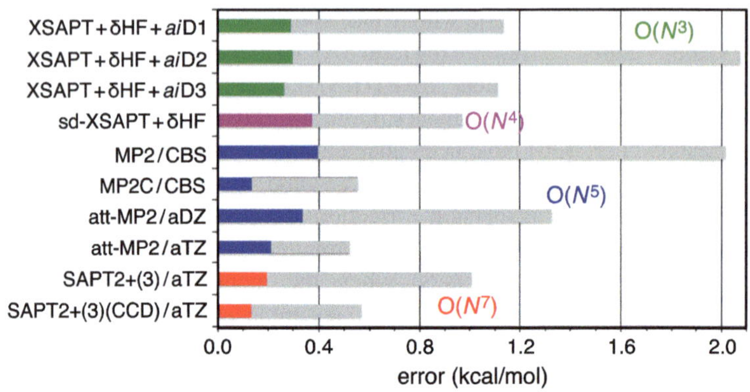

Fig. 3 Error statistics for total interaction energies in the S66 dataset, as compared to CCSD(T)/CBS benchmarks, for various post-SCF correlated methods. Colored bars indicate MAEs and gray bars are maximum absolute errors. The colors themselves suggest the formal scaling with system size, which is indicated at the right. XSAPT calculations employ ChElPG embedding charges with LRC-ω_{IE}PBE, meaning a IE-tuned version of LRC-ωPBE, and use the following basis sets: jun-cc-pVDZ for XSAPT + *ai*D1, hpTZVPP for XSAPT+*ai*D2 and +*ai*D3, and 6-31G(d,2p) for scaled-dispersion (sd-)XSAPT. The MP2-based methods used augmented Dunning basis sets. (Adapted from Ref. [37])

only MP2-based method with reasonable accuracy for this dataset is "attenuated" (att-)MP2 [133], which removes the long-range portion of the Coulomb operator from the post-SCF calculation and whose success for noncovalent interactions likely rests on error cancellation.

The issue for XSAPT+*ai*D in large complexes is dispersion nonadditivity arising from the presence of polarizable centers that screen the $C_{6,ab}$ (and higher-order) dispersion coefficients within a molecule, relative to what would be predicted based on isolated-atom C_6 coefficients using the combination rule in Eq. (26). Dobson has called this "type-B" dispersion nonadditivity [134] and to account for it, Grimme's D3 and D4 corrections incorporate three-body (triatomic) dispersion corrections of the Axilrod–Teller–Muto (ATM) "triple-dipole" form [3]. This is a different usage of "many-body" as compared to the discussion in Sect. 1.2.3, where the "bodies" were monomers. Here they are atoms, reflecting the atom–atom pairwise starting point in Eq. (25).

The particular version of the ATM correction that is used in Grimme's models is [1, 129]

$$E_{\text{3B-ATM}}^{(\text{Grimme})} = -\sum_{a}\sum_{b>a}\sum_{c>b}\frac{g_{abc}\; C_{9,abc}\, f_{3\text{B}}(\bar{R}_{abc})}{R_{ab}^3 R_{ac}^3 R_{bc}^3} \tag{27}$$

where g_{abc} is a geometric factor and $f_{3\text{B}}(\overline{R}_{abc})$ is a damping function that depends on the mutual three-body distance, $\overline{R}_{abc}$ [38]. The requisite C_9 coefficients can be estimated from the C_6 coefficients [129],

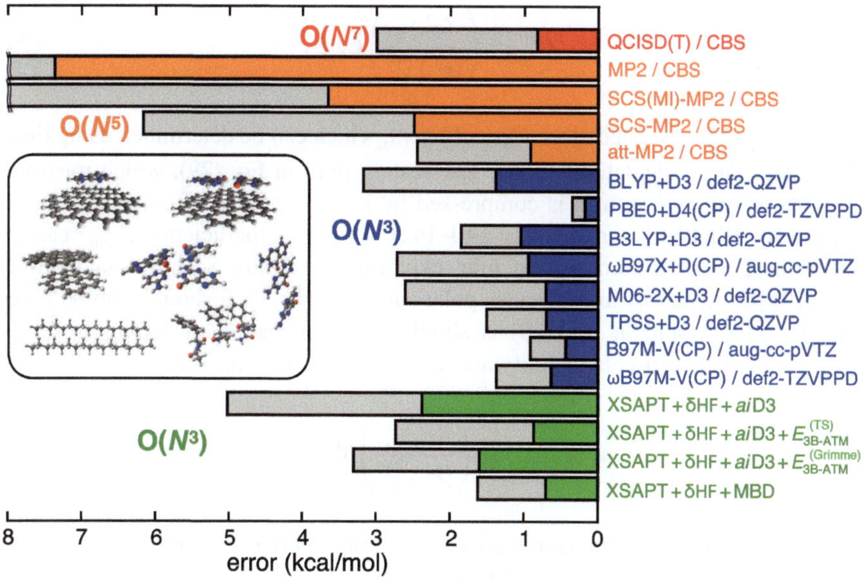

Fig. 4 MAEs (colored bars) and maximum absolute errors (gray bars) for the L7 dataset, assessed against CCSD(T)/CBS benchmarks. (The L7 complexes are shown in the inset.) Triple-ζ DFT results are CP-corrected and all XSAPT calculations use the def2-TZVPPD basis set with LRC-ω_{GDD}PBE. Each method is color-coded according to its formal asymptotic scaling with system size. Maximum errors for MP2/CBS and SCS(MI)-MP2/CBS are both 10 kcal/mol [38]. (Adapted with permission from Ref. [42]; copyright 2021 American Chemical Society)

$$C_{9,abc} = (C_{6,ab}\, C_{6,bc}\, C_{6,ac})^{1/2}. \tag{28}$$

A version with a modified form of damping has been used by Tkatchenko and Scheffler [38, 135–137]. When either this form ($E_{\text{3B-ATM}}^{(\text{TS})}$) or else $E_{\text{3B-ATM}}^{(\text{Grimme})}$ in Eq. (27) is combined with XSAPT+*ai*D3, results for the L7 complexes are greatly improved as documented in Fig. 4.

Nevertheless, the significance of nonadditive dispersion for large vdW complexes prompted a search for a less *ad hoc* way to describe this phenomenon. To that end, we have explored the use of XSAPT in conjunction with the MBD model developed for DFT by Tkatchenko and coworkers [4, 122–124]. It builds upon an earlier dispersion model developed by Tkatchenko and Scheffler [135], which is sometimes called "vdW-TS" in the condensed matter DFT literature [138]. The theory behind both MBD and its predecessor has been reviewed recently [1]. Briefly, the vdW-TS method determines an *in situ* (or "effective") value for $C_{6,a}$ by means of a volume-scaling relationship,

$$C_{6,a}^{\text{eff}} = \left(\frac{\langle r^3 \rangle_a}{\langle r^3 \rangle_a^{\text{free}}} \right) C_{6,a}^{\text{free}} \, , \tag{29}$$

where $C_{6,a}^{\text{free}}$ is the isolated atom C_6 coefficient, which can be determined using time-dependent DFT calculations [139]. The scaling ratio in Eq. (29), which measures how the volume of atom a is compressed by its molecular environment, is determined by Hirshfeld partition [1, 140–142]. Pairwise coefficients $C_{6,ab}$ can be determined using a combination rule, exploiting a volume scaling analogous to Eq. (29) for the atoms-in-molecules polarizabilties $\overline{\alpha}_a^{\text{eff}}$ [1]. It should also be noted that the relationship in Eq. (29) is simply a model, based on a long-presumed relationship between polarizability and volume [143]. Recent work suggests that

$$\overline{\alpha}_a^{\text{TS}} = \left(\frac{\langle r^4 \rangle_a}{\langle r^4 \rangle_a^{\text{free}}} \right) \overline{\alpha}_a^{\text{free}} \tag{30}$$

may be the more fundamental relationship, leading to improvements in the vdW-TS method [144]. Thus, the model may be improvable.

In the MBD extension of the vdW-TS method, imaginary-frequency atom-in-molecule polarizabilities, consistent with the dispersion coefficients $C_{6,a}$ in Eq. (29), are determined according to

$$\overline{\alpha}_a^{\text{TS}}(\mathrm{i}\omega) = \frac{\overline{\alpha}_a^{\text{free}}}{1 + (\omega/\omega_a)^2} \tag{31}$$

where the characteristic excitation frequency ω_a can be obtained from $C_{6,a}$ [1, 135]. These atomic polarizabilities could be used to compute pairwise dispersion coefficients using the Casimir–Polder relation [3, 28, 145, 146],

$$C_{6,ab} = \frac{3\hbar}{\pi} \int_0^\infty \overline{\alpha}_a(\mathrm{i}\omega) \, \overline{\alpha}_b(\mathrm{i}\omega) \, d\omega \, . \tag{32}$$

Instead, many-body dispersion effects are introduced via mutual (self-consistent) dipole screening of the atomic polarizabilities $\overline{\alpha}_a^{\text{TS}}$, which are mapped onto a $3N_{\text{atoms}}$-dimensional quantum harmonic oscillator Hamiltonian [122, 123]. Eigenmodes of that Hamiltonian provide information about collective (multiatom) contributions to dispersion [147, 148].

This approach was originally developed as an *ad hoc*, density-dependent dispersion correction for DFT [122–124], using range separation to make sure that the model accounts for long-range dispersion only (and to avoid short-range polarization catastrophes [149]), while the XC functional is tasked with the description of short-range correlation effects including dispersion. For this reason, the model described above has been called "MBD@rsSCS" [4], meaning "range-separated self-consistent screening," although we will simply call it MBD.

For use with SAPT, we require a version of MBD that describes dispersion at all length scales. This will ultimately be combined with an XC functional like PBE that is largely free of dispersion (even in the short range) [1], made asymptotically correct in LRC-ωPBE form. To that end, "effectively screened" C_8 coefficients were added to the original MBD model [40], which is a dipole polarizability (C_6) model. This effectively screened dipole–quadrupole (esDQ) dispersion model augments the original MBD, and the complete dispersion model to be used with SAPT is [40, 41]

$$E_{\text{disp}}^{\text{MBD+esDQ}} = E_{\text{disp}}^{\text{MBD}} - s_8 \sum_a \sum_{b>a} f_8(R_{ab}) \frac{C_{8,ab}}{R_{ab}^8} \,. \tag{33}$$

Coefficients $C_{8,ab}$ can be obtained from the $\{C_{6,a}\}$ computed for the MBD model [40].

The SAPT(KS)+MBD method consists in adding the dispersion model of Eq. (33) to the nondispersion SAPT(KS) terms in Eq. (13). XSAPT+MBD augments this with XPol charge embedding and uses LRC-ω_{GDD}PBE to compute the monomer wavefunctions. Results in Fig. 5 for the S66 dimers demonstrate that the accuracy of XSAPT+MBD is very similar to that of XSAPT+*ai*D3, without any three-body dispersion corrections, since type-B dispersion nonadditivity is insignificant for these small complexes. (See Table 1 for the corresponding numerical data.) For the L7 complexes, however, XSAPT+MBD is appreciably more accurate than XSAPT+*ai*D3 even when a nonadditive dispersion correction is included in the latter (Fig. 4). At present, XSAPT+MBD is the most accurate version of XSAPT, at least for charge-neutral complexes. (Performance for ions is discussed in Sect. 2.1.4.) Additional accuracy benchmarks are provided in Sect. 2.1, but we first introduce a few more aspects of the XSAPT methodology.

1.3 XSAPT Energy Decomposition

Having established the XSAPT formalism, we next discuss its inherent EDA scheme, focusing first on the accuracy of the energy components (Sect. 1.3.1). Then, we describe how the SAPT induction energy can be separated into polarization and CT contributions (Sect. 1.3.2). Finally, we discuss how many-body effects manifest in the EDA for systems with more than two monomers (Sect. 1.3.3).

1.3.1 Comparison to High-Level SAPT

Because SAPT2+(3) affords total interaction energies in good agreement with CCSD(T) benchmarks [25–27, 57, 58], we take its energy components as

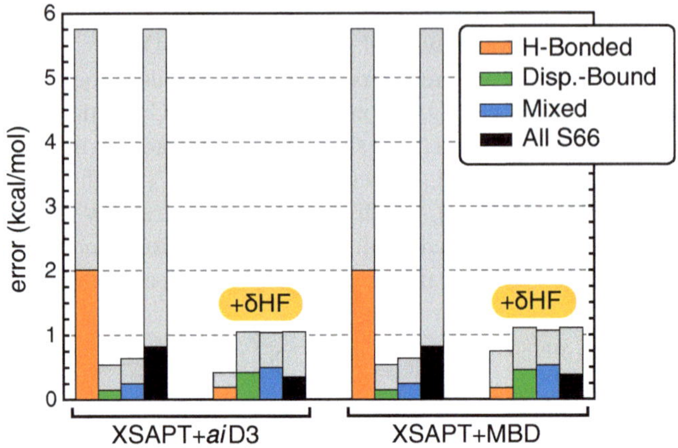

Fig. 5 Performance of XSAPT/def2-TZVPPD versus CCSD(T)/CBS benchmarks for the S66 dataset, using CM5 embedding charges. Results with and without the δHF correction are shown separately. Colored bars indicate mean absolute errors and gray bars indicate maximum errors. (Data are from Ref. [55] and the figure is reprinted with permission from Ref. [42]; copyright 2021 American Chemical Society)

benchmarks for other EDAs [1, 77]. Figure 6 plots those benchmarks against SAPT (HF)+MBD, SAPT(KS)+MBD, and XSAPT+MBD energy components, using the S22 dataset of small-molecule dimers [150, 151]. These three models allow us to compare the effects of monomer electron correlation [SAPT(HF) versus SAPT(KS)] and charge embedding [SAPT(KS) versus XSAPT]. Results are partitioned into three standard subsets defined by the ratio of E_{elst} to E_{disp} [152], as indicated in Fig. 6. Electrostatics-dominated complexes, defined as cases where $|E_{\text{elst}}| \geq 2|E_{\text{disp}}|$, correspond to hydrogen bonding in this dataset.

Broadly speaking, there is reasonable agreement among these methods (and with the benchmarks) for E_{elst} and E_{exch}, although the SAPT(HF) method slightly exaggerates electrostatic interactions for two of the hydrogen-bonded complexes where those interactions are largest, yet still affords sub-kcal/mol accuracy (MAE = 0.7 kcal/mol). However, that method significantly underestimates exchange repulsion in many of the hydrogen-bonded dimers, and the MAE is 3.6 kcal/mol for that component with a maximum error of 6.1 kcal/mol [77]. The two methods that use LRC-ω_{GDD}PBE wavefunctions for the monomers are much more accurate for these two energy components.

Induction energies are rather small except for the hydrogen-bonded dimers. Here, XSAPT's use of charge embedding clearly leads to better accuracy as compared to the SAPT(KS)+MBD method that includes monomer correlation but not self-consistent embedding (Fig. 6c). Considering only the hydrogen-bonded dimers, the MAE for XSAPT+MBD is 0.2 kcal/mol with respect to SAPT2+(3) benchmarks versus 1.5 kcal/mol for SAPT(KS)+MBD and 2.0 kcal/mol for SAPT(HF)+MBD. This is notable, given that all three methods include the δHF correction, yet there is still value in self-consistent embedding.

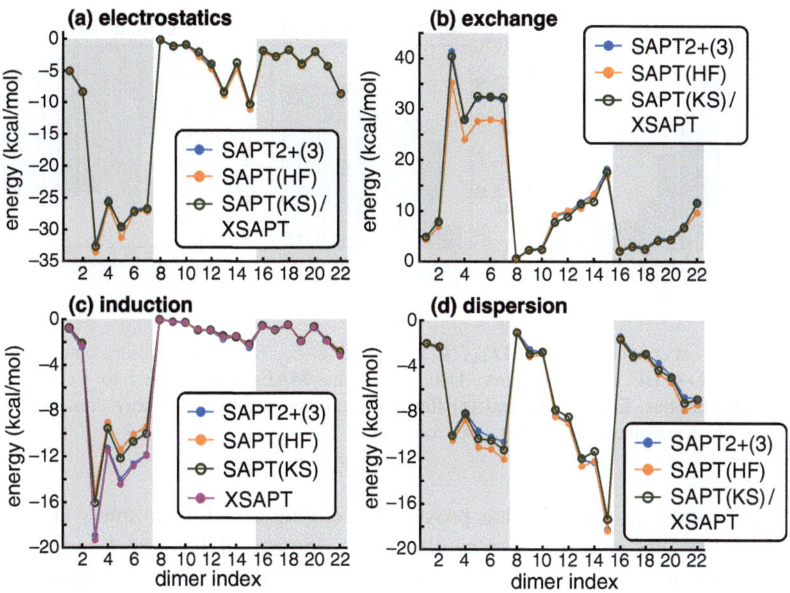

Fig. 6 Comparison of energy components for the S22 dimers: (**a**) E_{elst}, (**b**) E_{exch}, (**c**) E_{ind}, and (**d**) E_{disp}. SAPT(HF)+MBD, SAPT(KS)+MBD, and XSAPT+MBD calculations were performed using the aug-cc-pVQZ basis set and include the δHF correction, with LRC-ω_{GDD}PBE used for the latter two methods and CM5 charge embedding for XSAPT. The SAPT(KS)+MBD and XSAPT+MBD results are indistinguishable (and thus plotted together), except for E_{ind} in (**c**). Alternating shaded regions delineate the three standard subsets of S22: hydrogen-bonded dimers on the left, dispersion-bound dimers in the middle, and dimers with mixed influence interactions on the right. (Reproduced with permission from Ref. [77]; copyright 2022 American Chemical Society)

Finally, MBD dispersion energies are in very good agreement with the benchmarks regardless of which orbitals are used for the monomer wavefunctions (Fig. 6d). All together, this means that XSAPT+MBD reproduces all of the energy component accurately. Notably, results in Fig. 6 are converged with respect to basis set (aug-cc-pVQZ), so this accuracy does not rest on error cancellation as is the case for a method like SAPT0(HF). Basis-set convergence is systematic, as shown for E_{int} and several components in Fig. 7 using Karlsruhe basis sets. Other energy components exhibit similar or better convergence behavior, and the minimally augmented def2-ma-TZVP basis set affords converged results for all components [77]. This basis set, which was introduced in Ref. 77 as a proper subset of def2-TZVPD, differs from other minimally augmented extensions of def2-TZVP [153].

1.3.2 Separating Polarization and Charge Transfer

For chemists, the "induction energy" in the XSAPT-EDA of Eq. (15) may be unfamiliar; missing are the more familiar contributions from polarization and

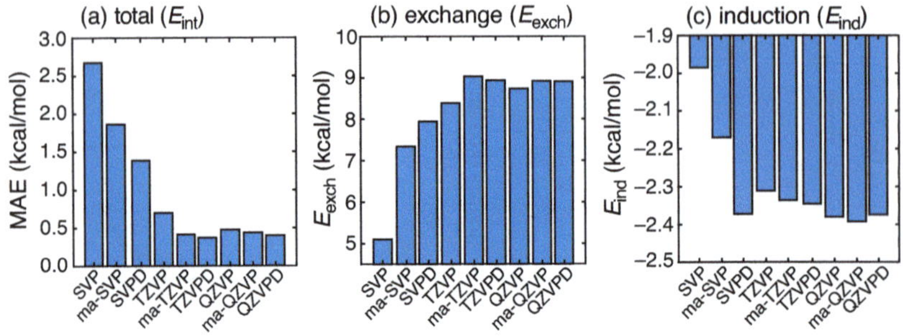

Fig. 7 Basis-set convergence of (**a**) E_{int}, (**b**) E_{exch}, and (**c**) E_{ind} for the S66 dimers, computed at the XSAPT+MBD+δ HF level of theory. Data in (**a**) are MAEs with respect to CCSD(T)/CBS interaction energies. Data in (**b**) and (**c**) illustrate convergence of the energy components. All basis sets are Karlsruhe "def2." (Data are from Ref. [77])

intermolecular CT. The latter has proven challenging to define uniquely [154–157], a debate that has played out recently in the context of halogen bonding [157–165]. Some authors, citing the Hellman–Feynman theorem [166], insist that all intermolecular forces—including polarization, CT, and even dispersion—are fundamentally electrostatic in origin [166–170]. This is not precisely *wrong*, insofar as they are all governed by a Coulombic Hamiltonian, but it ignores the idea of emergent complexity in many-body systems [171].

Within SAPT-based methods, both polarization and CT are contained within the induction energy as indicated in Eq. (3). Figure 8 illustrates the inherent difficulty in separating the two. Fundamentally, polarization is orbital relaxation in response to a perturbing influence, from a nearby nonbonded partner in this case. Mathematically, orbital relaxation can be cast as a unitary transformation from the isolated monomer MOs to the relaxed orbitals obtained in the presence of the partner, and such a transformation can always be written in terms of singly substituted Slater determinants expressed in the original, unrelaxed basis [172]. In a compact basis set, where the atomic orbital basis functions centered on monomer A have little overlap with those centered on B, one can readily identify the A-centered excitations that relax A's MOs in the presence of B, distinguishing them from A $\rightarrow$ B excitations. That dichotomy could provide a basis for separating E_{pol} from E_{CT}. In larger basis sets, however, functions centered on one monomer extend significantly over the other monomer, and the distinction between orbital relaxation and CT is obscured.

As a result, orbital-based definitions of CT are notoriously unstable with respect to expansion of the atom-centered Gaussian basis set [173–175]. Within SAPT, it has been suggested to define E_{CT} as the difference between $E_{ind}^{(2)}$ computed in the monomer- and dimer-centered basis sets [176]; this has been called the second-order CT energy, $E_{CT}^{(2)}$:

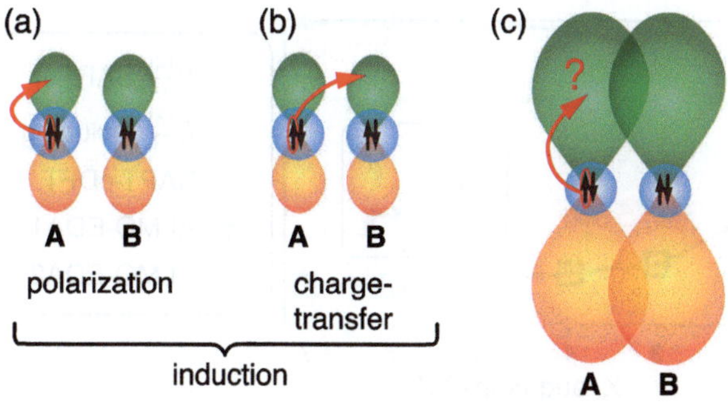

Fig. 8 Cartoon depiction of single excitations that contribute to induction. For compact atom-centered basis sets, these can be easily grouped into excitations that contribute to either (**a**) polarization (in this case, orbital relaxation of monomer A due to the presence of B) or else (**b**) CT (in this case, A→B). For larger basis sets, however, the distinction is muddied, as in (**c**). (Adapted with permission from Ref. [173]; copyright 2016 American Chemical Society)

$$E_{\text{CT}}^{(2)} = E_{\text{ind}}^{(2)}[\text{dimer basis}] - E_{\text{ind}}^{(2)}[\text{monomer basis}]. \quad (34)$$

However, $E_{\text{CT}}^{(2)} \to 0$ in the CBS limit, where monomer and dimer basis sets should afford the same wavefunctions. A numerical demonstration is provided in Fig. 9 using the prototypical Lewis acid/base system $H_3N\cdots BH_3$ (borazane) and extending previously reported data [173] to basis sets up to aug-cc-pV7Z. Alternatively, one may isolate E_{CT} via a regularization procedure that removes the Coulomb driving force that attracts electrons on A to monomer B [174].

A simpler procedure is based on constrained (c)DFT [159, 173, 175], a general technique in which the KS equation is solved in the presence of a constraint that forces a stipulated group of atoms to integrate to a user-specified number of electrons [177]. (There is some variability in how the constraint is implemented and our approach is documented elsewhere [178, 179].) Using cDFT, one can define a CT-free reference state in which the A···B complex is assembled in space at the geometry of the dimer, and its KS-DFT energy is computed under the constraint that both monomers contain an integer number of electrons. Thus, orbitals are polarized but cannot transfer any (net) charge. Relaxing this constraint defines E_{CT} (as the cDFT → DFT relaxation energy), which is subtracted from a SAPT-based calculation of E_{ind} to define E_{pol} [159, 173, 175]. For $H_3N\cdots BH_3$, Fig. 9 shows that this definition is exquisitely stable with respect to basis set, with a very minor change in E_{CT} between double- and triple-ζ basis sets but no change thereafter. This is consistent with the fact that DFT densities tend to be *almost* converged in double-ζ basis sets and well converged in triple-ζ basis sets.

It is worth noting that E_{CT} as defined by the second-generation ALMO-EDA2 method [13] does appear to have a well-defined CBS limit, as a result of a new definition of E_{CT} as compared to the first-generation ALMO-EDA1 [180]. Even for the second-generation EDA, however, convergence is slow and not complete until

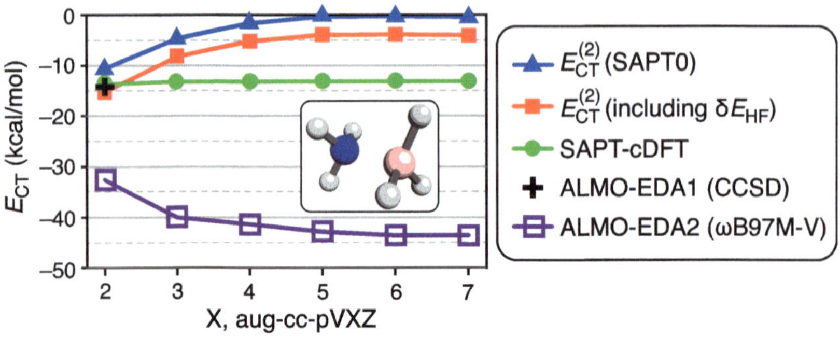

Fig. 9 Various definitions of the CT energy applied to $H_3N\cdots BH_3$ using aug-cc-pVXZ basis sets. (Angular momentum functions beyond g were removed from aug-cc-pV5Z and larger basis sets.) The quantity $E_{CT}^{(2)}$ is defined in Eq. (34) and might be evaluated with or without the δHF correction

aug-cc-pV6Z, with an 11 kcal/mol difference between double- and sextuple-ζ. That is less variability, however, as compared to ALMO-EDA1, where aug-cc-pV6Z results for the same borazane complex do not appear to have converged [173]. The SAPT-cDFT value of E_{CT} is considerably smaller than the ALMO-EDA2 value, which has been observed previously and attributed to a failure by cDFT to fully inhibit CT [156]. An alternative explanation is that the decision to use a pseudocanonicalized dimer basis for SAPT (Sect. 1.2.1) puts some intermolecular CT effects into the monomer wavefunctions [32], classified as electrostatics. In contrast, ALMO-EDA uses the eponymous, fragment-localized ALMO basis, leading to larger CT energies. This hypothesis is at least consistent with the fact that Eq. (34) affords smaller CT energies as compared to ALMO-EDA2, even in double-ζ basis sets where $|E_{CT}^{(2)}|$ is largest.

The SAPT-cDFT definition of E_{CT} affords what we believe to be a correct periodic trend for complexes $M^{+}\cdots(C_6H_6)$ involving alkali cations (Fig. 10a). Specifically, we expect the magnitude of E_{CT} to decrease rapidly with ion–molecule distance, due to exponential decay of the monomer wavefunctions. For the optimized geometries that are used in Fig. 10a, that distance is largest for K^+ and smallest for Li^+, and the SAPT-cDFT scheme predicts that $|E_{CT}(Li^+)| > |E_{CT}(Na^+)| > |E_{CT}(K^+)|$. The same cannot be said of several other EDA schemes. For this reason, we judge the cDFT-based approach to be reliable. For $M^{+}\cdots(H_2O)$ complexes, the same periodic trend is not observed (using either SAPT-cDFT or the ALMO-EDA2 scheme), as $|E_{CT}(K^+)|$ is slightly larger than $|E_{CT}(Na^+)|$, although $|E_{CT}(Li^+)|$ remains larger than both. However, the magnitude of the CT energies is smaller than it is for $M^{+}\cdots(C_6H_6)$ so something close to the anticipated group trend is still observed.

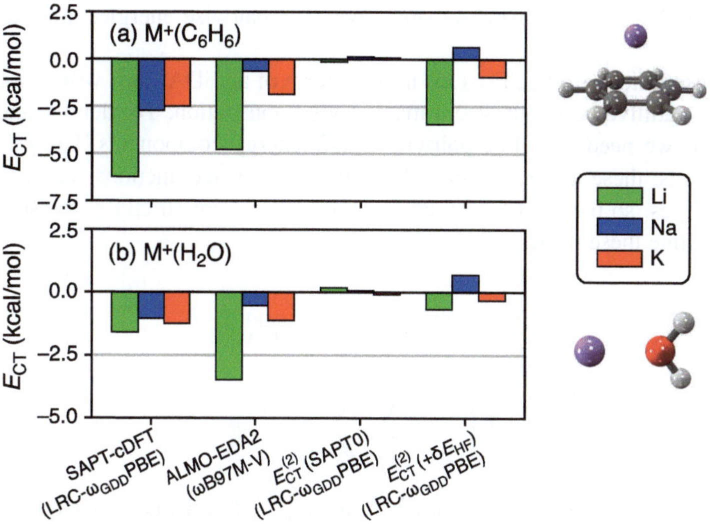

Fig. 10 CT energies for complexes of alkali cations with (**a**) C_6H_6 and (**b**) H_2O, computed according to various schemes. Geometries were optimized at the MP2/def2-SVPD level, and the ion–molecule distance increases in the order $Li^+ < Na^+ < K^+$. EDA calculations were performed using aug-cc-pVTZ for $M^+(C_6H_6)$ and aug-cc-pVQZ for $M^+(H_2O)$

1.3.3 Many-Body Systems

The XSAPT methodology was originally envisaged as "XPol + SAPT" [32–34], using a combination of XPol to describe many-body polarization in a cluster or molecular liquid and pairwise-additive SAPT0 for the remaining components of the intermolecular interaction energy [32]. Alternative dispersion models quickly became the focus [34], and there have been many applications to dimer systems. The methodology is efficient enough that a relatively large ion–water cluster, $X^-(H_2O)_n$, can be handled by treating $(H_2O)_n$ as a single monomer [77, 179], and results up to $X^-(H_2O)_{70}$ are used in Sect. 2.2.2 to establish how the cost scales with system size. Nevertheless, the methodology is formulated for many-body systems and that formalism is described next.

By definition, the XSAPT energy for a collection of N monomers is given by a pairwise sum [33, 36, 37],

$$E_{\text{int}}^{\text{XSAPT,MB}} = \sum_{A}^{N} \sum_{B>A}^{N} E_{\text{int},AB}^{\text{XSAPT,2B}}(A, B, \ldots, N), \tag{35}$$

where $E_{\text{int},AB}^{\text{XSAPT,2B}}(A, B, \ldots, N)$ is the two-body XSAPT-EDA in Eq. (15), applied to the dimer AB. The additional notation suggests that this quantity is a function of all N monomers, because the dimer SAPT calculations employ XPol monomer wavefunctions that are computed in the presence of embedding charges for the remaining $N-2$ monomers. As such, even the pairwise energies are many-body properties.

It is useful to rearrange Eq. (35) into the form of an EDA. This will contain some pairwise-additive parts but also a many-body contribution. To identify and isolate the latter, we need to define pairwise SAPT energy components. Unlike XSAPT components, these are based on isolated monomer wavefunctions with no embedding charges, so the total SAPT energy components are strictly pairwise-additive. Let us define these to be

$$E_{\text{elst}}^{\text{SAPT}} = \sum_{A}^{N} \sum_{B>A}^{N} E_{\text{elst},AB}^{\text{SAPT}}, \tag{36a}$$

$$E_{\text{exch}}^{\text{SAPT}} = \sum_{A}^{N} \sum_{B>A}^{N} E_{\text{exch},AB}^{\text{SAPT}}, \tag{36b}$$

etc., with analogous definitions for $E_{\text{ind}}^{\text{SAPT}}$ and $E_{\text{disp}}^{\text{SAPT}}$. For the latter, we might still use a hybrid dispersion model; the "SAPT" superscripts in Eq. (36) only signify that there is no charge embedding. Lastly, let us define

$$\delta E_{\text{HF}}^{\text{total}} = \sum_{A}^{N} \sum_{B>A}^{N} \delta E_{\text{HF},AB}. \tag{37}$$

For generality, the following EDA expressions will include the dimer corrections $\delta E_{\text{HF},AB}$ but these might be omitted for nonpolar systems, in the interest of cost, or evaluated using a smaller basis set as compared to the monomer wavefunctions and SAPT corrections [77].

Using the notation introduced above, a many-body generalization of the two-body XSAPT-EDA in Eq. (15) can be expressed as [36, 37]

$$E_{\text{int}}^{\text{XSAPT,MB}} = E_{\text{elst}}^{\text{SAPT}} + E_{\text{exch}}^{\text{SAPT}} + E_{\text{disp}}^{\text{SAPT}} + E_{\text{ind}}^{\text{SAPT}} + \delta E_{\text{HF}}^{\text{total}} + \Delta E_{\text{ind}}^{\text{pairwise}} + E_{\text{int}}^{\text{MB}} \tag{38}$$

where

$$\Delta E_{\text{ind}}^{\text{pairwise}} = \sum_{A}^{N} \sum_{B>A}^{N} \left[E_{\text{int},AB}^{\text{XSAPT}}(AB) - E_{\text{int},AB}^{\text{SAPT}} \right] \tag{39}$$

is a pairwise induction correction. Equation (38) is a trivial reformulation of Eq. (35), since the terms $-E_{\text{int},AB}^{\text{SAPT}}$ within the summation in Eq. (39) exactly cancel $E_{\text{elst}}^{\text{SAPT}} + E_{\text{exch}}^{\text{SAPT}} + E_{\text{disp}}^{\text{SAPT}} + E_{\text{ind}}^{\text{SAPT}}$. The many-body contribution in Eq. (38) can be expressed explicitly as

$$E_{\text{int}}^{\text{MB}} = \sum_{A}^{N} \sum_{B>A}^{N} \left[E_{\text{int},AB}^{\text{XSAPT,2B}}(AB\cdots N) - E_{\text{int},AB}^{\text{XSAPT,2B}}(AB) \right]. \tag{40}$$

The first term in this summand represents the *AB* energy computed using embedding charges for the full *N*-body system, whereas the second term is a self-consistently polarized dimer computed using only charges on the dimer *AB*. In other words, $E_{\text{int},AB}^{\text{XSAPT,2B}}(AB)$ is a dimer XSAPT calculation that uses charge embedding, to be contrasted with a dimer SAPT calculation $E_{\text{int},AB}^{\text{SAPT}}$, which does not. Energy components of the latter are what appear in Eq. (36).

In practical terms, the energy decomposition in Eq. (38) requires three calculations per dimer if the δHF correction is included. One is $\delta E_{\text{HF},AB}$ in Eq. (37), while the other two are XSAPT(KS) with embedding charges and a SAPT(KS) calculation without them. It can be argued that there is some double-counting of induction effects if the pairwise $\delta E_{\text{HF},AB}$ corrections are included, as some higher-order induction is partially (though approximately) captured by the self-consistent XPol procedure [36, 37]. In the formalism developed here, the δHF corrections are held apart as a separate term, and accuracy benchmarks both with and without this correction are presented in Sect. 2.1. Note that the $\delta E_{\text{HF},AB}$ corrections are decoupled from the (X)SAPT-SCF equations, so there is no difficulty in using different basis sets to evaluate them. Previous work has shown that def2-SVPD affords similar accuracy to def2-QZVPD for the δHF correction, at a small fraction of the cost [77].

Rewriting Eq. (38) using Eqs. (36) and (39), we obtain

$$\begin{aligned} E_{\text{int}}^{\text{XSAPT,MB}} &= \sum_{A}^{N} \sum_{B>A}^{N} \left(E_{\text{elst},AB}^{\text{SAPT}} + E_{\text{exch},AB}^{\text{SAPT}} + E_{\text{disp},AB}^{\text{SAPT}} + E_{\text{ind},AB}^{\text{SAPT}} + \delta E_{\text{HF},AB} \right) \\ &\quad + \Delta E_{\text{ind}}^{\text{pairwise}} + E_{\text{int}}^{\text{MB}}. \end{aligned} \tag{41}$$

Assuming that we use HF wavefunctions for the monomers, the definition of δE_{HF} (Eq. 10) greatly simplifies the term in parentheses in Eq. (41), leaving

$$E_{\text{int}}^{\text{XSAPT(HF),MB}} = \underbrace{\sum_{A}^{N} \sum_{B>A}^{N} \left(\Delta E_{\text{int},AB}^{\text{HF}} + E_{\text{disp},AB}^{\text{SAPT}} \right)}_{\text{HF+D}} + \Delta E_{\text{ind}}^{\text{pairwise}} + E_{\text{int}}^{\text{MB}}. \tag{42}$$

Here, $\Delta E_{\text{int},AB}^{\text{HF}}$ is the HF interaction energy for the dimer *AB* computed in the supramolecular way. Equation (42) reveals that the many-body XSAPT (HF) interaction energy can be interpreted as a form of pairwise HF+D plus additional induction, both pairwise and nonadditive.

In the case of XSAPT(KS), intramonomer correlation changes how the monomers polarize. We can write an expression similar to Eq. (42) by considering KS orbitals in Eq. (41), expressing the result as

$$E_{\text{int}}^{\text{XSAPT(KS),MB}} = \sum_{A}^{N} \sum_{B>A}^{N} \left(\Delta E_{\text{int},AB}^{\text{dlKS}} + E_{\text{disp},AB}^{\text{SAPT}} \right) + \Delta E_{\text{ind}}^{\text{pairwise}} + E_{\text{int}}^{\text{MB}} \tag{43}$$

where

$$\Delta E_{\text{int}}^{\text{dlKS}} = \Delta E_{\text{int}}^{\text{CP-HF}} + \underbrace{E_{\text{elst}}^{\text{KS}} + E_{\text{exch}}^{\text{KS}} + E_{\text{ind}}^{\text{KS}} - E_{\text{elst}}^{\text{HF}} - E_{\text{exch}}^{\text{HF}} - E_{\text{ind}}^{\text{HF}}}_{\Delta E_{\text{KS}}} . \tag{44}$$

The term ΔE_{KS} that is defined in Eq. (44) represents the difference brought about in the HF interaction energy by including electron correlation in the zeroth-order wavefunctions used for subsequent intermolecular energy corrections. This implies that the quantity $\Delta E_{\text{int}}^{\text{dlKS}}$ defined in Eq. (44) can be interpreted as a "dispersionless" KS interaction energy, which is added to SAPT(KS) dispersion and XPol polarization in Eq. (43).

The many-body EDA in Eq. (38) has been discussed in prior work [36, 37] and was used in a recent benchmark study of small-molecule trimers [181], where nonadditive dispersion and induction effects computed using XSAPT+MBD were compared against those obtained from CCSD(T) calculations and from SAPT(DFT) augmented with three-body corrections taken from wavefunction theory. Consistent with the discussion in Sect. 1.2.3, these nonadditive dispersion effects are quite small, as shown in Fig. 11a. Even so, they are better described by XSAPT+MBD than by SAPT(DFT), based on comparison to CCSD(T) benchmarks.

As a final note, we observe that second-order perturbation theory is sufficient to connect three monomers. Three-body induction couplings for XPol + SAPT0 have been derived and implemented [33] but only tested in a few cases [33, 37, 181]. Recent three-body benchmarks suggest that the XPol treatment affords an accurate treatment of nonadditive polarization for small-molecule trimers, which is degraded by the inclusion of the three-body induction couplings [181]. This is demonstrated in Fig. 11b, where XSAPT nonadditive induction energies computed *without* the three-body induction couplings are in reasonable agreement with SAPT (DFT) results whose three-body corrections are taken from wavefunction-based SAPT [181]. In contrast, XSAPT induction energies that include the three-body induction couplings are much larger. It is suggested that this overestimation likely results from neglect of the complementary nonadditive exchange-repulsion interactions [181], which have not been implemented. As such, these three-body induction couplings are not considered further, and they were not included in the many-body EDA scheme described above. See Ref. [37] for a version of the XSAPT-EDA that does include those couplings.

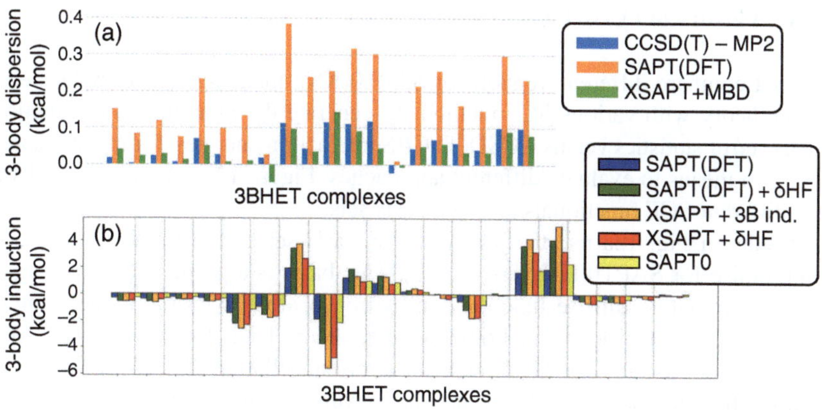

Fig. 11 Three-body results for a small-molecule dataset of trimers ("3BHET"), including (**a**) dispersion energies and (**b**) induction energies. XSAPT+MBD and SAPT(DFT) calculations employ the aug-cc-pVTZ basis set, whereas CCSD(T) and MP2 results are extrapolated to the CBS limit. The dataset consists of trimers of molecules ranging in size up to phenol and bromobenzene, along with some ions (NH_4^+ and Cl^-). (Adapted with permission from Ref. [181]; copyright 2023 the PCCP Owner Societies)

2 Performance Benchmarks

Throughout Sect. 1.2, we introduced data documenting the accuracy of various approximations, as a way to motivate the historical development of XSAPT. With the complete formalism now assembled, additional accuracy benchmarks are provided in Sect. 2.1. In Sect. 2.2, we document the cost and effective computational scaling with system size.

2.1 Accuracy

To provide a realistic overview of a family of XSAPT methods, we divide this discussion into several parts. Sect. 2.1.1 compares different XC functionals used in SAPT0(KS) and SAPT(KS) methods. This choice has important consequences for accuracy, and some poor choices that should not be emulated can nevertheless be found in the literature. Basis-set demands are discussed in Sect. 2.1.2, and because they are relatively forgiving (by *ab initio* standards), calculations can be performed in systems composed of large monomers. This capability is illustrated with examples in Sect. 2.1.3. Finally, Sect. 2.1.4 provides an overview of XSAPT for ions, which is one place where there is certainly room for future improvement.

2.1.1 Flavors of (X)SAPT

We first compare different versions of SAPT0(KS), meaning second-order perturbation theory with various SCF methods to generate the monomer wavefunctions. We use error statistics versus CCSD(T)/CBS interaction energies, for the S66 dataset [73], as a means to evaluate different approaches. Figure 12 summarizes results for four different SCF functionals and two basis sets.

Results in Fig. 12a use the jun-cc-pVDZ basis set that is found to be a good compromise for SAPT0(HF) [26, 27, 77]. This basis is a proper subset of aug-cc-pVDZ, eliminating diffuse functions on hydrogen along with the highest angular momentum set of diffuse functions on other atoms [74]. Its success for SAPT0 (HF) calculations (and for MP2) rests on error cancellation [27, 75], as is evident from the fact that both the MAEs and the maximum errors increase when aug-cc-pVTZ is used instead (Fig. 12b). Nevertheless, MAEs are no larger than 1.1 kcal/mol for each of the traditional subsets of S66 including hydrogen-bonded dimers, dispersion-dominated complexes, and dimers with mixed influence interactions. Interestingly, opposite basis-set behavior is obtained with and without the δHF correction. At the SAPT0(HF)/jun-cc-pVDZ level, the errors for hydrogen-bonded

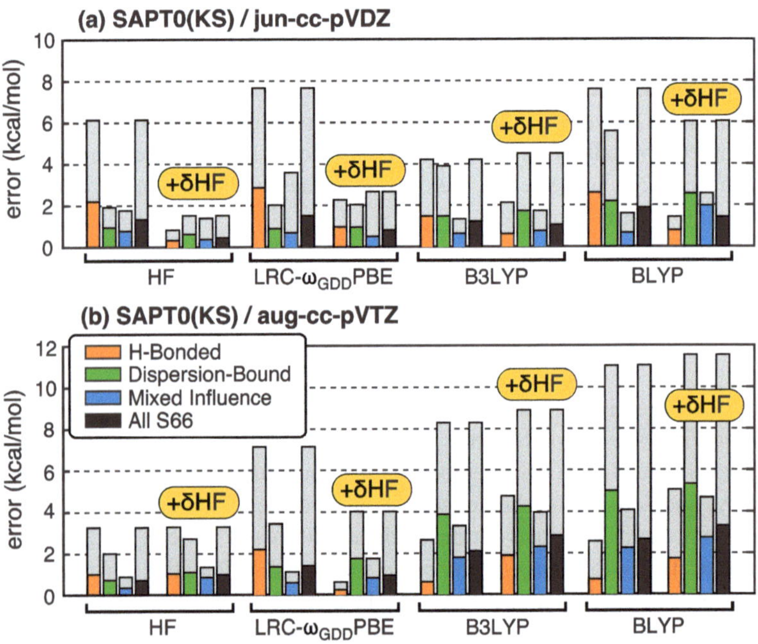

Fig. 12 Performance of second-order perturbation theory for interaction energies in the S66 dataset and its subsets, using either (**a**) the jun-cc-pVDZ basis set or (**b**) the aug-cc-pVTZ basis set. SAPT0 (KS) results are compared using various SCF methods, as indicated, either with or without the δHF correction. Colored bars indicate MAEs with respect to CCSD(T)/CBS benchmarks and gray bars are maximum absolute errors. (Data are from Ref. [55])

dimers are unacceptably large (up to 6 kcal/mol) when this correction is omitted, yet for SAPT0(HF)/aug-cc-pVTZ calculations, the δHF correction has little statistical effect. This is a curious result that went unnoticed in comprehensive benchmarking of various levels of SAPT approximations [26, 27], because the δHF correction was always included at the SAPT0 level. This observation points to an additional source of error cancelation.

Interestingly, error statistics for S66 are not improved, and are actually slightly worsened, by the use of LRC-ω_{GDD}PBE for the monomers. The δHF correction remains essential, and results using jun-cc-pVDZ are reasonable when it is included. Differences with respect of HF-based SAPT0 may lie in the imbalance between $E_{\text{ind}}^{(2)}$ based on KS orbitals and the infinite-order induction correction that is contained in δE_{HF}, which subtracts out second-order corrections based on HF theory (Eq. 10). This is done to avoid contaminating E_{ind} with other energy components that are present in a supramolecular KS-DFT calculation [55].

While SAPT0(LRC-ω_{GDD}PBE)/jun-cc-pVDZ errors are perhaps acceptable, dispersion-dominated outliers are larger when the aug-cc-pVTZ basis set is used; see Fig. 12b. For second-order SAPT0 calculations, there seems to be no reason to recommend using a SCF method other than HF theory. That is especially true when XC functionals lacking correct asymptotic behavior are inserted into the SAPT0 formalism, as demonstrated by results using BLYP and B3LYP in Fig. 12.

In a review of different EDAs targeted at drug discovery [108], the SAPT0 (B3LYP) method was included as the sole representative of the entire SAPT formalism. The present comparison demonstrates that this approach is a poor substitute even for conventional HF-based SAPT0, even if the δHF correction were included. In the larger aug-cc-pVTZ basis set, the reason is obvious: significant overestimation of dispersion-dominated interaction energies. This is the net result of a second-order perturbation theory formalism that overestimates dispersion, combined with a functional that lacks correct asymptotic behavior and affords smaller HOMO/LUMO gap as compared to HF theory [182]. Within a perturbative formulation of E_{disp}, smaller gaps manifest as dispersion energies that are too attractive. GGA functionals narrow the gaps even further as compared to global hybrids like B3LYP, with correspondingly worse results for SAPT0(KS) calculations, as documented in Fig. 12. Functionals that lack correct asymptotic behavior should not be used in SAPT0(KS) calculations.

These problems are alleviated through the use of MBD. Figure 13 juxtaposes SAPT0(HF) and SAPT0(LRC-ω_{GDD}PBE) results from Fig. 12b with the corresponding results when MBD replaces $E_{\text{disp}}^{(2)}$. The SAPT(LRC-ω_{GDD} PBE) +MBD method affords very good results in a high-quality basis set (aug-cc-pVTZ), demonstrating that its favorable accuracy does not rely on a limited basis set for error cancellation. That said, note that SAPT(HF)+MBD results are somewhat worse than SAPT0(HF) when the δHF correction is included in both. The damping function that defines the MBD + esDQ model in Eq. (33) was only fitted for use with LRC-ωPBE, which may result in slightly degraded performance for HF monomers.

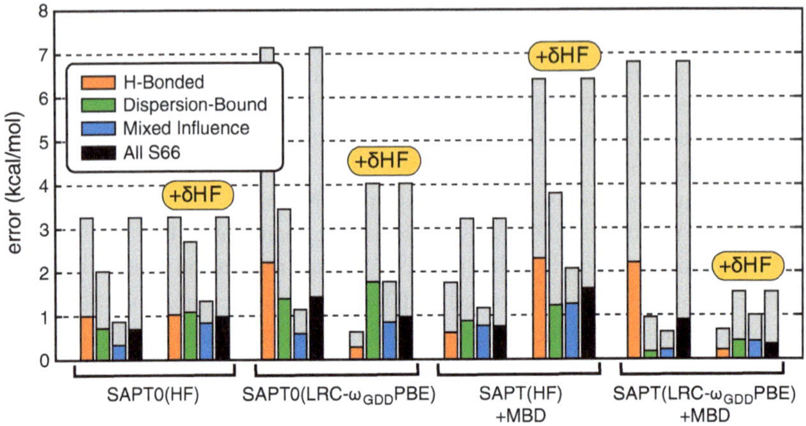

Fig. 13 MAEs (in color) and maximum absolute errors (in gray) for the S66 dimers, comparing SAPT(KS)-based interaction energies to CCSD(T)/CBS results. Methods labeled "SAPT0" use $E^{(2)}_{\text{disp}}$ (Eq. 8a), whereas the two SAPT(KS)+MBD methods on the right use the MBD model (Eq. 33). (All calculations use the aug-cc-pVTZ basis set and the data are from Ref. [77])

2.1.2 Basis Sets

One advantage of introducing dispersion models in place of perturbation theory is a reduction in basis-set demands, as *ai*D models are entirely independent of the basis set and MBD exhibits mild basis-set dependence [77]. Extensive benchmarking of the XSAPT+MBD energy components has been reported recently [77], and Fig. 14 summarizes how the choice of basis set impacts the accuracy of total interaction energies computed using SAPT(LRC-ω_{GDD}PBE)+δHF+MBD.

Errors incurred using double-ζ basis sets are unacceptably large except for jun-cc-pVDZ or aug-cc-pVDZ. This likely results from an imbalance in the convergence behavior of $E^{\text{MBD}}_{\text{disp}}$ as compared to other energy components; indeed, these errors are reduced dramatically in triple-ζ basis sets. The minimally augmented def2-ma-TZVP basis set [77], which is a proper subset of def2-TZVPD [183], is essentially converged with respect to its parent basis set but considerably more efficient, as documented below. Comparing def2-ma-TZVP and def2-QZVPD results in Fig. 14a, we see that differences in the MAEs are well within the intrinsic accuracy of the method itself, although there is some reduction in the outliers when the larger basis set is used. The same can be said for the comparison between jun-cc-pVTZ and aug-cc-pVQZ results in Fig. 14b. This suggests using minimally augmented triple-ζ basis sets for XSAPT+MBD calculations.

Of these, def2-ma-TZVP is by far the least expensive, adding only modest overhead upon def2-TZVP as documented in Fig. 15. In contrast, the upgrade from def2-ma-TZVP to def2-TZVPD improves the MAE for the S66 dimers by ≤ 0.1 kcal/mol but nearly doubles the cost for the π-stacked uracil dimer that is used

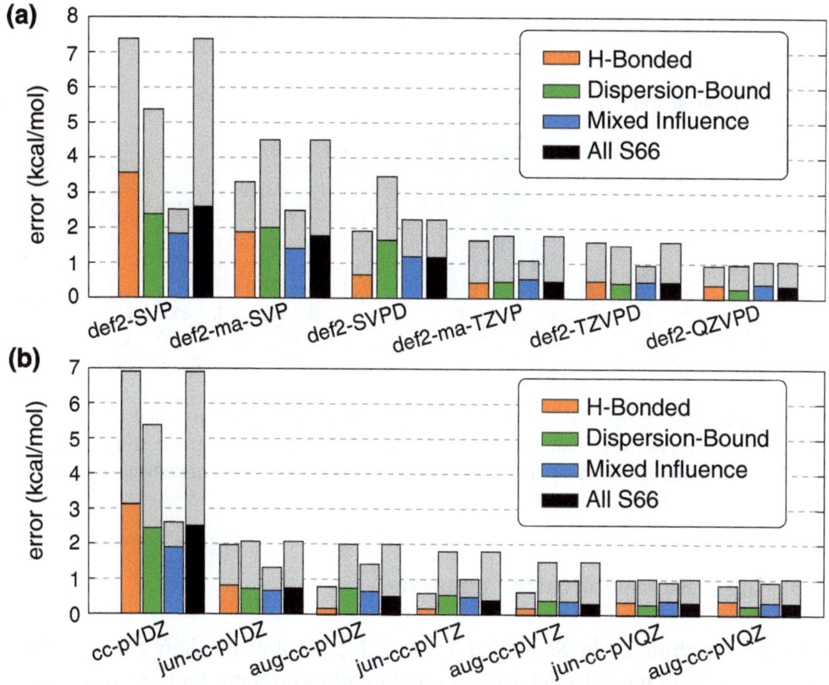

Fig. 14 Errors across the S66 dataset, for SAPT(LRC-ω_{GDD}PBE)+δHF+MBD calculations using (**a**) Karlsruhe and (**b**) Dunning basis sets. Colored bars represent MAEs and gray bars are maximum absolute errors, with respect to CCSD(T)/CBS benchmarks. (Data are from Ref. [77])

for timing benchmarks. The def2-QZVPD basis set improves the MAE by < 0.1 kcal/mol with respect to def2-TZVPD, although it does reduce some of the outliers, but increases the cost by nearly $5\times$. These disparities are exacerbated in Dunning basis sets, where the full complement of diffuse functions in aug-cc-pVTZ reduces the MAE by < 0.1 kcal/mol, as compared to jun-cc-pVTZ results, at $1.4\times$ the cost. Quadruple-ζ basis sets provide very limited improvements over aug-cc-pVTZ at prohibitive cost. We recommend def2-ma-TZVP unless one is pushing for absolute convergence to the CBS limit, in which case Dunning basis sets provide slightly more consistent results.

Convergence data for energy components were discussed above (see Fig. 7). With the exception of dispersion, all energy components exhibit similar convergence behavior and have nearly reached the CBS limit using def2-ma-TZVP [77]. The dispersion energy $E_{\text{disp}}^{\text{MBD}}$ converges even faster because MBD depends on the density but not directly on the wavefunction. This should be contrasted with the slow convergence of SAPT0 (Fig. 1), to an undesirable answer. Relatively uniform convergence of the energy components means that the XSAPT EDA does not benefit significantly from error cancellation, provided that basis sets of triple-ζ quality are used.

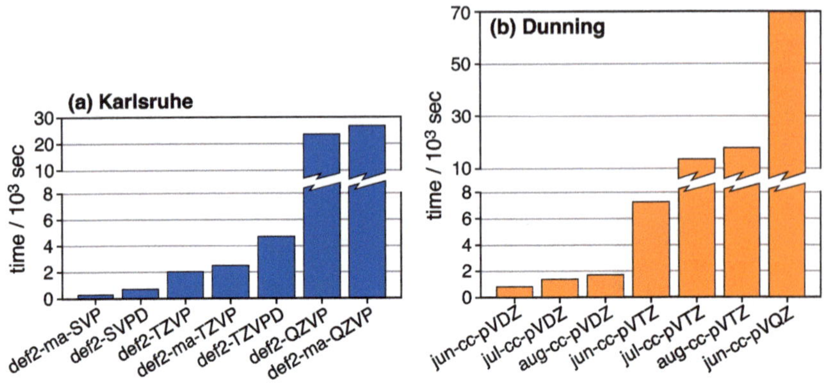

Fig. 15 Total CPU times for a SAPT(LRC-ω_{GDD} PBE) calculation on uracil dimer using (**a**) Karlsruhe basis sets or (**b**) Dunning basis sets. All calculations were performed using 12 processors of a Dual Intel Xeon 8268 node with 64 Gb of memory. Note the change in vertical scale beginning at 8000 s, in both panels

2.1.3 Large Supramolecular Complexes

Accuracy benchmarks in Sects. 2.1.1 and 2.1.2 employ the S66 dataset of small dimers in order to have decent statistics and good throughput on numerous variants. Where XSAPT+MBD truly shines, however, is in applications to large complexes. This was previewed in Fig. 4, demonstrating a MAE below 1 kcal/mol for the L7 dataset, which is comparable to the best available DFT methods for noncovalent interactions, while affording an EDA that is less ambiguous as discussed in Sect. 3.2.

Here, we consider some additional large-system benchmarks starting with another set of calculations for the L7 complexes, the results of which are shown in Fig. 16. Data for MP2 and SAPT0(HF) emphasize the failure of second-order perturbation theory for systems involving large, conjugated monomers. In contrast, XSAPT+*ai*D3 and (especially) XSAPT+MBD perform extremely well for these systems. This is true even in comparison to state-of-the-art semiempirical methods designed for noncovalent interactions, exemplified in Fig. 16 by the HF-3c [184] and PBEh-3c [185] methods. These are designed for geometry optimization of large molecules, and they complement XSAPT by providing an affordable means to obtain geometries. Reliable interaction energies can then be computed using XSAPT-based methods.

Figure 17 presents E_{int} benchmarks for a set of large host–guest complexes common to the S12L and S30L datasets [186, 187]. Unlike other benchmarks discussed in this chapter, which come primarily from CCSD(T)/CBS calculations, these are obtained from experimental association energies (ΔG°), back-corrected to gas-phase interaction energies using theoretical estimates of the vibrational entropy and the change in solvation energy upon complexation. These benchmarks come with estimated uncertainties of ± 3 kcal/mol, with the solvation correction as the largest source of error [186, 187].

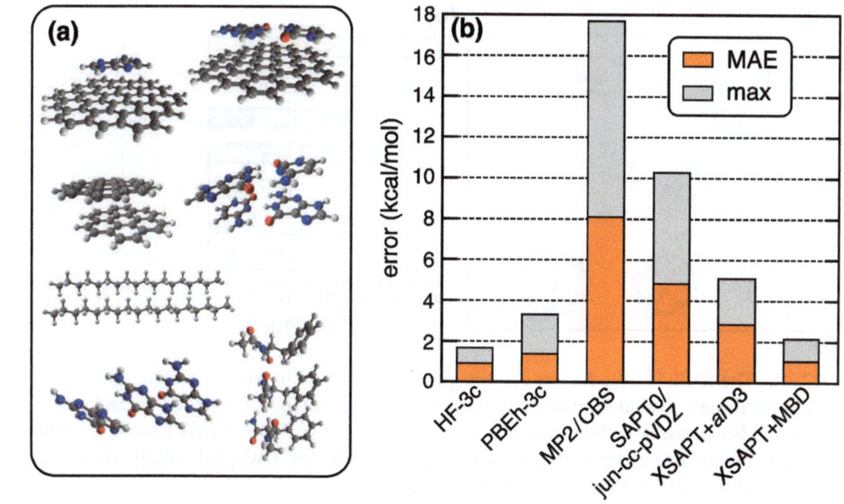

Fig. 16 (**a**) Dimers in the L7 dataset and (**b**) error statistics for L7 versus CCSD(T)/CBS benchmarks. XSAPT+*ai*D3/def2-TZVPP and XSAPT+MBD/def2-TZVPPD calculations use LRC-ω_{GDD}PBE and ChElPG embedding charges. None of these calculations employs the δHF correction. (Data are from Ref. [38])

MP2-based methods are effectively useless for these systems and even B97M-V, which is usually an outstanding semilocal functional for noncovalent interactions [188], exhibits a MAE of 6 kcal/mol and a maximum error of 11 kcal/mol. Notably, these are CP-corrected B97M-V/aug-cc-pVTZ calculations that should lie near the CBS limit [59]. In complexes of this size, BSSE effects can reach 50 kcal/mol or more in double-ζ basis sets [1, 59].

As with the large-monomer L7 complexes, the XSAPT+*ai*D3 method requires three-body (triatomic) dispersion corrections to achieve reasonable accuracy for this S12L∩S30L dataset, whereas those effects are inherent in XSAPT+MBD. The latter method exhibits accuracy comparable to B97M-V(CP)/aug-cc-pVTZ, in a more modest triple-ζ basis set and without the need for CP correction.

2.1.4 Performance for Ions

Early versions of XSAPT used ChElPG embedding charges for the XPol procedure [32, 33]. These are designed to approximate the molecular electrostatic potential and seem fit to purpose but engender significant computational overhead (Fig. 2). In contrast, CM5 charges [120] are designed only to reproduce the molecular dipole moment but can be evaluated much more efficiently using DFT quadrature grids [41]. Although implemented for reasons of efficiency, CM5 embedding charges also turn out to improve the accuracy, especially for complexes involving ions and ion

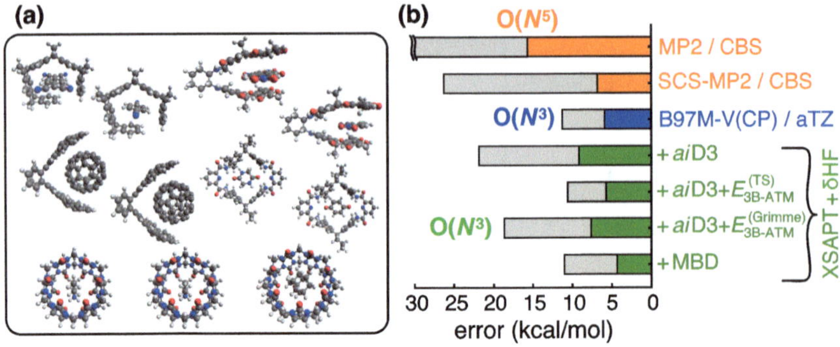

Fig. 17 (**a**) The S12L∩S30L dataset of host–guest complexes. (**b**) Mean absolute errors (colored bars) and maximum absolute errors (gray bars) for various methods, compared to back-corrected experimental benchmarks with estimated uncertainties of ± 3 kcal/mol. XSAPT+δHF calculations use ChElPG embedding charges and the def2-TZVPPD basis set. (Adapted with permission from Ref. [42]; copyright 2021 American Chemical Society)

pairs, for which ChElPG embedding charges afford errors up to 15 kcal/mol that are reduced to <4 kcal/mol using CM5 charges [41].

Table 2 provides an accuracy assessment for XSAPT+δHF+*ai*D3 and XSAPT +δHF+MBD calculations on ion–molecule complexes, using the AHB21 and CHB6 datasets [189] that consist of anions and cations, respectively. For AHB21, the benchmarks range up to $|E_{\text{int}}| = 66$ kcal/mol, and ten of the 21 dimers have absolute interaction energies larger than 20 kcal/mol. (For comparison, $|E_{\text{int}}| < 20$ kcal/mol for all of the S66 dimers.) For CHB6, the benchmarks range from -20 kcal/mol to -39 kcal/mol. As such, these are more challenging tests and errors are correspondingly larger than what we have seen for neutral complexes.

For the anion–molecule dataset, MAEs with respect to CCSD(T)/CBS benchmarks are $\lesssim 3$ kcal/mol when the def2-ma-TZVP basis set is used, although maximum errors are about 8.5 kcal/mol. Independent of basis set, the largest error is incurred for the very challenging Cl^-(HCl) complex consisting of a proton that is shared equally between two Cl^- ions yet must be assigned to one monomer or the other for the purpose of (X)SAPT calculations. For the XSAPT+δHF+MBD calculations, this outlier is noticeably reduced by additional diffuse functions so in that case there is a reason to prefer def2-ha-TZVP over def2-ma-TZVP, which is not surprising for a calculation involving an anion.

The CHB6 dataset consists of $M^+(C_6H_6)$ and $M^+(H_2O)$ complexes for alkali metal cations: M = Li, Na, and K. As compared to the anions, XSAPT+δHF+*ai*D3 calculations result in smaller errors for these cationic complexes, with MAEs below 2 kcal/mol and maximum errors <4 kcal/mol. However, the MBD model affords excessively large dispersion energies for the two complexes involving K^+, e.g., $E_{\text{disp}}^{\text{MBD+esDQ}} = -84$ kcal/mol for $K^+(C_6H_6)$, as compared to $E_{\text{disp}}^{ai\text{D3}} = -4$ kcal/mol for the same complex. These very large MBD dispersion energies manifest as significant

Table 2 Error statistics for the AHB21 and CHB6 datasets of ionic complexes using XSAPT with CM5 embedding charges

Basis set	Error (kcal/mol)[a]							
	XSAPT+δHF+*ai*D3[b]				XSAPT+δHF+MBD			
	AHB21		CHB6		AHB21		CHB6	
	MAE	Max	MAE	Max	MAE	Max	MAE[c]	Max[c]
def2-ma-SVP	3.7	11.9	1.8	3.4	3.2	8.3	10.6	19.0
def2-ha-SVP	2.9	8.4	1.2	2.6	2.4	6.1	11.8	21.0
def2-SVPD	2.5	8.1	1.5	2.9	2.8	12.1	12.4	22.0
def2-ma-TZVP	2.6	8.6	1.7	3.9	3.2	8.5	11.6	23.0
def2-ha-TZVP	2.7	8.2	1.3	3.7	1.8	5.0	10.9	23.0
def2-TZVPD	2.5	8.4	1.3	3.6	1.7	4.8	10.9	23.0

[a] With respect to CCSD(T)/CBS benchmarks from Ref. [189]
[b] Data from Ref. [55]
[c] $K^+(H_2O)$ and $K^+(C_6H_6)$ removed from the dataset

overbinding of the K^+ complexes by XSAPT+δHF+MBD, with errors of 23 kcal/mol for $K^+(H_2O)$ and 80 kcal/mol for $K^+(C_6H_6)$, using def2-ma-TZVP. As such, these two complexes have been removed from the error statistics in Table 2 in order to provide a more representative perspective or the other cation–molecule dimers. Even so, errors remain considerably larger than they are for XSAPT+δHF+*ai*D3 applied to the CHB6 dataset.

The source of these errors remains a topic for future work and development, but we can speculate. It has been demonstrated that MBD is sometimes vulnerable to polarization catastrophes for short-range interactions [149], as one might anticipate in strong interactions with ions, although the reason why cations are more problematic is unclear. Especially for ions [190], the model is sensitive to the input polarizabilities [149], which are controlled by the oscillator frequency ω_a in Eq. (31). Alternative models for obtaining these polarizabilities have been suggested [191], which are available in the `libMBD` library that is interfaced to Q-Chem [192], but these have not yet been adapted for use with XSAPT. At a minimum, iterative determination of Hirshfeld volumes [193–195], to assign excess charge to appropriate atoms in a molecular density, is probably necessary to make MBD generally applicable to ions [190, 194, 196] and the ordinary Hirshfeld partition is known to produce unrealistically large polarizabilities for cations [196]. However, it is unclear that this should improve the description of complexes involving monatomic ions, such as those in the CHB6 dataset. A more rigorous treatment of dipole–quadrupole (and higher-order) polarizability contributions to MBD, as compared to the MBD + esDQ model in Eq. (33), might improve the results for strong, short-range interactions. Clearly, there remains room for improvement to the performance of XSAPT-based methods for ions.

2.2 Cost

In this section, we document the cost for XSAPT calculations and how it scales with system size. All calculations were performed using the Q-Chem program, which remains the only platform in which XSAPT is implemented [45, 46]. Timing benchmarks were performed using a pre-release copy of Q-Chem v. 6.3, in order to document improved multithreading for conventional SAPT0 calculations. The resolution-of-identity (RI) technique leads to considerable reduction in the cost of this $O(N^5)$ method [76, 78, 197, 198] and the SAPT0 timings reported in Sect. 2.2.1 use this technique. Additional parallelization improvements were made in Q-Chem v. 6.0, as documented in Fig. 2, so that multithreaded performance should be significantly improved as compared to Q-Chem v. 5.4. Further improvements in parallel efficiency since that release are documented in Sect. 2.2.1.

These timing benchmarks use an SCF convergence threshold of 10^{-5} E_h in the electronic energy gradient $\|\mathbf{FP}-\mathbf{PF}\|$ [199] and a screening threshold of 10^{-12} a.u. (The latter is appropriate for medium-size systems with diffuse basis functions [200].) Timing benchmarks are run on dedicated nodes.

2.2.1 Parallel Performance

Figure 18 documents the parallel efficiency of the new RI-SAPT0 code, running on a single compute node using OpenMP parallelization. Even for the smallest basis set that we consider (cc-pVDZ), the scaling across an entire 48-processor node is reasonably good. The parallel efficiency exceeds 70% in most cases (Fig. 18b) and is considerably improved relative to the old code. Memory usage is also improved, enabling cc-pVQZ calculations (1345 basis functions) that were not previously feasible on our hardware.

We next examine parallel efficiencies for XSAPT+*ai*D3, which has a much different cost profile. For these tests, we have extended upon data for $Cl^-(H_2O)_n$ clusters that were published elsewhere [77], adding new basis sets and updated timings. These $Cl^-(H_2O)_n$ clusters were extracted from a molecular dynamics simulation of Cl^-(aq) [201], with $\langle n \rangle = 28.4 \pm 2.4$ water molecules per snapshot, and the data presented in Table 3 are averages across 51 snapshots, treating $(H_2O)_n$ as a single monomer. For Pople basis sets, parallel efficiencies approach or exceed 80% on 26 processors, although they are lower (46–67%) for Karlsruhe basis sets. The difference likely lies in the use of compound ("sp") shells for Pople basis sets, which Q-Chem's integrals code handles efficiently [200].

Although Pople basis sets have fallen out of favor for modern electronic structure calculations (for good reason in certain cases [202–205]), they can yield high-quality results for DFT calculations if sufficient polarization functions are included, beyond the usual 6-31G(d) or 6-311G(d,p) [200]. In the present context, Pople basis sets can reproduce induction energies obtained in much larger Karlsruhe or Dunning basis sets, to an acceptable degree, if additional polarization functions are included

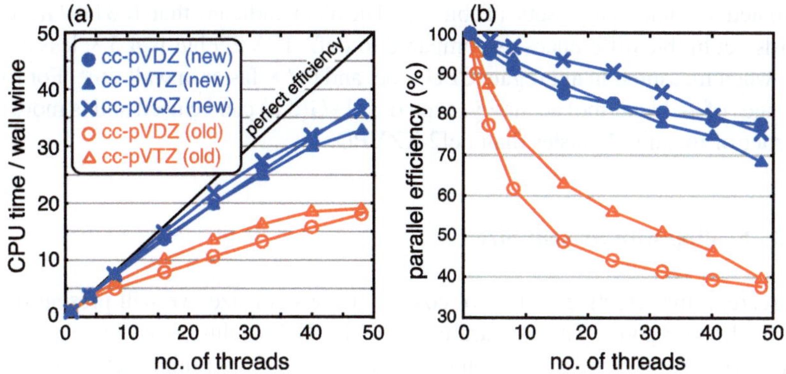

Fig. 18 Strong-scaling data for multithreaded RI-SAPT0 calculations on a guanine–cytosine base pair, $(C_5H_5N_5O)\cdots(C_4H_5N_3O)$, using either the original SAPT0 code (Q-Chem versions up to v. 6.2) or the new code (pre-release Q-Chem v. 6.3). (**a**) Speedup measured by the ratio of CPU time to wall time ($t_{\mathrm{CPU}}/t_{\mathrm{wall}}$). (**b**) Parallel efficiency, defined as $t_{\mathrm{CPU}}/t_{\mathrm{wall}}/n_{\mathrm{threads}}$ and expressed as a percentage. All calculations were run on a single Dual Intel Xeon 8268 node with 48 processors and 192 Gb memory

Table 3 Error statistics and timing data for XSAPT+*ai*D3 calculations on $Cl^-(H_2O)_n^a$

Basis set	E_{int} Error (kcal/mol)[b]		Time (sec)[c]		
	MAE[d]	Max	CPU	Wall	Parallel efficiency[e]
6-31+G(d)	1.9 ± 0.8	3.8	1226	57	82%
6-311+G(d,p)	1.3 ± 0.5	2.4	2834	140	78%
6-311+G(2df,2p)	0.1 ± 0.1	0.4	7813	355	85%
6-311+G(3df,2pd)	0.3 ± 0.1	0.6	16,564	761	84%
def2-ma-SVP	3.1 ± 0.7	4.4	2678	154	67%
def2-SVPD	1.2 ± 0.2	1.5	11,063	928	46%
def2-ma-TZVP	1.3 ± 0.3	1.9	9982	586	66%
def2-TZVPD	–	–	34,812	2371	56%

[a] Data represent averages over 51 snapshots with $\langle n \rangle = 28.4 \pm 2.4$. The $(H_2O)_n$ cluster is treated as a single monomer
[b] With respect to the def2-TZVPD value
[c] Using 26 processors (Dell Intel CPU Max 9470 HBM)
[d] Uncertainty represents one standard deviation
[e] PE = $t_{\mathrm{CPU}}/t_{\mathrm{wall}}/n_{\mathrm{proc}}$

[77]. For example, XSAPT+δHF+MBD induction energies computed using the 6-311++G(3df,2pd) basis set exhibit slightly smaller errors, as compared to SAPT2+(3)/aug-cc-pVTZ benchmarks [151], than when the def2-QZVPD basis set is used [77]. Errors are small in both cases, but the Pople basis set is dramatically more efficient.

This is a rather general observation [77, 200]. To see this, we take XSAPT+MBD/def2-TZVPD results as converged and examine interaction energies for $Cl^-(H_2O)_n$

computed in other basis sets (Table 3). The data indicate that 6-311+G(2df,2p) affords negligible differences as compared to def2-TZVPD but is 4.5 × less expensive, when measured in aggregate CPU time, and 6.7 × faster in wall time. For rough estimates of ~ 1 kcal/mol accuracy, even 6-311+G(d,p) can be used and is more than an order of magnitude faster than def2-TZVPD.

2.2.2 Scaling with System Size

To ascertain the effective scaling of cost versus system size, we will present timing data for homologous systems starting with $Cl^-(H_2O)_n$ clusters of increasing size, treating $(H_2O)_n$ as a single monomer. Figure 19 presents timing data for XSAPT +MBD calculations. The quantity that is plotted is the total CPU time aggregated across all 20 processors that were used, not the time to completion. The latter is much shorter, as may be inferred from the reasonably good parallel efficiencies that are documented in Table 3, but it is total CPU time that captures the true cost of the calculation [206]. Fits of the total time t to the function $t = \alpha n^\beta$ reveal that the scaling is effectively cubic, for both the XPol-SCF procedure (which contributes 80% of the total cost, for the largest systems examined) and for the post-SCF SAPT and MBD corrections (comprising the remaining 20% of the cost).

For completeness, we also present timings for RI-SAPT0(HF)/jun-cc-pVDZ calculations on $(\text{guanine})_n$ stacks extracted from an idealized structure of B-DNA. Timing data are presented in Fig. 20, for calculations in which $(\text{guanine})_{n-1}$ is treated as a single monomer. For the post-SCF steps, we obtain an effective scaling of $O(N^{4.2})$ that is better than the theoretical result. However, the SCF time constitutes a significant fraction of the overall time for the system sizes considered (up to $n = 6$), resulting in an effective scaling of $O(N^{3.1})$ for the entire calculation. For the same

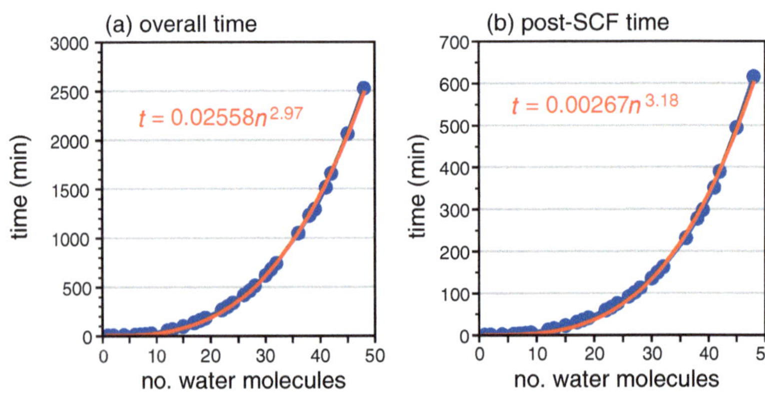

Fig. 19 Timing data for XSAPT+MBD/def2-ma-TZVP calculations on $Cl^-(H_2O)_n$ clusters of increasing size, treating $(H_2O)_n$ as a single monomer and examining (**a**) the total compute time and (**b**) the post-SCF time. Fits to the function $t = \alpha n^\beta$ are indicated. Calculations were performed on 20 processors of a single Dual Intel Xeon 8268 node and the δHF correction is not included

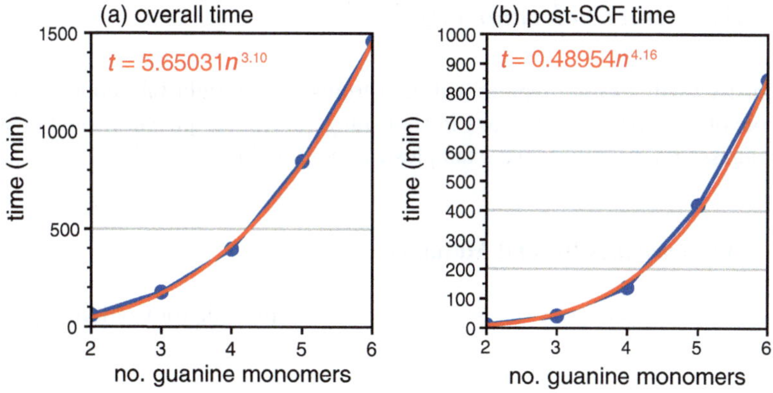

Fig. 20 Timing data for RI-SAPT0/jun-cc-pVDZ calculations on (guanine)$_n$, including (**a**) the overall time, aggregated across all processors, and (**b**) the post-SCF time. The system is treated as a dimer of guanine and (guanine)$_{n-1}$. Calculations were performed using 40 processors of a single Dual Intel Xeon 8268 node

systems and basis set, XSAPT+MBD calculations afford effective scaling of $\mathcal{O}(N^{2.6})$ for both the overall time and the post-SCF time, demonstrating that this approach does actually reduce the effective scaling, relatively to RI-SAPT0, even when super-theoretical scaling is obtained. Moreover, the memory footprint of XSAPT+MBD is much smaller.

3 Illustrative Applications

This section explores the use of XSAPT+MBD as a tool for understanding the physics of intermolecular interactions, taking advantage of the dimer EDA in Eq. (1) and relying on the sub-kcal/mol accuracy that was established in Sect. 2.1 and elsewhere. We first establish some basics about short-range repulsion, which originates in the Pauli principle and not with Coulomb interactions, as discussed in Sect. 3.1. Then, the MBD dispersion model is compared directly to dispersion energies obtained from a popular DFT-based EDA scheme [13], in Sect. 3.2. That discussion highlights the fact that $E_{\text{disp}}^{\text{MBD+esDQ}}$ in Eq. (33) agrees very well with SAPT2+(3) dispersion energies, even though it was not parameterized using these data (unlike the *ai*D models). That discussion also highlights ambiguities in the definition of E_{disp} in the DFT-based EDA. Having established the veracity of the XSAPT+MBD energy decomposition, we apply it to analyze π–π interactions in Sect. 3.3. Finally, hydrogen bonding in $F^{-}(H_2O)$ is examined in Sect. 3.4, using the SAPT-cDFT decomposition to isolate E_{CT} and thereby elucidate the role of CT in the anion–water hydrogen bond.

3.1 Basic Intermolecular Physics

Chemistry textbooks are replete with half-truths and outright falsehoods regarding the role of multipolar interactions in short-range molecular physics, some of which are addressed here. Additional examples can be found in Ref. [43].

3.1.1 Electrostatics Beyond Multipoles

In the interest of pedagogy, let us introduce the electrostatic (or Coulomb) potential

$$\tilde{\varphi}\,(\mathbf{R}) = \int \frac{\tilde{\rho}\,(\mathbf{r})}{\|\mathbf{r}-\mathbf{R}\|}\,d\mathbf{r} \tag{45}$$

for a continuous charge distribution $\tilde{\rho}(\mathbf{r})$. We use $\tilde{\rho}(\mathbf{r})$ to indicate a molecular charge distribution that includes both nuclei and electrons. If $\rho(\mathbf{r})$ is the charge distribution for the electrons only, which is positive-valued by convention, then

$$\tilde{\rho}(\mathbf{r}) = -\rho(\mathbf{r}) + \sum_{a}^{\text{nuclei}} Z_a\,\delta(\mathbf{r}-\mathbf{R}_a). \tag{46}$$

Given charge densities $\tilde{\rho}_{\mathrm{A}}(\mathbf{r})$ and $\tilde{\rho}_{\mathrm{B}}(\mathbf{r})$ for isolated molecules A and B, along with the corresponding electrostatic potentials $\tilde{\varphi}_{\mathrm{A}}(\mathbf{r})$ and $\tilde{\varphi}_{\mathrm{B}}(\mathbf{r})$, the total electrostatic interaction in the SAPT scheme (Eq. 16a) is equivalent to

$$E_{\text{elst}} = \int \tilde{\varphi}_{\mathrm{A}}(\mathbf{R})\,\tilde{\rho}_{\mathrm{B}}(\mathbf{R})\,d\mathbf{R} = \int \tilde{\varphi}_{\mathrm{B}}(\mathbf{R})\,\tilde{\rho}_{\mathrm{A}}(\mathbf{R})\,d\mathbf{R}. \tag{47}$$

For interpretative purposes, we might approximate one or both of the electrostatic potentials using a multipole expansion. For example,

$$\tilde{\varphi}_{\mathrm{A}}(R) = \frac{Q_{\mathrm{A}}}{R} + \frac{\boldsymbol{\mu}_{\mathrm{A}}\cdot\mathbf{R}}{R^3} + \frac{\mathbf{R}^{\dagger}\boldsymbol{\Theta}_{\mathrm{A}}\mathbf{R}}{R^5} + \cdots \tag{48}$$

where Q_{A}, $\boldsymbol{\mu}_{\mathrm{A}}$, and $\boldsymbol{\Theta}_{\mathrm{A}}$ are the total charge, the dipole moment vector, and the quadrupole moment tensor of molecule A, respectively. One could make a multipole expansion for the density $\tilde{\rho}_{\mathrm{B}}(\mathbf{R})$ and then recast Eq. (47) into a form involving multipole–multipole interactions [207]. In doing so, it is important to remember that Eq. (48) is an asymptotic expansion that is valid only for a test charge that is located at a distance R that is large compared to the details of the molecular charge distribution, $\tilde{\rho}_{\mathrm{A}}(\mathbf{r})$. In fact, the multipole expansion is divergent as $R \to 0$ [22, 208–210]. More importantly, it is readily demonstrable that nonbonded close-contact distances (including typical equilibrium geometries of vdW complexes) are not usually in the asymptotic regime where the electrostatic potential is well described (or even *qualitatively* described) by leading-order multipoles. An example is shown in Fig. 21 for different geometries of the benzene dimer.

Exact electrostatic interactions, computed using Eq. (47) without any multipole approximation, are attractive in the typical range of intermolecular π-stacking separations (3.4–3.8 Å), and they are attractive in the edge-to-face (C–H$\cdots\pi$) configuration also [211–213]. This has implications for understanding π-stacking [42, 148, 214–217], as discussed in Sect. 3.3. For now, we simply note that an atom-centered multipole approximation to E_{elst} (Fig. 21a), which should converge faster than the single-center expansion in Eq. (48), nevertheless does not agree with exact calculations for the cofacial arrangement until the face-to-face separation R is much larger than typical π-stacking distances. A leading-order approximation, using only the molecular quadrupole moments of the two benzene monomers, does not agree with exact electrostatics until even larger face-to-face separations (Fig. 21b). At typical π-stacking separations, such as $R = 3.35$ Å for graphite [218] or $R = 3.4$ Å for the parallel-displaced benzene dimer [212], exact electrostatics is attractive but quadrupolar approximations are repulsive [148, 211, 213, 215].

This breakdown in multipolar approximations is sometimes ascribed to *charge penetration* effects, meaning that multipole approximations no longer describe E_{elst} when $\rho_{\mathrm{A}}(\mathbf{r})$ and $\rho_{\mathrm{B}}(\mathbf{r})$ interpenetrate. Another interesting example of this effect can be found in the corannulene dimer, $(C_{20}H_{10})_2$. The corannulene monomer consists of six hexagonal rings surrounding a central pentagonal ring, the latter of which lends curvature to the structure, which has been called a "buckybowl" [219–221]. This and

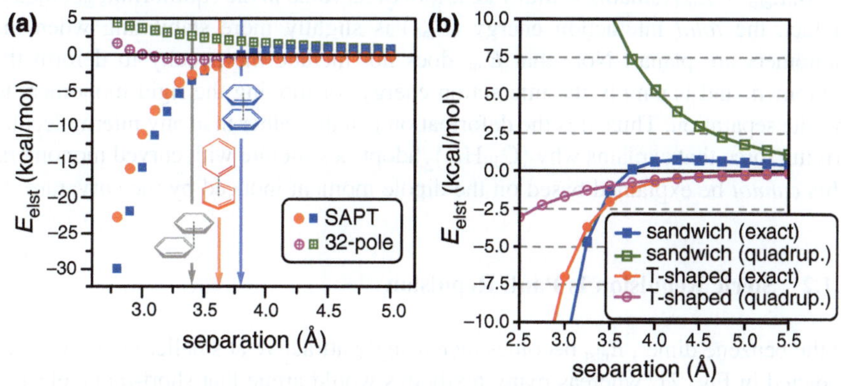

Fig. 21 (**a**) Electrostatic interaction energies for the cofacial (solid blue symbols) and T-shaped (solid red symbols) geometries of $(C_6H_6)_2$, computed at the HF/jun-cc-pVDZ level as a function of intermolecular separation. (For the T-shaped structure, the distance coordinate is the H-to-face separation.) Open symbols with crosses correspond to an atom-centered distributed multipole calculation at the HF/6-311G(d,p) level, including interactions up to octupole–quadrupole. Arrows indicate the minimum-energy intermolecular separations for the three structures that are shown. (Adapted with permission from Ref. [211]; copyright 2016 American Chemical Society.) (**b**) Exact versus quadrupolar electrostatics for $(C_6H_6)_2$, computed at the LRC-ω_{GDD}PBE/def2-TZVPD level along the same separation coordinates

other concave hydrocarbons form concentric stacks in the solid state [221–224]. (A five-member ring is not required to induce curvature [225].) The π-electron density at the bottom of these buckybowls must be compensated by electropositive terminal C–H bonds, as it is in benzene as well but the buckybowls break planar symmetry and may develop dipole moments. These range from $\mu = 2.07$ D in $C_{20}H_{10}$, based on a gas-phase measurement [226], to $\mu > 10$ D in larger buckybowls, based on calculations [227, 228]. As a result, this phenomenon has been called "flexo-electric" π-stacking [228], with the implication that the driving force for this solid-state stacking phenomenon is to align the dipole moments of the monomers. In view of exact electrostatic calculations for $(C_6H_6)_2$, however, this hypothesis should be regarded with suspicion.

To investigate this in detail, we used XSAPT+MBD calculations to examine intermolecular interactions in $(C_{20}H_{10})_2$ in a sequence of calculations in which we gradually flatten each monomer at fixed intermolecular separation, starting at the equilibrium dimer structure (with concave monomers) and ending at a structure in which both monomers are planar (so $\mu = 0$) [148]. Results of the energy decomposition along this coordinate are plotted in Fig. 22. Monomer densities $\rho_A(\mathbf{r})$ and $\rho_B(\mathbf{r})$ are also shown, using an isocontour value ($\rho = 0.001$ a.u.) that is typical for exhibiting molecular size [146]. These plots demonstrate that the monomer densities significantly interpenetrate at the equilibrium stacking distance.

Despite diminishing (and ultimately vanishing) monomer dipole moments along the planarization coordinate, electrostatic interactions remain significant. What is plotted in Fig. 22 is actually polarized electrostatics ($E_{\text{elst}} + E_{\text{ind}}$), but this sum is dominated by E_{elst} at all geometries. Even when the monomers are planar and $\mu = 0$, $E_{\text{elst}} + E_{\text{ind}}$ remains within 1 kcal/mol of its value in the equilibrium geometry. In fact, the *total* interaction energy (E_{int}) is slightly more stabilizing when the monomers are planar. Note that E_{int} does not include the penalty to deform the monomers and is simply the interaction energy for moving the rigid monomers to infinite separation. Thus, it is the deformation penalty rather than any intermolecular driving force that explains why $(C_{20}H_{10})_2$ adopts a structure with curved monomers. This *cannot* be explained based on the dipole moment induced by the curvature.

3.1.2 Steric Repulsion Is Pauli Repulsion

In the benzene dimer, E_{elst} becomes increasingly attractive at smaller values of R, as depicted in Fig. 21, whereas many textbooks would argue that short-range electrostatics should be repulsive due to interpenetration of the electron clouds on the two monomers. To debunk this myth, we consider an even simpler system: Ar_2. Energy components for Ar_2 are plotted as a function of R in Fig. 23a [43]. Note that $E_{\text{elst}}(R) < 0$ at all values of R, indicating a strictly attractive electrostatic interaction, even for values of R that are much smaller than the equilibrium separation.

The balance of forces that enters $E_{\text{elst}}(R)$ is rather subtle. To appreciate how much so, it is necessary to decompose the total electrostatic interaction,

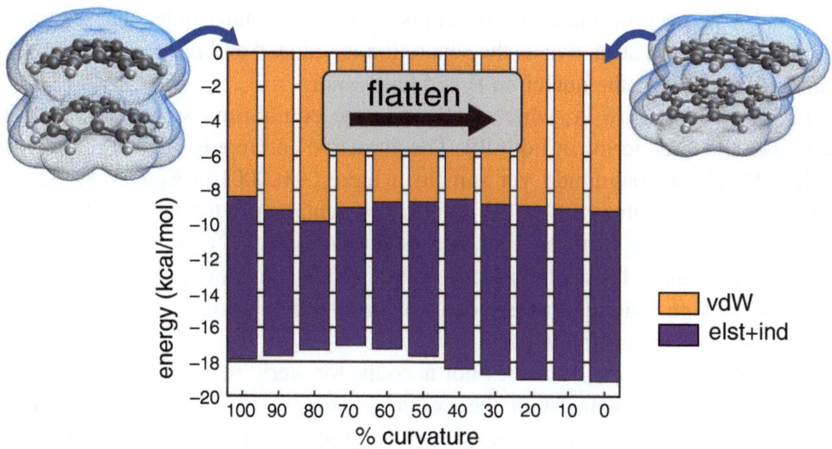

Fig. 22 XSAPT+δHF+MBD/def2-TZVPPD results for $(C_{20}H_{10})_2$ along a coordinate in which the monomer geometries are gradually flattened at fixed intermolecular separation, starting from the equilibrium geometry on the left and ending at a geometry in which both monomers are planar, on the right. The quantities $E_{\text{vdW}} = E_{\text{exch}} + E_{\text{disp}}$ and $E_{\text{elst}} + E_{\text{ind}}$ are provided along this coordinate. Monomer electron densities at the starting and ending geometries are plotted using an isocontour $\rho = 0.001$ a.u. (Data are from Ref. [148])

$$E_{\text{elst}} = E_{\text{elst}}^{\text{ee}} + E_{\text{elst}}^{\text{nn}} + E_{\text{elst}}^{\text{en}} \tag{49}$$

into electron–electron repulsion ($E_{\text{elst}}^{\text{ee}} > 0$), nuclei–nuclei repulsion ($E_{\text{elst}}^{\text{nn}} > 0$), and electron–nuclei attraction ($E_{\text{elst}}^{\text{en}} < 0$). Written in atomic units, the expressions for these terms are

$$E_{\text{elst}}^{\text{nn}} = \sum_{a \in \text{A}} \sum_{b \in \text{B}} \frac{Z_a Z_b}{R_{ab}} \tag{50}$$

for atomic numbers Z_a and Z_b,

$$E_{\text{elst}}^{\text{ee}} = \int \frac{\rho_{\text{A}}(\mathbf{r}_1)\, \rho_{\text{B}}(\mathbf{r}_2)}{\|\mathbf{r}_1 - \mathbf{r}_2\|}\, d\mathbf{r}_1\, d\mathbf{r}_2 \tag{51}$$

for the electron–electron repulsion, and

$$E_{\text{elst}}^{\text{en}} = -\sum_{a \in \text{A}} \int \frac{Z_a \rho_{\text{B}}(\mathbf{r})}{\|\mathbf{r} - \mathbf{R}_a\|}\, d\mathbf{r} - \sum_{b \in \text{B}} \int \frac{Z_b \rho_{\text{A}}(\mathbf{r})}{\|\mathbf{r} - \mathbf{R}_b\|}\, d\mathbf{r} \tag{52}$$

for the attractive contribution. Signs in Eq. (52) reflect the convention that electron densities are output from electronic structure programs as strictly positive quantities, representing probability densities of negatively charged electrons [43]. Although

$E_{\text{elst}}^{\text{nn}}$ and $E_{\text{elst}}^{\text{ee}}$ become increasingly repulsive at shorter intermolecular separation, $E_{\text{elst}}^{\text{en}}$ becomes more attractive as the separation is reduced and the electrons in $\rho_{\text{A}}(\mathbf{r})$ gain better access to the nuclei on B and *vice versa*.

For Ar_2 at $R = 3.76$ Å, which is close to the equilibrium separation, Fig. 23b quantifies the three terms in Eq. (49). The attractive and repulsive contributions are each $\sim 90\ E_h$ in magnitude yet sum to a mere $-0.000164\ E_h = -35.9\ \text{cm}^{-1}$, which is the total value of E_{elst} [43]. In this particular case, the increased attraction of $E_{\text{elst}}^{\text{en}}$ at short range wins out over increased repulsion of $E_{\text{elst}}^{\text{ee}} + E_{\text{elst}}^{\text{nn}}$, although the effect is very small. It may be larger in other systems, as in $(C_6H_6)_2$ (see Fig. 21), although the sign is not guaranteed one way or the other and examples can be found where $E_{\text{elst}} > 0$ at short range [213].

Thus, Coulomb repulsion does not account for steric repulsion. Instead, short-range repulsion originates with the exchange energy (E_{exch}), as seen for Ar_2 in Fig. 23a. This is *Pauli repulsion*, originating from the antisymmetry requirement. Within the SAPT formalism, that requirement is not built into the direct-product wavefunction $|\psi_0^{\text{A}}\rangle|\psi_0^{\text{B}}\rangle$. This is more than a semantic or a model-dependent distinction, however. Results above demonstrate that interpenetration of two charge densities $\tilde{\rho}_{\text{A}}(\mathbf{r})$ and $\tilde{\rho}_{\text{B}}(\mathbf{r})$ need not lead to a repulsive interaction, when the electron–nuclear interactions are considered. Thus, a distinct physical phenomenon is needed to explain steric repulsion.

One may conceptualize Pauli repulsion in the following way. If the isolated monomer wavefunctions ψ_0^{X} (for X = A or B) are computed by means of a variational principle, as in any SCF calculation, then ψ_0^{X} will adopt whatever shape minimizes its energy in isolation. Introduction of another molecule brings with it an antisymmetry constraint, so that the energy of the constrained system must be higher than that of the unconstrained system, for any finite intermolecular

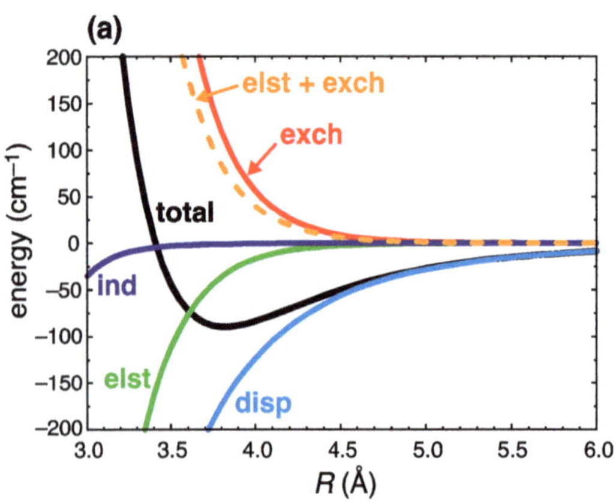

Fig. 23 (**a**) Total interaction potential $E_{\text{int}}(R)$ for Ar_2 (in black) and its SAPT decomposition (in color), computed at the SAPT0/aug-cc-pVTZ level. (Adapted with permission from Ref. [43]; copyright 2021 American Chemical Society.) (**b**) Isocontours $\rho = 0.001$ a.u. of the HF electron density and components of E_{elst}, computed for $R = 3.76$ Å

separation. In other words, the antisymmetry constraint forces a deformation of the wavefunction away from the minimum-energy form that it adopted in isolation and that deformation is associated with an energy penalty (steric or Pauli repulsion). An alternative but equivalent view is that steric repulsion represents an orthogonalization penalty [229]. Since the MOs for molecules A and B are computed in isolation, they will not be mutually orthogonal and must be deformed (orthogonalized) in order to construct a common Slater determinant that satisfies the Pauli principle (antisymmetry requirement) for the composite AB system.

It is worth noting that some EDA schemes choose to orthogonalize the MOs and construct an antisymmetric Slater determinant $|\psi_0^A \psi_0^B\rangle$ prior to computing electrostatic interactions [9]. This differs from the direct-product $|\psi_0^A\rangle\,|\psi_0^B\rangle$ that is used in SAPT. Where electrostatic interactions are computed using an antisymmetrized reference state, the antisymmetrized electrostatics should be compared to $E_{\text{elst}} + E_{\text{exch}}$ from SAPT. For Ar_2, Fig. 23a shows that this sum is also repulsive, as the Pauli repulsion (E_{exch}) more than compensates for the attractiveness of E_{elst}.

3.1.3 Dispersion Is Size Extensive

There is a tendency to describe dispersion as a weak interaction. (Sometimes, the phrase "van der Waals interactions" is used to mean dispersion, but we avoid that terminology.) Dismissing dispersion as a weak effect is prevalent in the literature on drug discovery and noncovalent enzyme inhibition, where "vdW interactions" are sometimes described as a sub-kcal/mol phenomenon [230]. This is true for the rare-gas dimers that are often used to introduce London dispersion, as evident from the Ar_2 data in Fig. 23. However, the magnitude of E_{disp} scales with system size and need not be small in large molecular systems, even as compared to covalent bond energies of ~ 100 kcal/mol. Assuming typical values for atomic C_6 coefficients, a rough estimate suggests that two 100-atom organic molecules at 5 Å separation might exhibit a mutual dispersion interaction $E_{\text{disp}} \sim -60$ kcal/mol [5]. Entropic effects tend to reduce interaction energies through entropy/enthalpy compensation, such that $|\Delta G^\circ_{\text{bind}}| < 20$ kcal/mol even for ~ 100-atom host–guest complexes for which $|\Delta U^\circ_{\text{bind}}|$ exceeds 100 kcal/mol [186].

Nevertheless, in order to understand the balance of intermolecular forces it is important to get the interaction energies right. Qualitative effects of dispersion on organic reactivity have been discussed [6], aided by *in silico* computational modeling in which dispersion interactions can be turned off in order to isolate their effects. This has led to the fascinating idea of "steric attraction" [6, 7, 231–235], in which a bulky substituent also contributes substantially to the polarizability, enhancing E_{disp} to the point that it may outcompete E_{exch} and the larger substituent becomes the less repulsive one. This phenomenon is only possible because dispersion is size extensive.

To illustrate how the magnitude of the dispersion interaction generally increases with molecular size, Fig. 24 shows two examples relevant to drug discovery, namely, a DNA intercalation ligand (the antitumor agent ellipticine) and also the protease inhibitor indinavir in a model of the HIV protease binding site. These two systems have become standard benchmarks for noncovalent quantum chemistry [236, 237]. In both cases, the XSAPT+MBD interaction energy lies rather close to a benchmark CCSD(T)/CBS value [236]. The interesting observation in the present context is the magnitude of the dispersion energy. For the DNA intercalation complex, which is an obvious example of π-stacking where dispersion might be expected to play a prominent role, the XSAPT+MBD dispersion energy is -71 kcal/mol [41]. Meanwhile, the HIV–indinavir system has no obvious π–π interactions, yet the dispersion contribution is nearly twice as large, at -135 kcal/mol [42]. The model system for HIV–indinavir contains 323 atoms as compared to 157 atoms for the DNA–ellipticine model, and the dispersion energy is more attractive in the larger model system.

3.2 Comparison of Dispersion Models

Having established that SAPT calculations can reveal fundamental physics regarding the origins of intermolecular forces, and that the magnitude of dispersion increases with system size, we next want to consider the dispersion interactions

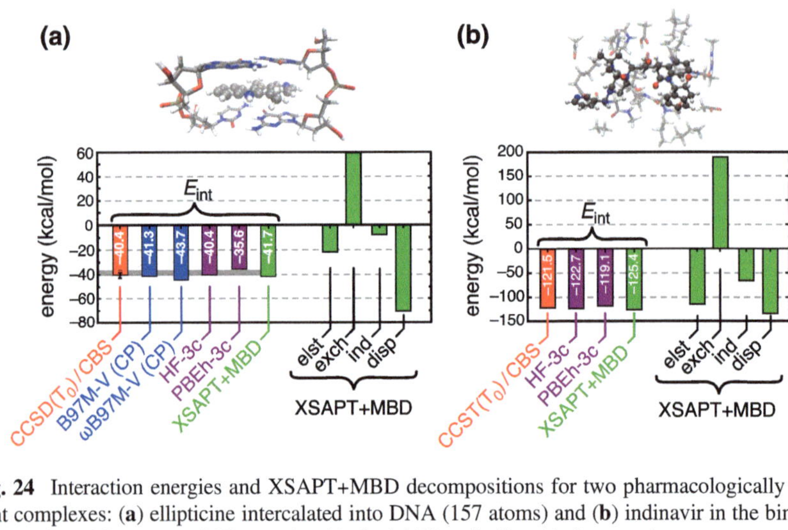

Fig. 24 Interaction energies and XSAPT+MBD decompositions for two pharmacologically relevant complexes: (**a**) ellipticine intercalated into DNA (157 atoms) and (**b**) indinavir in the binding pocket of HIV-2 protease (323 atoms), with the inhibitor (indinavir) shown using opaque atoms. In (**a**), CP-corrected DFT/def2-TZVPPD results are provided for two different functionals, whereas XSAPT+MBD calculations use the def2-hpTZVPP basis set for both systems; these calculations are from Ref. [41]. Semiempirical HF-3c and PBEh-3c results are from Ref. [42]. (Adapted with permission from Ref. [42] but updated to reflect new CCSD(T_0)/CBS benchmarks from Ref. [236]; copyright 2021 American Chemical Society)

more carefully. The second-order expression obtained by London is [43, 145, 238, 239]

$$E^{(2)}_{\text{disp,AB}} = -\sum_{m>0}\sum_{n>0} \frac{|\langle \psi^{\text{A}}_m \psi^{\text{B}}_n | (\hat{H}_{\text{AB}} - \hat{H}_{\text{A}} - \hat{H}_{\text{B}}) | \psi^{\text{A}}_0 \psi^{\text{B}}_0 \rangle|^2}{(\varepsilon^{\text{A}}_m - \varepsilon^{\text{A}}_0)(\varepsilon^{\text{B}}_n - \varepsilon^{\text{B}}_0)} . \tag{53}$$

This corresponds to the dispersion energy in SAPT0 (Eq. 8a). As discussed previously, it is typically much too attractive, e.g., by 80–100% for π-stacked benzene dimer [75, 240]. For larger systems, errors in second-order perturbation theory are even more egregious [38, 71].

For that reason, we substitute a dispersion model in XSAPT+MBD, which performs extremely well as compared to dispersion energies obtained from SAPT2+(3) calculations, as discussed previously in the context of Fig. 6. We consider that SAPT2+(3) constitutes a benchmark level of theory for the energy components because total interaction energies computed at that level agree with CCSD(T) results to sub-kcal/mol accuracy [25–27, 57, 58]. Good agreement between XSAPT+MBD and SAPT2+(3) energy components thus lends credence to the idea that the former does not benefit significantly from error cancellation, so that individual energy components within the SAPT formalism can be swapped out for better models, e.g., substituting the non-perturbative MBD model for dispersion.

It is interesting to compare these benchmark-quality dispersion energies to those obtained from the popular ALMO-EDA2 scheme [13], which is DFT-based. We select the PBE0+D3 functional for this assessment, as it performs well for total interaction energies on a variety of noncovalent datasets [1]. ALMO-EDA2 provides a means to separate induction from the total (CP-corrected) interaction energy [82], and the remaining energy components are grouped together as the "frozen density" part of ΔE_{int}, which is

$$\Delta E_{\text{FRZ}} = E_{\text{elst}} + E_{\text{exch}} + E_{\text{disp}} . \tag{54}$$

To isolate the dispersion energy, it is recommended [13] to compute ΔE_{FRZ} using both the target functional (PBE0+D3 in this case) and also with a "dispersionless" functional that is assumed to describe dispersion poorly or not at all (e.g., HF theory). The difference between ΔE_{FRZ} computed with either functional defines the ALMO-EDA2 dispersion energy [13]:

$$E^{\text{ALMO-EDA2}}_{\text{disp}} = \Delta E^{\text{target}}_{\text{FRZ}} - \Delta E^{\text{disp'less}}_{\text{FRZ}} . \tag{55}$$

If the target functional is a hybrid then HF theory is suggested for computing $\Delta E^{\text{disp'less}}_{\text{FRZ}}$ [13].

Application of this procedure to the S22 dataset, however, leads to ALMO-EDA2 dispersion energies that are systematically less attractive than SAPT2+(3) values, as documented in Fig. 25. These sizable discrepancies may arise because the recipe in

Eq. (55) assumes that every correlation effect is dispersion, when the "dispersionless" functional is selected to be HF exchange (only). Correlation corrections to electrostatics or Pauli repulsion might reduce the attractiveness of what is counted as dispersion. To investigate this, we experimented with using HF exchange plus PBE correlation (the HF-PBE functional) as the "dispersionless" functional in Eq. (55) [1]. This affords dispersion energies that are even farther from SAPT2+(3)-values, as shown also in Fig. 25.

Figure 26 provides an alternative presentation of these data, plotting $E_{\rm disp}^{\rm MBD}$ and $E_{\rm disp}^{\rm ALMO\text{-}EDA2}$ versus $E_{\rm disp}^{\rm SAPT2+(3)}$. A linear fit of the MBD data affords a slope of ≈ 1.0 with an intercept of -0.4 kcal/mol, which we regard as good agreement with the benchmarks. ALMO-EDA2 dispersion energies also correlate reasonably well with SAPT2+(3) benchmarks, although less so when HF-PBE is used as the "dispersionless" functional. Given that removing the semilocal correlation effects from ALMO-EDA2 dispersion seems to worsen agreement with benchmark dispersion energies, it is unclear whether ALMO-EDA2 energy components can be

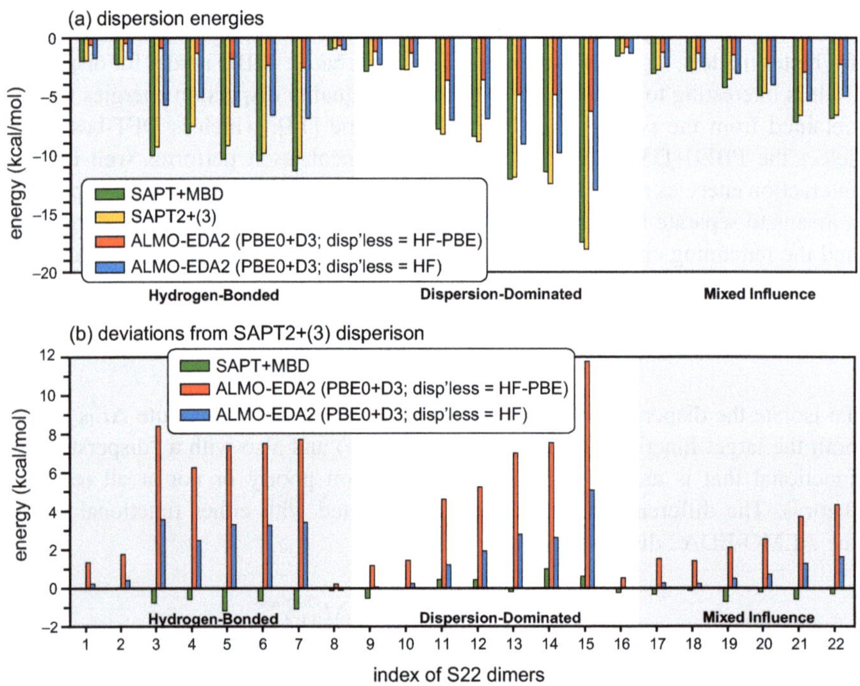

Fig. 25 (**a**) Dispersion energies for the S22 dataset computed using various methods. (**b**) Differences between the SAPT(KS)+MBD or ALMO-EDA2 dispersion energies and SAPT2+(3)/aug-cc-pVTZ benchmarks from Ref. [151]. ALMO-EDA2 dispersion energies are computed using Eq. (55) with PBE0+D3 as the target functional and either HF or HF-PBE as the dispersionless functional. (Adapted with permission from Ref. [1]; copyright 2024 Elsevier)

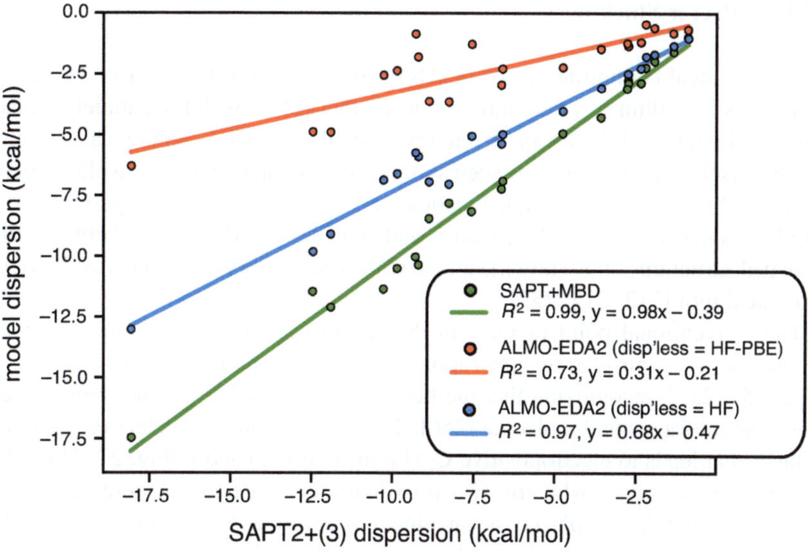

Fig. 26 Correlation between dispersion energies computed using either SAPT(LRC-ω_{GDD}PBE)+MBD or else ALMO-EDA2 applied to PBE0+D3 (on the vertical axis), and benchmarks computed at the SAPT2+(3)/aug-cc-pVTZ level (on the horizontal axis). For ALMO-EDA2, two different "dispersionless" XC functionals are examined for separation of E_{disp} from ΔE_{FRZ} via Eq. (55). These are the same data as in Fig. 25. (Adapted with permission from Ref. [1]; copyright 2024 Elsevier)

compared directly to SAPT values. Consult Ref. [1] for some additional comparisons between dispersion models.

3.3 Arene Stacking Interactions

In a series of publications, we have used XSAPT+MBD calculations to explore the fundamental nature of π–π interactions [42, 148, 215–217], which have also been investigated using conventional SAPT [212–214, 241–243]. Although the benzene dimer has long been the archetypal model system for studying such interactions, it is not a good model for larger polycyclic aromatic hydrocarbon (PAH) dimers [148], for reasons discussed in Sect. 3.3.2. Nevertheless, understanding substituent effects on stacking energies in dimers of benzene and small heterocycles remains an important guide for synthetic chemistry [244–250]. Below, we use XSAPT+MBD to explore the origins of parallel-displaced π-stacking (Sect. 3.3.1) and then address an old question of whether π–π interactions are a unique type of interaction that is distinct from "ordinary" dispersion (Sect. 3.3.3).

3.3.1 Offset π-Stacking

Three canonical configurations of $(C_6H_6)_2$ are depicted in Fig. 27a–c. These three geometries lie within ≈ 1 kcal/mol of one another [251], with the parallel-displaced structure being slightly lower in energy than the cofacial "sandwich" structure, which is itself a saddle point between symmetry-equivalent parallel-displaced geometries [212, 240]. The T-shaped geometry is essentially iso-energetic with the parallel-displaced π-stacked structure, and may be a saddle point between tilted T-shaped structures (not shown), albeit with a negligibly small energy barrier of ~ 0.2 kcal/mol [212, 215, 252]

The conventional point of view in the organic chemistry literature is that π–π interactions are controlled by quadrupole–quadrupole interactions between the arene rings [253–255], recognizing that the benzene monomer has a rather large quadrupole moment for a small molecule [256]. This results from the π-electron density on its faces that leads to electropositive C–H bonds, as depicted in Fig. 27. The offset-stacking phenomenon, whereby two proximal arenes are much more likely to be found in a parallel-displaced arrangement than in a face-to-face-configuration [257, 258], is also typically explained in terms of quadrupolar electrostatics [253, 254]. The argument is that quadrupole–quadrupole interactions are repulsive in the face-to-face configuration but attractive in the edge-to-face arrangement, which is certainly true according to multipolar electrostatic calculations (Fig. 21) but misrepresents exact electrostatics.

This quadrupolar electrostatic viewpoint on π-stacking exemplifies how electrostatic considerations dominate the thinking in organic chemistry and how leading-order multipolar interactions are often used as surrogates for electrostatics. Not only does the multipolar picture disagree with exact electrostatics at π-stacking distances (predicting even the wrong sign for E_{elst} at $R = 3.4$ Å separation), it also cannot explain the origin of parallel-displaced π-stacking in the $(C_6H_6)\cdots(C_6F_6)$. heteromolecular dimer, which is predicted by CCSD(T) calculations [259–261] and illustrated in Fig. 27. For a symmetric top molecule, the quadrupole moment

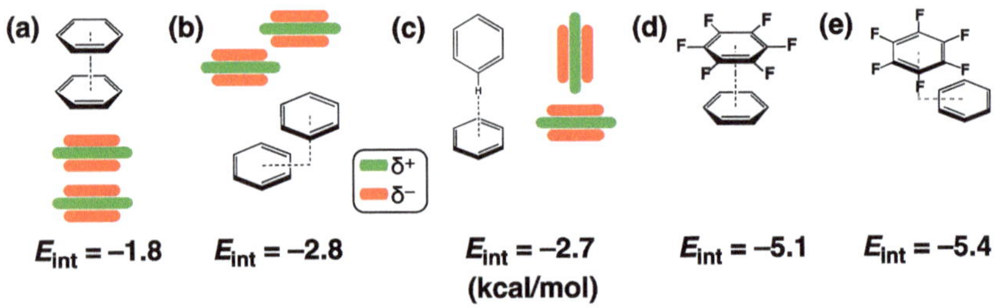

Fig. 27 Three canonical configurations of $(C_6H_6)_2$ and $(C_6F_6)\cdots(C_6H_6)$: (**a**) eclipsed-cofacial ("sandwich") isomer of $(C_6H_6)_2$, (**b**) slip-stacked isomer of $(C_6H_6)_2$, (**c**) T-shaped isomer of $(C_6H_6)_2$, (**d**) cofacial isomer of $(C_6H_6)\cdots(C_6F_6)$, and (**e**) offset-stacked isomer of $(C_6H_6)\cdots(C_6F_6)$. Cartoon charge distributions are shown for the benzene dimer structures. CCSD(T)/CBS interaction energies are from Ref. [251] for $(C_6H_6)_2$ and from Ref. [260] for $(C_6H_6)\cdots(C_6F_6)$

tensor has only one independent component (call it Θ_{zz}) [207, 262], which is opposite in sign but nearly equal in magnitude in C_6F_6 and C_6H_6: $\Theta_{zz}(C_6F_6) = +9.50 \pm 0.51$ B versus $\Theta_{zz}(C_6H_6) = -8.69 \pm 0.51$ B [256]. As a result, the quadrupole–quadrupole interaction is *attractive* in face-to-face orientations of $(C_6H_6)\cdots(C_6F_6)$ [215]. Nevertheless, accurate calculations suggest that $(C_6H_6)\cdots(C_6F_6)$ also exhibits offset π-stacking [215, 259–261].

We pause here for an aside on nomenclature and attribution. In the supramolecular organic chemistry literature, the quadrupolar electrostatic idea described above is often used synonymously with the term "Hunter–Sanders model" [217]. Although widely repeated, this misrepresents force-field models developed to explain the configurational preferences of $(C_6H_6)_2$ and other systems [257, 263–265]. Even so, Hunter himself would eventually come to describe π–π interactions in quadrupolar terms [253]. The idea that π-stacking has a quadrupolar electrostatic basis remains widespread in the synthetic chemistry literature, as discussed elsewhere [217] and evidenced by recent reviews [254, 255]. In view of this continuing misconception, we have argued that cartoon charge distributions such as those as in Fig. 27a–c are misleading because they do not adequately capture the change in quadrupolar interactions for $(C_6H_6)\cdots(C_6F_6)$ [217].

While it may be ubiquitous, the quadrupolar picture of π-stacking is also irredeemably incorrect. Charge penetration leads to a qualitative breakdown in the multipole approximation at typical vdW contact distances, including typical face-to-face π-stacking separations (3.4–3.8 Å) as seen in Fig. 21. Not only is the multipole expansion divergent as $R \rightarrow 0$ [22, 208–210], but leading-order molecular multipole approximations and even distributed (atom-centered) multipole expansions may afford the wrong sign for E_{elst} at these intermolecular distances.

To make this explicit, Fig. 28 juxtaposes two-dimensional contour plots of exact and quadrupolar electrostatics for $(C_6H_6)_2$ [217]. One of the coordinates in these plots is the intermolecular distance R, representing the face-to-face separation for parallel arrangements of the dimer and face-to-center distance for perpendicular orientations. The other coordinate is a lateral displacement that generates slip stacking in the cofacial arrangement, while in perpendicular orientations it moves the dimer between T-shaped and L-shaped geometries. The exact E_{elst} calculation in Fig. 28 uses SAPT electrostatics, computed at the LRC-ω_{GDD}PBE/def2-TZVPD level in order to be consistent with XSAPT+MBD calculations to follow, while the quadrupolar calculation uses the quadrupole tensor $\boldsymbol{\Theta}$ for C_6H_6 that is obtained at the same level of theory.

Consistent with organic chemistry lore, quadrupole–quadrupole repulsion in the face-to-face orientation would push the system away from the sandwich geometry and into an offset-stacked arrangement (Fig. 28a), while quadrupolar attraction in the perpendicular orientation favors a T-shaped geometry over an L-shaped one (Fig. 28d). However, these results are inconsistent with exact electrostatic calculations. For the face-to-face arrangement, contours of (exact) E_{elst} are relatively flat along the lateral displacement coordinate (Fig. 28b), at least for typical π-stacking distances, indicating that electrostatics is largely ambivalent toward offset-stacking.

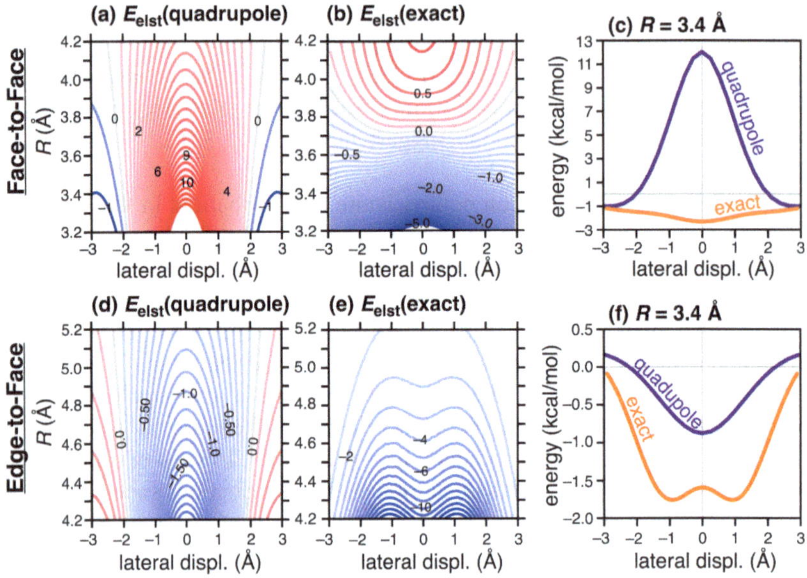

Fig. 28 Electrostatic interaction profiles for $(C_6H_6)_2$. Top row: parallel orientation, plotting (**a**) the quadrupole–quadrupole interaction along the face-to-face distance (R) and lateral displacement (slip-stacking) coordinates; (**b**) exact electrostatics along the same two coordinates, computed using full charge densities; and (**c**) one-dimensional slices through (**a**) and (**b**) at $R = 3.4$ Å, corresponding to the parallel-displaced minimum-energy geometry. Bottom row: perpendicular (edge-to-face) configuration, plotting (**d**) quadrupolar electrostatics along the center-to-face distance R and lateral displacement coordinates; (**e**) exact electrostatics; and (**f**) one-dimensional slices through (**a**) and (**b**) at $R = 5.0$ Å, corresponding to the T-shaped saddle point. Contours are drawn in intervals of (**a**) 0.5 kcal/mol, (**b**) 0.1 kcal/mol, (**d**) 0.1 kcal/mol, and (**e**) 1.0 kcal/mol. (Most data are from Ref. [217], replotted here with some additional calculations)

This is emphasized in Fig. 28c, which plots one-dimensional slices at the minimum-energy stacking distance, $R = 3.4$ Å. If anything, exact electrostatics favors the sandwich configuration rather than the offset-stacked geometry, albeit very weakly. Meanwhile in the edge-to-face geometry, exact electrostatics favors an L-shaped geometry with a 1 Å offset, which is seldom observed [257], rather than a T-shaped geometry with no offset.

The quadrupolar electrostatic explanation is clearly deficient, and we turn to XSAPT+MBD to explain the physical origins of slip stacking. In Fig. 29, we examine lateral displacement potentials for the parallel and perpendicular arrangements of $(C_6H_6)_2$. The total interaction potential exhibits a T-shaped minimum for the edge-to-face benzene dimer (Fig. 29a) and symmetric parallel-displaced minima in the face-to-face arrangement (Fig. 29b). The same critical points are exhibited by the *vdW potential,*

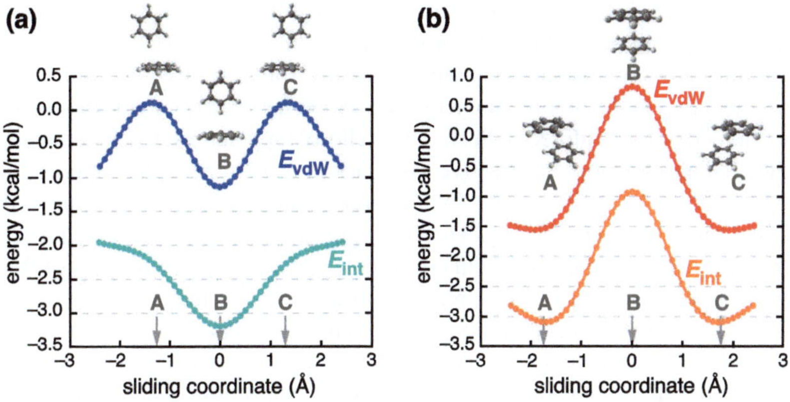

Fig. 29 Total interaction potentials versus vdW potentials ($E_{\text{vdW}} = E_{\text{exch}} + E_{\text{disp}}$) for sliding one benzene monomer past another in either (**a**) an edge-to-face orientation or (**b**) a face-to-face configuration. Arrows at the bottom of each panel indicate the position coordinate of the structures that are shown. Intermolecular distances are consistent with either (**a**) the T-shaped structure of $(C_6H_6)_2$, with a 5.0 Å face-to-center distance, or (**b**) the parallel-displaced minimum of $(C_6H_6)_2$, with a 3.4 Å face-to-face stacking distance. All calculations were performed using XSAPT+δHF+MBD/def2-TZVPPD. (Adapted from Ref. [215] under a CC BY 3.0 license)

$$E_{\text{vdW}} = E_{\text{exch}} + E_{\text{disp}}. \tag{56}$$

This terminology recognizes that van der Waals' modifications to the ideal gas equation contained the effects of both attractive interactions (E_{disp}) and finite molecular size (E_{exch}). The data in Fig. 29 suggest that E_{vdW} is qualitatively sufficient to explain the origin of offset π-stacking in $(C_6H_6)_2$. It arises from a competition between Pauli repulsion, which favors an offset in order to reduce density overlap, and dispersion that favors spatially proximal atoms. Electrostatic interactions are not required for this geometric preference to emerge.

That is not to say that there is no role for electrostatics in π-stacking or that the vdW potential alone sets the stacking parameters. Although the attractive electrostatic contours for cofacial benzene dimer (Fig. 28b) have little influence on whether slip stacking occurs or not, they are essential for obtaining the correct face-to-face separation because they provide a stabilizing interaction at short range. Absent E_{elst}, one would obtain a stacking distance $R_{\text{min}} \approx 3.9$ Å for the parallel-displaced configuration [217], significantly larger than the correct value $R_{\text{min}} = 3.4$ Å.

To examine this in a system where electrostatics plays an even more important role, energy profile contours were computed for the nitrobenzene dimer and plotted in Fig. 30. By itself, the vdW potential affords a minimum-energy separation of $R_{\text{min}} \approx 3.8$ Å, whereas the correct value (considering all energy components) is about 3.4 Å. This difference is directly attributable to electrostatic interactions that are attractive at these distances, contrary to the quadrupolar electrostatic picture. In the case of $(C_6H_5NO_2)_2$, short-range electrostatics is attractive regardless of whether the monomer dipole moments are aligned parallel or antiparallel; see Fig. 30c for the parallel configuration and Fig. 30f for the antiparallel arrangement.

Although the slip-stacking phenomenon arises strictly from E_{vdW}, electrostatics-driven reduction in the minimum-energy separation has general implications for π–π interactions that are explored in Sect. 3.3.3. Those implications depend critically on the fact that electrostatics is attractive at π-stacking distances, a phenomenon that is not captured—even qualitatively—by low-order multipole expansions. This is demonstrated for $(C_6H_5NO_2)_2$ in Fig. 31, which presents the results of an approximate electrostatic calculation that uses only the dipole and quadrupole moments of the two monomers. These multipolar electrostatic surfaces are qualitatively different from the exact ones. In the parallel-dipole arrangement, the multipolar approximation predicts strong electrostatic repulsion, in geometries consistent with π-stacking (Fig. 31a), whereas short-range electrostatics is always attractive when it is computed using full monomer charge densities (Fig. 30c). For the antiparallel-dipole arrangement, the multipolar approximation changes sign near zero displacement (Fig. 31b), which is also completely inconsistent with an exact calculation where there is no repulsive region at all (Fig. 30f).

A recent survey of protein crystal structures confirms this viewpoint [258]. It updates an early census of neighboring phenylalanines by Hunter et al. [257] but demonstrates that the "vdW picture" of π-stacking (driven by E_{vdW}) better explains the prevalence of parallel-displaced cofacial stacking. As we speculated earlier [215], the vdW picture provides a much more satisfactory explanation for why offset stacking is so pervasive in protein crystal structures, across disparate

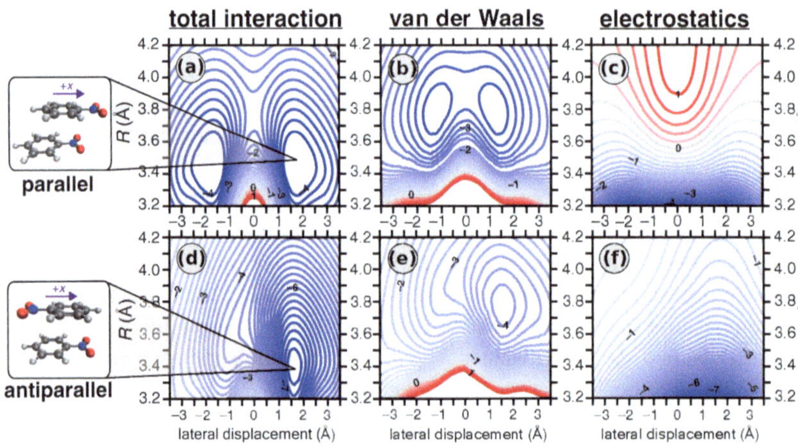

Fig. 30 Potential energy scans for $(C_6H_5NO_2)_2$ in configurations where the monomer dipole moments are either (**a**)–(**c**) parallel-aligned, or else (**d**)–(**f**) antiparallel, as shown in the structures on the left. The vertical coordinate is the cofacial separation R and the horizontal coordinate represents lateral displacement along the dipolar axis. Contours are drawn every 0.2 kcal/mol with blue and red indicating negative and positive values, respectively; selected contours are labeled in kcal/mol. (Adapted with permission from Ref. [217]; copyright 2025 American Chemical Society)

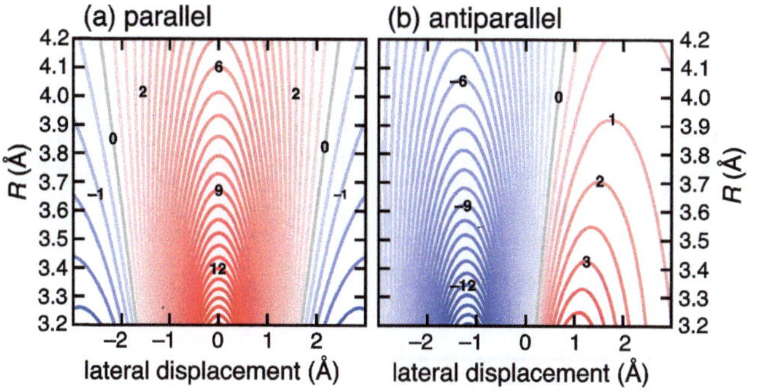

Fig. 31 Multipolar electrostatic potentials for cofacial $(C_6H_5NO_2)_2$ in orientations where the monomer dipole moments are either (**a**) parallel or (**b**) antiparallel, as in Fig. 30. In either case, the electrostatic potential is approximated using monomer dipole and quadrupole moments computed at a level of theory that is consistent with XSAPT+MBD electrostatics. The corresponding exact results, using full charge densities, can be found in Fig. 30c for the parallel configuration and in Fig. 30f for the antiparallel arrangement. (Adapted with permission from Ref. [217]; copyright 2025 American Chemical Society)

electrostatic environments. The reason is that E_{vdW} consists of only short-range intermolecular forces, which are omnipresent for arene–arene interactions regardless of the electrostatic environment. In contrast, it is more difficult to rationalize how an explanation based on longer-range electrostatic interactions could persist across the myriad electrostatic environments that are encountered within different proteins.

3.3.2 π–π Interactions in Larger Systems

Examples considered in Sect. 3.3.1 exploit XSAPT's energy decomposition but don't require its reduced scaling. A larger example along the same lines is benzene atop a $C_{94}H_{24}$ graphene nanoflake, for which XSAPT+MBD results are shown in Fig. 32 [42]. Here, eclipsed-cofacial stacking (analogous to the sandwich structure of benzene dimer) represents a saddle point between symmetry-equivalent slip-stacked minima, where an offset of approximately 1.4 Å lowers the energy by about 1 kcal/mol.

The vdW explanation for offset stacking persists in more exotic systems as well. For example, we have examined slip stacking in models of the covalent organic framework (COF) known as "COF1" [266], models for which are illustrated in Fig. 33a [216]. COFs are layered materials with potential applications in catalysis, but those applications require that the pore structure be open. Although the structures are likely amorphous or polycrystalline, characterized by microdomains, there has been much discussion regarding whether solvent molecules can occlude the pores [267–269] or whether even the solvent-free structure might be characterized by slippage between the layers that could block the pores [270, 271]. Using the nomenclature of graphene [272, 273], the question is whether these materials exhibit

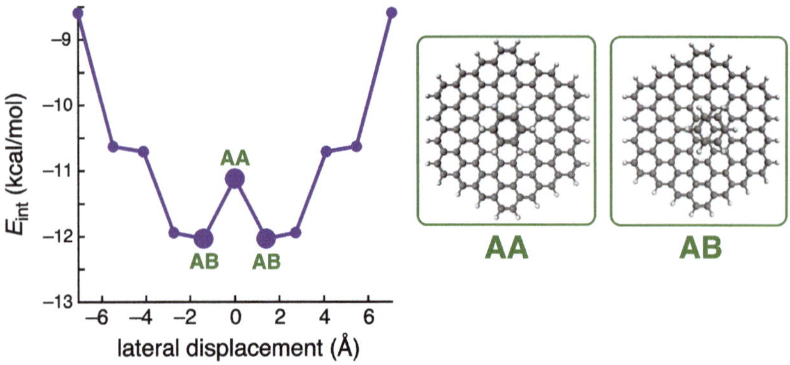

Fig. 32 Total interaction energy (E_{int}) for lateral displacement of C_6H_6 atop a $C_{94}H_{24}$ graphene nanoflake, in a cofacial orientation, computed at the XSAPT+MBD/def2-TZVPPD level. Critical points are labeled using the AA/AB nomenclature for graphene, as illustrated on the right. (Data are from Ref. [148])

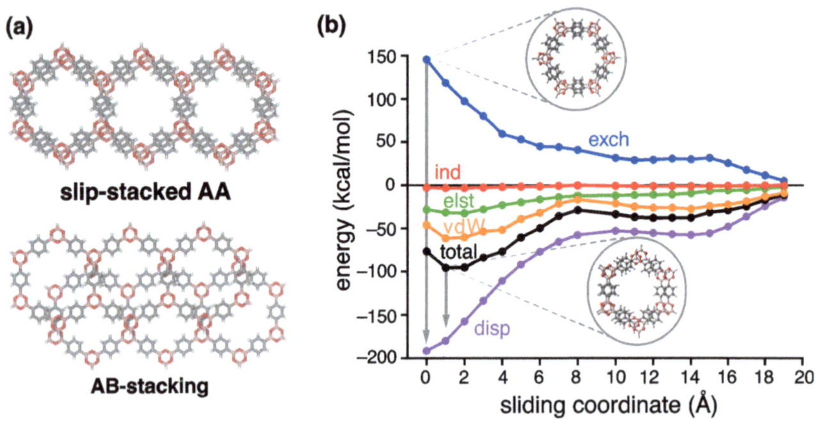

Fig. 33 (**a**) Three-unit dimer model of COF1, showing a slip-stacked version of AA-stacking (upper structure) as well as an AB-stacked structure in which the pores are occluded. (**b**) XSAPT +MBD/def2-ma-TZVP interaction potential and energy components, for lateral displacement of a one-unit dimer model at a fixed separation of 3.28 Å, corresponding to the parallel-displaced value of R_{min}.

"AA stacking" (perfectly cofacial, with pores open) or "AB stacking" (with alternating offset layers that will occlude the pores) [267–269, 274–280].

These possibilities are illustrated in Fig. 33a and can be investigated with XSAPT +MBD, whose low cost is crucial because the smallest monomer unit of COF1 is a 12-member ring composed of six-membered rings, containing 102 atoms. The XSAPT+MBD/def2-ma-TZVP calculations presented in Fig. 33b, for the dimer of that fundamental unit, are updates from previous calculations using def2-SVPD [216], taking advantage of improved memory management in the new code to improve the basis-set quality. These energy profiles indicate a small lateral offset of $\lesssim$ 1.5 Å, much smaller than the pore size. This is more consistent with benzene dimer and other small π-stacked systems than it is with true AB-stacking. As seen in Fig. 33b, the slight offset in the $(\mathrm{COF1})_2$ model originates mostly from E_{vdW}, although E_{elst} makes a nontrivial contribution to the well depth.

3.3.3 Are π–π Interactions Unique?

We next use XSAPT+MBD to consider a question that has been asked repeatedly in the literature [281–283], namely, do π–π interactions constitute a unique form of dispersion, different from that encountered in other molecules? CCSD(T) calculations on the coronene dimer $(C_{24}H_{12})_2$ and perhydrocoronene dimer $(C_{24}H_{36})_2$, as models of graphene and graphane, respectively, suggest that E_{int} is more than 30% larger per carbon atom in the [24]graphene model as compared to [24]graphane [284]. This effect is not captured in B97+D/6-31G(d,p) calculations, which predict similar interaction energies per carbon atom for much larger [n]graphene and [n] graphane models [285].

Close examination of stacking energies for acene dimers and their saturated perhydroacene analogues suggests an explanation for enhanced stacking energies in [n]graphenes [148]. Figure 34 compares stacking energies for the acene and perhydroacene dimers, where the former are considered in both cofacial (parallel-displaced) and perpendicular (T-shaped) geometries. For $n \geq 2$ rings, the π-stacked acene is systematically more stable than either the C–H$\cdots\pi$ (T-shaped) acene dimer or the stacked perhydroacene dimer, by an amount that increases with the number of rings. In contrast, all three structures have very similar interaction energies for $n = 1$ rings (i.e., the benzene and cyclohexane dimers), demonstrating that results for benzene dimer are not representative of larger PAHs. These differences are confirmed in calculations of C_6H_6 atop $C_{96}H_{24}$, which exhibit a clear preference for cofacial π-stacking over perpendicular C–H$\cdots\pi$ structures [148].

Remarkably, stacking energies for the perhydroacene dimers with one to five cyclohexane rings are a nearly identical match to data for T-shaped acene dimers with the same number of carbon atoms [148]. This suggests that there is something special about cofacial π–π interactions specifically. To understand what it might be, we consider isosurfaces of the reduced density gradient (RDG) [286],

$$\mathrm{RDG}(\mathbf{r}) = \left(\frac{1}{2(3\pi^2)^{1/3}} \right) \frac{\| \hat{\nabla}\rho(\mathbf{r}) \|}{\rho(\mathbf{r})^{4/3}}, \tag{57}$$

which has been used extensively as an indicator of noncovalent interactions [286–291]. This works because noncovalent interactions are characterized by relatively large oscillations in the density (manifesting as large values of $\|\hat{\nabla}\rho(\mathbf{r})\|$) in regions where the density itself is small, due to disturbances in the tails of the monomer electron densities originating in the presence of the other monomer.

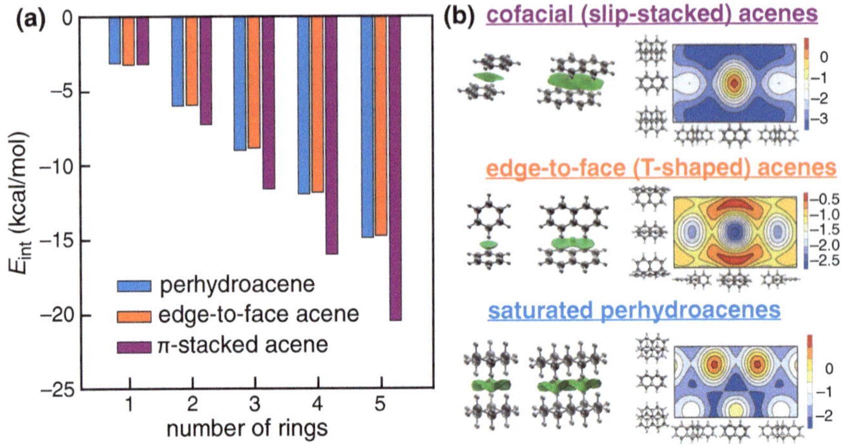

Fig. 34 (**a**) Interaction energies computed using XSAPT+δHF+MBD/def2-TZVPPD for perhydroacene dimers and for acene dimers, with the latter considered in either a parallel-displaced π-stacked or a T-shaped edge-to-face arrangement. (Adapted with permission from Ref. [148]; copyright 2020 the PCCP Owner Societies.) (**b**) Illustration of these three systems for $n = 1$ and $n = 2$ rings including isosurfaces of RDG(**r**) (in green) and lateral displacement potential energy profiles. The energy scales in (**b**) are in kcal/mol. (Adapted with permission from Ref. [42]; copyright 2021 American Chemical Society)

Isosurfaces of RDG(**r**) for the acene and perhydroacene dimers with $n = 1$ or 2 rings are plotted in Fig. 34b [148]. We observe that the RDG isosurface is corrugated for both the T-shaped acene dimer and the perhydroacene dimer, with stronger interactions when the C–H bonds of one monomer point into the middle of the ring of its partner. In contrast, the RDG isosurfaces for the cofacial acene dimers are relatively flat, meaning that strong interactions persist upon lateral displacement.

From this analysis, we conclude that π-stacking is unique among other dispersion-driven interactions, specifically due to the planarity of the arene moiety that facilitates exceptionally strong interactions when two arenes are arranged in a cofacial geometry but considerably weaker interactions (at least for PAHs) in the T-shaped geometry. The difference is that the cofacial arrangement allows for closer approach of the two arenes, which can then access short-range electrostatic interactions that are attractive in the cofacial geometry. Thus, the fact that exact electrostatics differs in sign from quadrupolar electrostatics for cofacial arenes is a crucial aspect of what makes π-stacking special, different from an ordinary dispersion interaction. Two arenes in a T-shaped geometry still interact via dispersion forces but are kept farther apart due to Pauli repulsion of the C–H$\cdots\pi$ arrangement and do not access the attractive short-range electrostatic interaction, leading to systematically weaker interaction energies as compared to those in cofacial geometries. Interactions in [n] graphane models are weaker for the same reason.

Note that this explanation for what is special about π-stacking invokes planarity but not aromaticity. For arenes, the former obviously derives from the latter, yet our analysis suggests that exceptionally strong stacking interactions should be available to systems that are planar but not aromatic. In fact, this has been observed [255, 292–295]. Our mental image is that the dispersion interactions associated with the π-electron density (or the electron density in some otherwise planar moiety) are served up on a two-dimensional platter, resembling a pizza peel, which can be rotated into an optimal coplanar configuration but whose shape cannot be deformed. If it can be, as in the perhydroacenes, then the stronger stacking interactions are lost. We refer to this as the "pizza-π" model of stacking interactions [148]. Strong π–π interactions turn out to be more about the pizza (molecular planarity) than they are about the π (aromaticity).

3.4 Charge Transfer in Anion–Water Hydrogen Bonds

Partial covalency of hydrogen bonds has long been a topic of debate [154]. For anion–water hydrogen bonds, that debate is settled (in our view) by the observation of strong redshifts in gas-phase vibrational spectra of hydrated anion clusters [296–298]. These are the signature of $n \rightarrow \sigma^*_{\mathrm{OH}}$ charge donation from the anion into antibonding MOs of the hydrogen-bonded O–H moiety [299, 300], which is then considerably redshifted, whereas the "free" or non-hydrogen-bonded O–H vibration is not [296–298]. Quantitative values for the fraction of an electron that is transferred vary greatly depending on the theoretical method that is used but are on the order of a few milli-electrons in the neutral water dimer [154], up to perhaps ≈ 0.2 electrons for $F^-(H_2O)$ [300].

Here, we examine the role of CT in the $F^-(H_2O)$ hydrogen bond using SAPT0+δHF in conjunction with the cDFT-based procedure for separating polarization and CT that was introduced in Sect. 1.3.2 [178]. Figure 35b plots various energy components as the ion is swept radially around the H_2O molecule. Note that the C_{2v} geometry of $F^-(H_2O)$, in which the ion sits at the positive end of water's dipole moment vector, is actually a saddle point between two symmetry-equivalent minima corresponding to quasi-linear hydrogen bonds. Halide–water coordination is monodentate rather than bidentate, consistent with the interpretation of cluster ion vibrational spectroscopy [296–298], solution-phase neutron diffraction [301], and x-ray absorption experiments [302].

In contrast to the dipolar picture of anion–water interactions that remains common in textbooks [303], electrostatics alone actually favors the hydrogen-bonded structure over the C_{2v} geometry. That said, E_{exch} is maximally repulsive in the hydrogen-bonded configuration because the ion–water distance is smaller as compared to that in the C_{2v} geometry. (Note that the plots in Fig. 35 represent relaxed scans, meaning that the geometry of the complex is optimized at each fixed angle that is used to generate the one-dimensional energy profile. The small cusps that can be seen at the C_{2v} geometry are artifacts of these relaxed scans that disappear if the H_2O deformation energy is included, as explained in Ref. [178].) A plot of

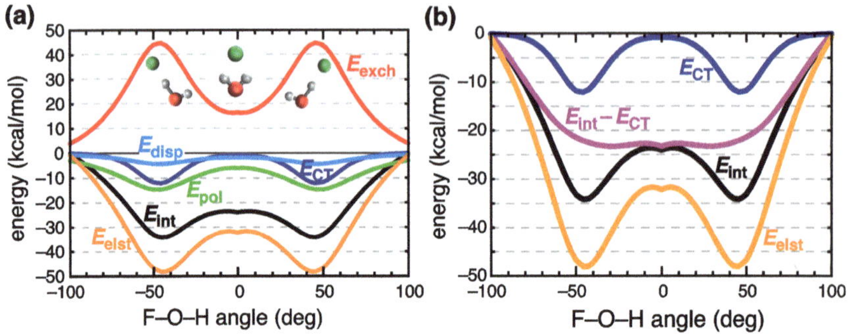

Fig. 35 (**a**) Energy components for $F^-(H_2O)$ computed at the SAPT0+δHF/jun-cc-pVDZ level, along a relaxed scan in which the F–O–H angle is varied while the other degrees of freedom are relaxed at the MP2/aug-cc-pVTZ level. Three representative geometries are depicted, corresponding to right and left hydrogen bonds with a C_{2v}-symmetric saddle point between them. (**b**) A closer look at some energy component profiles. (Reproduced with permission from Ref. [178]; copyright 2021 American Chemical Society)

$E_{int} - E_{CT}$ is nearly flat for angular displacements to either side of the C_{2v} geometry (Fig. 35b), indicating that energy components other than CT are ambivalent toward the position of the ion, so long as it resides within the angle subtended by the two O–H bonds. In contrast, E_{CT} turns on sharply for quasi-linear $F^- \cdots H$–O geometries, implying that CT is primarily responsible for stabilizing the hydrogen-bonded configuration relative to the ion–dipole arrangement. Quasi-linearity is a defining feature of the hydrogen bond [304, 305] and arises from the strong angle dependence of E_{CT}, which can certainly be rationalized in terms of overlap between the MOs n(F) and σ^*(O–H), but emerges here without explicit MO considerations. The same conclusions are valid for complexes $X^-(H_2O)$ with larger halide ions [178].

Encouraged by this successful description, we turned to SAPT(HF)+MBD calculations to analyze anion–water interactions for inorganic anions in aqueous solution [179]. This originates in an attempt to determine whether the solvation environment of simple anions (including halides, ClO_x^-, NO_x^-, and cyanates) is different in bulk water versus the liquid/vapor interface, and a provocative suggestion that it might not be [201]. Figure 36 shows results in the form of box-and-whisker plots representing ensemble averages of SAPT(HF)+MBD interaction energies and energy components for anion–water clusters extracted from a simulation [179]. (Each cluster contains approximately two solvation shells of water molecules that are treated as a single monomer.) For a variety of surface-active monovalent anions [306–311], we find little difference in E_{int} between the bulk and interfacial environments (Fig. 36a) [179]. Interestingly, the CT energies are also quite similar (Fig. 36b), with the possible exception of I^-, ClO^-, and SCN^-, although differences are modest even in those cases. Since E_{int} and E_{CT} are similar, it follows that polarization energies are not very different between bulk and interfacial environments [179].

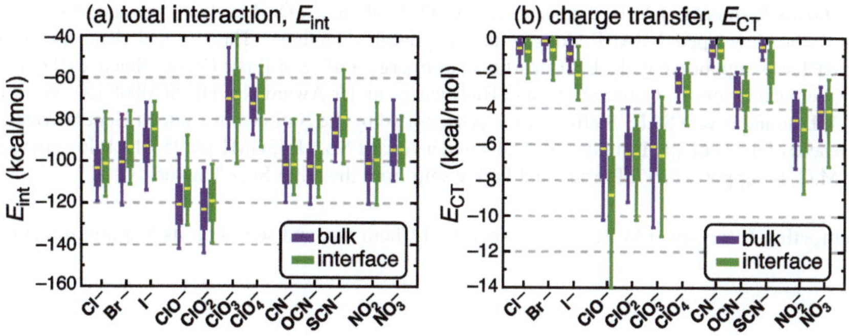

Fig. 36 Box-and-whisker plots showing mean values, standard deviations, and extremal values of (**a**) total interaction energies and (**b**) CT energies for ion–water clusters extracted from a molecular dynamics simulation. Electronic structure calculations were performed at the SAPT(HF)+MBD/6-311+G(d,p) level, and each snapshot from the simulation was binned according to whether the ion was in a bulk solvation environment or else was present at the air/water interface. (Adapted from Ref. [179] under a CC BY 4.0 license)

4 Summary

The SAPT formalism, and XSAPT with it, has quite a few moving parts that can make it intimidating to beginners, and setting up the calculations requires a variety of decisions on the user's part. In this lengthy overview of XSAPT, we have taken the time to explain carefully how those various parts work and what decisions need to be made. Philosophically, we suggest that the need for choices is a feature rather than a bug. Compared to DFT-based EDAs, which are subject to the vagaries of XC functional choice and an ambiguous definition of dispersion, (X)SAPT-based energy decomposition is well defined and improvable, in principle. Clean separability between energy components means that it is possible to test the effect that specific decisions have on individual energy components and thereby gauge the reliability of one's conclusions based on their sensitivity to this menu of choices.

That said, our preferred choice for complexes not involving ions is XSAPT+MBD/def2-ma-TZVP, with monomer wavefunctions obtained using the LRC-ω_{GDD}PBE functional, tuned in a monomer-specific way, and with CM5 embedding charges. The latter perform well for ions, and we recommend XSAPT+*ai*D3 as a starting point for ionic complexes. These methods are available in the Q-Chem code [46], with improved multithreading capabilities in recent releases.

Exemplary applications discussed herein demonstrate how the XSAPT formalism can be brought to bear on important problems in physical (and physical organic) chemistry, exposing certain fallacies regarding intermolecular interactions that persist in textbooks and in research literature. We hope that XSAPT-based calculations can start to change conventional thinking by furnishing qualitative insight based on quantitative calculations.

Acknowledgements J.M.H. thanks his former Ph.D. students, Dr. Leif Jacobson and Prof. Ka Un Lao, who developed XSAPT along with the present coauthors. The original development of XSAPT was supported by the US Department of Energy, Office of Basic Energy Sciences, Division of Chemical Sciences, Geosciences, and Biosciences under Award No. DE-SC0008550. Work by K.-Y.L. to improve XSAPT's efficiency was supported by the National Institute of General Medical Sciences of the National Institutes of Health under Award No. 1R43GM148095-01A1. Recent work by M.G. is supported by a Presidential Fellowship from the Ohio State University.

Competing Interests J.M.H. is part owner of Q-Chem Inc. and serves on its board of directors.

References

1. Gray M, Herbert JM (2024) Density functional theory for van der Waals complexes: Size matters. Annu Rep Comput Chem 20:1–61. https://doi.org/10.1016/bs.arcc.2024.03.001
2. Grimme S (2011) Density functional theory with London dispersion corrections. Wiley Interdiscip Rev Comput Mol Sci 1:211–228. https://doi.org/10.1002/wcms.30
3. Grimme S, Hansen A, Brandenburg JG, Bannwarth C (2016) Dispersion-corrected mean-field electronic structure methods. Chem Rev 116:5105–5154. https://doi.org/10.1021/acs.chemrev.5b00533
4. Hermann J, DiStasio Jr. RA, Tkatchenko A (2017) First-principles models for van der Waals interactions in molecules and materials: Concepts, theory, and applications. Chem Rev 117: 4714–4758. https://doi.org/10.1021/acs.chemrev.6b00446
5. Grimme S (2014) Dispersion interaction and chemical bonding. In: Frenking G, Shaik S (eds) The chemical bond: chemical bonding across the periodic table, chap 16. Wiley-VCH, Weinheim, pp 477–499. https://doi.org/10.1002/9783527664658.ch16
6. Wagner JP, Schreiner PR (2015) London dispersion in molecular chemistry—Reconsidering steric effects. Angew Chem Int Ed Engl 54:12274–1229. https://doi.org/10.1002/anie.201503476
7. Rummel L, Schreiner PR (2024) Advances and prospects in understanding London dispersion interactions in molecular chemistry. Angew Chem Int Ed Engl 63:e202316364. https://doi.org/10.1002/anie.202316364
8. Kitaura K, Morokuma K (1976) A new energy decomposition scheme for molecular interactions within the Hartree-Fock approximation. Int J Quantum Chem 10:325–340. https://doi.org/10.1002/qua.560100211
9. Bickelhaupt FM, Baerends EJ (2000) Kohn-Sham density functional theory: Predicting and understanding chemistry. In: Lipkowitz KB, Boyd DB (eds) Reviews in Computational Chemistry, vol 15, chap 1. Wiley-VCH, New York, pp 1–86. https://doi.org/10.1002/9780470125922.ch1
10. Francisco E, Pendás AM (2017) Energy partition analyses: Symmetry-adapted perturbation theory and other techniques. In: de la Roza AO, DiLabio GA (eds) Non-Covalent Interactions in Quantum Chemistry and Physics, chap 2. Elsevier, Amsterdam, pp 27–64. https://doi.org/10.1016/B978-0-12-809835-6.00003-7
11. Zhao L, von Hopffgarden M, Andrada DM, Frenking G (2018) Energy decomposition analysis. Wiley Interdiscip Rev Comput Mol Sci 8:e1345. https://doi.org/10.1002/wcms.1345
12. Su P, Tang Z, Wu W (2020) Generalized Kohn-Sham energy decomposition analysis and its applications. Wiley Interdiscip Rev Comput Mol Sci 10:e1460. https://doi.org/10.1002/wcms.1460
13. Mao Y, Loipersberger M, Horn PR, Das A, Demerdash O, Levine DS, Veccham SP, Head-Gordon T, Head-Gordon M (2021) From intermolecular interaction energies to observable shifts to component contributions and back again: A tale of variational energy decomposition

analysis. Annu Rev Phys Chem 72:641–666. https://doi.org/10.1146/annurev-physchem-090419-115149
14. Thirman J, Head-Gordon M (2015) An energy decomposition analysis for second-order Møller–Plesset perturbation theory based on absolutely localized molecular orbitals. J Chem Phys 143:084124. https://doi.org/10.1063/1.4929479
15. Thirman J, Head-Gordon M (2017) Efficient implementation of energy decomposition analysis for second-order Møller–Plesset perturbation theory and application to anion-π interactions. J Phys Chem A 121:717–728. https://doi.org/10.1021/acs.jpca.6b11516
16. Bistoni G (2020) Finding chemical concepts in the Hilbert space: Coupled cluster analyses of noncovalent interactions. Wiley Interdiscip Rev Comput Mol Sci 10:e1442. https://doi.org/10.1002/wcms.1442
17. Andrés J, Ayers PW, Boto RA, Carbó-Dorca R, Chermette H, Cioslowski J, Contreras-García J, Cooper DL, Frenking G, Gatti C, Heidar-Zadeh F, Joubert L, Pendás AM, Matito E, Mayer I, Misquitta AJ, Mo Y, Pilmé J, Popelier PLA, Rahm M, Ramos-Cordoba E, Salvador P, Schwarz WHE, Shahbazian S, Silvi B, Solà M, Szalewicz K, Tognetti V, Weinhold F, Zins EL (2019) Nine questions on energy decomposition analysis. J Comput Chem 40:2248–2283. https://doi.org/10.1002/jcc.26003
18. Andrada DM, Foroutan-Nejad C (2020) Energy components in energy decomposition analysis (EDA) are path functions; why does it matter? Phys Chem Chem Phys 22:22459–22464. https://doi.org/10.1039/D0CP04016A
19. Poater J, Andrada DM, Solà M, Foroutan-Nejad C (2022) Path-dependency of energy decomposition analysis & the elusive nature of bonding. Phys Chem Chem Phys 24:2344–2348. https://doi.org/10.1039/D1CP04135E
20. Clark T (2023) How deeply should we analyze non-covalent interactions? J Mol Model 29:66. https://doi.org/10.1007/s00894-023-05460-4
21. Rocher-Casterline BE, Ch'ng LC, Mollner AK, Reisler H (2011) Communication: Determination of the bond dissociation energy (D_0) of the water dimer, $(H_2O)_2$, by velocity map imaging. J Chem Phys 134:211101. https://doi.org/10.1063/1.3598339
22. Jeziorski B, Moszynski R, Szalewicz K (1994) Perturbation theory approach to intermolecular potential energy surfaces of van der Waals complexes. Chem Rev 94:1887–1930. https://doi.org/10.1021/cr00031a008
23. Szalewicz K (2012) Symmetry-adapted perturbation theory of intermolecular forces. Wiley Interdiscip Rev Comput Mol Sci 2:254–272. https://doi.org/10.1002/wcms.86
24. Jansen G (2014) Symmetry-adapted perturbation theory based on density functional theory for noncovalent interactions. Wiley Interdiscip Rev Comput Mol Sci 4:127–144. https://doi.org/10.1002/wcms.1164
25. Hohenstein EG, Sherrill CD (2012) Wavefunction methods for noncovalent interactions. Wiley Interdiscip Rev Comput Mol Sci 2:304–326. https://doi.org/10.1002/wcms.84
26. Parker TM, Burns LA, Parrish RM, Ryno AG, Sherrill CD (2014) Levels of symmetry adapted perturbation theory (SAPT). I. Efficiency and performance for interaction energies. J Chem Phys 140:094106. https://doi.org/10.1063/1.4867135
27. Schriber J, Cheney D, Sherrill CD (2023) Levels of symmetry adapted perturbation theory (SAPT). II. Convergence of interaction energy components. ChemRxiv. https://doi.org/10.26434/chemrxiv-2023-ftt1v
28. Patkowski K (2020) Recent developments in symmetry-adapted perturbation theory. Wiley Interdiscip Rev Comput Mol Sci 10:e1452. https://doi.org/10.1002/wcms.1452
29. Gao J (1998) A molecular-orbital derived polarization potential for liquid water. J Chem Phys 109:2346–2354. https://doi.org/10.1063/1.476802
30. Xie W, Song L, Truhlar DG, Gao J (2008) The variational explicit polarization potential and analytical first derivative of energy: Towards a next generation force field. J Chem Phys 128: 234108. https://doi.org/10.1063/1.2936122

31. Wang Y, Mazack MJM, Truhlar DG, Gao J (2016) Explicit polarization theory. In: Cui Q, Meuwly M, Ren P (eds) Many-Body Effects and Electrostatics in Biomolecules, chap 2. Pan Stanford, Boca Raton, FL, pp 33–64. https://doi.org/10.1201/b21343-5
32. Jacobson LD, Herbert JM (2011) An efficient, fragment-based electronic structure method for molecular systems: Self-consistent polarization with perturbative two-body exchange and dispersion. J Chem Phys 134:094118. https://doi.org/10.1063/1.3560026
33. Herbert JM, Jacobson LD, Lao KU, Rohrdanz MA (2012) Rapid computation of intermolecular interactions in molecular and ionic clusters: Self-consistent polarization plus symmetry-adapted perturbation theory. Phys Chem Chem Phys 14:7679–7699. https://doi.org/10.1039/C2CP24060B
34. Lao KU, Herbert JM (2012) Accurate intermolecular interactions at dramatically reduced cost: XPol+SAPT with empirical dispersion. J Phys Chem Lett 3:3241–3248. https://doi.org/10.1021/jz301015p
35. Jacobson LD, Richard RM, Lao KU, Herbert JM (2013) Efficient monomer-based quantum chemistry methods for molecular and ionic clusters. Annu Rep Comput Chem 9:25–58. https://doi.org/10.1016/B978-0-444-62672-1.00002-9
36. Lao KU, Herbert JM (2013) An improved treatment of empirical dispersion and a many-body energy decomposition scheme for the explicit polarization plus symmetry-adapted perturbation theory (XSAPT) method. J Chem Phys 139:034107. https://doi.org/10.1063/1.4813523
37. Lao KU, Herbert JM (2015) Accurate and efficient quantum chemistry calculations of noncovalent interactions in many-body systems: The XSAPT family of methods. J Phys Chem A 119:235–253. https://doi.org/10.1021/jp5098603
38. Lao KU, Herbert JM (2018) Atomic orbital implementation of extended symmetry-adapted perturbation theory (XSAPT) and benchmark calculations for large supramolecular complexes. J Chem Theory Comput 14:2955–2978. https://doi.org/10.1021/acs.jctc.8b00058
39. Lao KU, Herbert JM (2018) A simple correction for nonadditive dispersion within extended symmetry-adapted perturbation theory (XSAPT). J Chem Theory Comput 14:5128–5142. https://doi.org/10.1021/acs.jctc.8b00527
40. Carter-Fenk K, Lao KU, Liu KY, Herbert JM (2019) Accurate and efficient *ab initio* calculations for supramolecular complexes: Symmetry-adapted perturbation theory with many-body dispersion. J Phys Chem Lett 10:2706–2714. https://doi.org/10.1021/acs.jpclett.9b01156
41. Liu KY, Carter-Fenk K, Herbert JM (2019) Self-consistent charge embedding at very low cost, with application to symmetry-adapted perturbation theory. J Chem Phys 151:031102. https://doi.org/10.1063/1.5111869
42. Carter-Fenk K, Lao KU, Herbert JM (2021) Predicting and understanding non-covalent interactions using novel forms of symmetry-adapted perturbation theory. Acc Chem Res 54:3679–3690. https://doi.org/10.1021/acs.accounts.1c00387
43. Herbert JM (2021) Neat, simple, and wrong: Debunking electrostatic fallacies regarding noncovalent interactions. J Phys Chem A 125:7125–7137. https://doi.org/10.1021/acs.jpca.1c05962
44. Patkowski K (2017) Benchmark databases of intermolecular interaction energies: Design, construction, and significance. Annu Rep Comput Chem 13:3–91. https://doi.org/10.1016/bs.arcc.2017.06.004
45. Shao Y, Gan Z, Epifanovsky E, Gilbert ATB, Wormit M, Kussmann J, Lange AW, Behn A, Deng J, Feng X, Ghosh D, Goldey M, Horn PR, Jacobson LD, Kaliman I, Khaliullin RZ, Kús T, Landau A, Liu J, Proynov EI, Rhee YM, Richard RM, Rohrdanz MA, Steele RP, Sundstrom EJ, Woodcock III HL, Zimmerman PM, Zuev D, Albrecht B, Alguire E, Austin B, Beran GJO, Bernard YA, Berquist E, Brandhorst K, Bravaya KB, Brown ST, Casanova D, Chang CM, Chen Y, Chien SH, Closser KD, Crittenden DL, Diedenhofen M, DiStasio Jr. RA, Do H, Dutoi AD, Edgar RG, Fatehi S, Fusti-Molnar L, Ghysels A, Golubeva-Zadorozhnaya A, Gomes J, Hanson-Heine MWD, Harbach PHP, Hauser AW, Hohenstein EG, Holden ZC, Jagau TC, Ji H, Kaduk B, Khistyaev K, Kim J, Kim J, King RA, Klunzinger P, Kosenkov D, Kowalczyk T, Krauter CM, Lao KU, Laurent AD, Lawler KV, Levchenko SV, Lin CY, Liu F,

Livshits E, Lochan RC, Luenser A, Manohar P, Manzer SF, Mao SP, Mardirossian N, Marenich AV, Maurer SA, Mayhall NJ, Neuscamman E, Oana CM, Olivares-Amaya R, O'Neill DP, Parkhill JA, Perrine TM, Peverati R, Prociuk A, Rehn DR, Rosta E, Russ NJ, Sharada SM, Sharma S, Small DW, Sodt A, Stein T, Stück D, Su YC, Thom AJW, Tsuchimochi T, Vanovschi V, Vogt L, Vydrov O, Wang T, Watson MA, Wenzel J, White A, Williams CF, Yang J, Yeganeh S, Yost SR, You ZQ, Zhang IY, Zhang X, Zhao Y, Brooks BR, Chan GKL, Chipman DM, Cramer CJ, Goddard III WA, Gordon MS, Hehre WJ, Klamt A, Schaefer III HF, Schmidt MW, Sherrill CD, Truhlar DG, Warshel A, Xu X, Aspuru-Guzik A, Baer R, Bell AT, Besley NA, Chai JD, Dreuw A, Dunietz BD, Furlani TR, Gwaltney SR, Hsu CP, Jung Y, Kong J, Lambrecht DS, Liang W, Ochsenfeld C, Rassolov VA, Slipchenko LV, Subotnik JE, Van Voorhis T, Herbert JM, Krylov AI, Gill PMW, Head-Gordon M (2015) Advances in molecular quantum chemistry contained in the Q-Chem 4 program package. Mol Phys 113:184–215. https://doi.org/10.1080/00268976.2014.952696

46. Epifanovsky E, Gilbert ATB, Feng X, Lee J, Mao Y, Mardirossian N, Pokhilko P, White AF, Coons MP, Dempwolff AL, Gan Z, Hait D, Horn PR, Jacobson LD, Kaliman I, Kussmann J, Lange AW, Lao KU, Levine DS, Liu J, McKenzie SC, Morrison AF, Nanda KD, Plasser F, Rehn DR, Vidal ML, You ZQ, Zhu Y, Alam B, Albrecht BJ, Aldossary A, Alguire E, Andersen JH, Athavale V, Barton D, Begam K, Behn A, Bellonzi N, Bernard YA, Berquist EJ, Burton HGA, Carreras A, Carter-Fenk K, Chakraborty R, Chien AD, Closser KD, Cofer-Shabica V, Dasgupta S, de Wergifosse M, Deng J, Diedenhofen M, Do H, Ehlert S, Fang PT, Fatehi S, Feng Q, Friedhoff T, Gayvert J, Ge Q, Gidofalvi G, Goldey M, Gomes J, González-Espinoza CE, Gulania S, Gunina AO, Hanson-Heine MWD, Harbach PHP, Hauser A, Herbst MF, Hernández Vera M, Hodecker M, Holden ZC, Houck S, Huang X, Hui K, Huynh BC, Ivanov M, Jász A, Ji H, Jiang H, Kaduk B, Kähler S, Khistyaev K, Kim J, Kis G, Klunzinger P, Koczor-Benda Z, Koh JH, Kosenkov D, Koulias L, Kowalczyk T, Krauter CM, Kue K, Kunitsa A, Kus T, Ladjánszki I, Landau A, Lawler KV, Lefrancois D, Lehtola S, Li RR, Li YP, Liang J, Liebenthal M, Lin HH, Lin YS, Liu F, Liu KY, Loipersberger M, Luenser A, Manjanath A, Manohar P, Mansoor E, Manzer SF, Mao SP, Marenich AV, Markovich T, Mason S, Maurer SA, McLaughlin PF, Menger MFSJ, Mewes JM, Mewes SA, Morgante P, Mullinax JW, Oosterbaan KJ, Paran G, Paul AC, Paul SK, Pavošević F, Pei Z, Prager S, Proynov EI, Rák A, Ramos-Cordoba E, Rana B, Rask AE, Rettig A, Richard RM, Rob F, Rossomme E, Scheele T, Scheurer M, Schneider M, Sergueev N, Sharada SM, Skomorowski W, Small DW, Stein CJ, Su YC, Sundstrom EJ, Tao Z, Thirman J, Tornai GJ, Tsuchimochi T, Tubman NM, Veccham SP, Vydrov O, Wenzel J, Witte J, Yamada A, Yao K, Yeganeh S, Yost SR, Zech A, Zhang IY, Zhang X, Zhang Y, Zuev D, Aspuru-Guzik A, Bell AT, Besley NA, Bravaya KB, Brooks BR, Casanova D, Chai JD, Coriani S, Cramer CJ, Cserey G, DePrince III AE, DiStasio Jr. RA, Dreuw A, Dunietz BD, Furlani TR, Goddard III WA, Hammes-Schiffer S, Head-Gordon T, Hehre WJ, Hsu CP, Jagau TC, Jung Y, Klamt A, Kong J, Lambrecht DS, Liang W, Mayhall NJ, McCurdy CW, Neaton JB, Ochsenfeld C, Parkhill JA, Peverati R, Rassolov VA, Shao Y, Slipchenko LV, Stauch T, Steele RP, Subotnik JE, Thom AJW, Tkatchenko A, Truhlar DG, Van Voorhis T, Wesolowski TA, Whaley KB, Woodcock III HL, Zimmerman PM, Faraji S, Gill PMW, Head-Gordon M, Herbert JM, Krylov AI (2021) Software for the frontiers of quantum chemistry: An overview of developments in the Q-Chem 5 package. J Chem Phys 155:084801. https://doi.org/10.1063/5.0055522

47. Moszynski R (2007) Theory of intermolecular forces: An introductory account. In: Sokalski WA (ed) Molecular Materials and Specific Interactions—Modeling and Design. Springer, New York, pp 1–152. https://doi.org/10.1007/1-4020-5372-X_1

48. Jeziorski B, Bulski M, Piela L (1976) First-order perturbation treatment of the short-range repulsion in a system of many closed-shell atoms or molecules. Int J Quantum Chem 10:281–297. https://doi.org/10.1002/qua.560100208

49. Schäffer R, Jansen G (2012) Intermolecular exchange-induction energies without overlap expansion. Theor Chem Acc 131:1235. https://doi.org/10.1007/s00214-012-1235-6

50. Schäffer R, Jansen G (2013) Single-determinant-based symmetry-adapted perturbation theory without single-exchange approximation. Mol Phys 111:2570–2584. https://doi.org/10.1080/00268976.2013.827253
51. Hirschfelder JO (1967) Perturbation theory for exchange forces, I. Chem Phys Lett 1:325–329. https://doi.org/10.1016/0009-2614(67)80007-1
52. Jeziorski B, Moszynski R, Ratkiewicz A, Rybak S, Szalewicz K, Williams HL (1993) SAPT: A program for many-body symmetry-adapted perturbation theory calculations of intermolecular interaction energies. In: Clementi E (ed) Methods and Techniques in Computational Chemistry: METECC-94, vol B, chap 3. STEF, Cagliari, pp 79–129
53. Szalewicz K, Patkowski K, Jeziorski B (2005) Intermolecular interactions via perturbation theory: From diatoms to biomolecules. In: Wales DJ (ed) Intermolecular Forces and Clusters II, Structure and Bonding, vol 116. Springer, Berlin, pp 43–117. https://doi.org/10.1007/430_004
54. Szalewicz K, Jeziorski B (2022) Physical mechanisms of intermolecular interactions from symmetry-adapted perturbation theory. J Mol Model 28:273. https://doi.org/10.1007/s00894-022-05190-z
55. Gray M, Herbert JM (2021) Simplified tuning of long-range corrected density functionals for symmetry-adapted perturbation theory. J Chem Phys 155:034103. https://doi.org/10.1063/5.0059364
56. Moszynski R, Jeziorski B, Ratkiewicz A, Rybak S (1993) Many-body perturbation theory of electrostatic interactions between molecules: Comparison with full configuration interaction for four-electron dimers. J Chem Phys 99:8856–8869. https://doi.org/10.1063/1.465554
57. Patkowski K, Szalewicz K, Jeziorski B (2006) Third-order interactions in symmetry-adapted perturbation theory. J Chem Phys 125:154107. https://doi.org/10.1063/1.2358353
58. Patkowski K, Szalewicz K, Jeziorski B (2010) Orbital relaxation and the third-order induction energy in symmetry-adapted perturbation theory. Theor Chem Acc 127:211–221. https://doi.org/10.1007/s00214-010-0748-0
59. Gray M, Bowling PE, Herbert JM (2022) Systematic examination of counterpoise correction in density functional theory. J Chem Theory Comput 18:6742–6756. https://doi.org/10.1021/acs.jctc.2c00883
60. Altun A, Neese F, Bistoni G (2019) HFLD: A nonempirical London dispersion-corrected Hartree–Fock method for the quantification and analysis of noncovalent interaction energies of large molecular systems. J Chem Theory Comput 15:5894–5907. https://doi.org/10.1021/acs.jctc.9b00425
61. Garcia J, Szalewicz K (2020) Ab initio extended Hartree–Fock plus dispersion method applied to dimers with hundreds of atoms. J Phys Chem A 124:1196–1203. https://doi.org/10.1021/acs.jpca.9b11900
62. Bistoni G, Altun A, Wang Z, Neese F (2024) Local energy decomposition analysis of London dispersion effects: From simple model dimers to complex biomolecular assemblies. Acc Chem Res 57:1411–1420. https://doi.org/10.1021/acs.accounts.4c00085
63. Conway A, Murrell JN (1974) The exchange energy between two neon atoms. Mol Phys 27: 873–878. https://doi.org/10.1080/00268977400100791
64. Cohen JS, Pack RT (1974) Modified statistical method for intermolecular potentials. Combining rules for higher van der Waals coefficients. J Chem Phys 61:2372–2382. https://doi.org/10.1063/1.1682318
65. Hepburn J, Scoles G, Penco R (1975) A simple but reliable method for the prediction of intermolecular potentials. Chem Phys Lett 36:451–456. https://doi.org/10.1016/0009-2614(75)80278-8
66. Ahlrichs R, Penco R, Scoles G (1977) Intermolecular forces in simple systems. Chem Phys 19: 119–130. https://doi.org/10.1016/0301-0104(77)85124-0
67. Douketis C, Scoles G, Marchetti S, Zen M, Thakkar AJ (1982) Intermolecular forces via hybrid Hartree–Fock–SCF plus damped dispersion (HFD) energy calculations. An improved spherical model. J Chem Phys 76:3057–3063. https://doi.org/10.1063/1.443345

68. Hobza P, Zahradník R (1983) Van der Waals molecules: Quantum chemistry, physical properties, and reactivity. Int J Quantum Chem 23:325–338. https://doi.org/10.1002/qua.560230128
69. Douketis C, Hutson JM, Orr BJ, Scoles G (1984) Anisotropic intermolecular forces from Hartree-Fock plus damped dispersion (HFD) calculations. Mol Phys 52:763–781. https://doi.org/10.1080/00268978400101541
70. Cybulski SM, Lytle ML (2007) The origin of the deficiency of the supermolecule second-order Møller-Plesset approach for evaluating interaction energies. J Chem Phys 127:141102. https://doi.org/10.1063/1.2795693
71. Nguyen B, Chen GP, Agee MM, Burow AM, Tang M, Furche F (2020) Divergence of many-body perturbation theory for noncovalent interactions of large molecules. J Chem Theory Comput 16:2258–2273. https://doi.org/10.1021/acs.jctc.9b01176
72. Szabo A, Ostlund NS (1977) The correlation energy in the random phase approximation: Intermolecular forces between closed-shell systems. J Chem Phys 67:4351–4360. https://doi.org/10.1063/1.434580
73. Řezác J, Riley KE, Hobza P (2011) S66: A well-balanced database of benchmark interaction energies relevant to biomolecular structures. J Chem Theory Comput 7:2427–2438. https://doi.org/10.1021/ct2002946
74. Papajak E, Zheng J, Xu X, Leverentz HR, Truhlar DG (2011) Perspectives on basis sets beautiful: Seasonal plantings of diffuse basis functions. J Chem Theory Comput 7:3027–3034. https://doi.org/10.1021/ct200106a
75. Sherrill CD (2009) Computations of noncovalent π interactions. In: Lipkowitz KB, Cundari TR (eds) Reviews in Computational Chemistry, vol 26, chap 1. Wiley-VCH, Weinheim, pp 1–38. https://doi.org/10.1002/9780470399545.ch1
76. Hohenstein EG, Sherrill CD (2010) Density fitting of intramonomer correlation effects in symmetry adapted perturbation theory. J Chem Phys 133:014101. https://doi.org/10.1063/1.3451077
77. Gray M, Herbert JM (2022) Comprehensive basis-set testing of extended symmetry-adapted perturbation theory and assessment of mixed-basis combinations to reduce cost. J Chem Theory Comput 18:2308–2330. https://doi.org/10.1021/acs.jctc.1c01302
78. Heßelmann A, Jansen G, Schütz M (2005) Density-functional theory symmetry-adapted intermolecular perturbation theory with density fitting: A new efficient method to study intermolecular interaction energies. J Chem Phys 122:014103. https://doi.org/10.1063/1.1824898
79. Lao KU, Herbert JM (2014) Symmetry-adapted perturbation theory with Kohn-Sham orbitals using non-empirically tuned, long-range-corrected density functionals. J Chem Phys 140:044108. https://doi.org/10.1063/1.4862644
80. Khaliullin RZ, Cobar EA, Lochan RC, Bell AT, Head-Gordon M (2007) Unravelling the origin of intermolecular interactions using absolutely localized molecular orbitals. J Phys Chem A 111:8753–8765. https://doi.org/10.1021/jp073685z
81. Williams HL, Mas EM, Szalewicz K, Jeziorski B (1995) On the effectiveness of monomer-, dimer-, and bond-centered basis functions in calculations of intermolecular interaction energies. J Chem Phys 103:7374–7391. https://doi.org/10.1063/1.470309
82. Khaliullin RZ, Head-Gordon M, Bell AT (2006) An efficient self-consistent field method for large systems of weakly-interacting fragments. J Chem Phys 124:204105. https://doi.org/10.1063/1.2191500
83. Rybak S, Jeziorski B, Szalewicz K (1991) Many-body symmetry-adapted perturbation theory of intermolecular interactions. H_2O and HF dimers. J Chem Phys 95:6576–6601. https://doi.org/10.1063/1.461528
84. Misquitta AJ, Bukowski R, Szalewicz K (2000) Spectra of Ar–CO_2 from *ab initio* potential energy surfaces. J Chem Phys 112:5308–5319. https://doi.org/10.1063/1.481120

85. Williams HL, Chabalowski CF (2001) Using Kohn-Sham orbitals in symmetry-adapted perturbation theory to investigate intermolecular interactions. J Phys Chem A 105:646–659. https://doi.org/10.1021/jp003883p
86. Jansen G, Heßelmann A (2001) Comment on "Using Kohn–Sham orbitals in symmetry-adapted perturbation theory to investigate intermolecular interactions". J Phys Chem A 105: 11156–11157. https://doi.org/10.1021/jp0112774
87. Misquitta AJ, Szalewicz K (2002) Intermolecular forces from asymptotically corrected density functional description of monomers. Chem Phys Lett 357:301–306. https://doi.org/10.1016/S0009-2614(02)00533-X
88. Heßelmann A, Jansen G (2002) First-order intermolecular interaction energies from Kohn-Sham orbitals. Chem Phys Lett 357:464–470. https://doi.org/10.1016/S0009-2614(02)00538-9
89. Heßelmann A, Jansen G (2002) Intermolecular induction and exchange-induction energies from coupled-perturbed Kohn-Sham density functional theory. Chem Phys Lett 362:319–325. https://doi.org/10.1016/S0009-2614(02)01097-7
90. Perdew JP, Burke K (1996) Comparison shopping for a gradient-corrected density functional. Int J Quantum Chem 57:309–319. https://doi.org/10.1002/(SICI)1097-461X(1996)57:3%3C309::AID-QUA4%3E3.0.CO;2-1
91. Tozer DJ, Handy NC (1998) Improving virtual Kohn–Sham orbitals and eigenvalues: Application to excitation energies and static polarizabilities. J Chem Phys 109:10180–10189. https://doi.org/10.1063/1.477711
92. Perdew JP, Parr RG, Levy M, Balduz, Jr. JL (1982) Density-functional theory for fractional particle number: Derivative discontinuities of the energy. Phys Rev Lett 49:1691–1694. https://doi.org/10.1103/PhysRevLett.49.1691
93. Levy M, Perdew JP, Sahni V (1984) Exact differential equation for the density and ionization energy of a many-particle system. Phys Rev A 30:2745–2748. https://doi.org/10.1103/PhysRevA.30.2745
94. Perdew JP (1985) What do the Kohn–Sham orbitals energies mean? How do atoms dissociate? In: Dreizler RM, da Providência J (eds) Density Functional Methods in Physics. NATO Advanced Science Institute Series, vol 123. Springer, pp 265–308. https://doi.org/10.1007/978-1-4757-0818-9_10
95. Perdew JP, Levy M (1997) Comment on "Significance of the highest occupied Kohn-Sham eigenvalue". Phys Rev B 56:16021–16028. https://doi.org/10.1103/PhysRevB.56.16021
96. van Leeuwen R, Baerends EJ (1994) Exchange-correlation potential with correct asymptotic behavior. Phys Rev A 49:2421–2431. https://doi.org/10.1103/PhysRevA.49.2421
97. Grüning M, Gritsenko OV, van Gisbergen SJA, Baerends EJ (2001) Shape corrections to exchange-correlation potentials by gradient-regulated seamless connection of model potentials for inner and outer region. J Chem Phys 114:652–660. https://doi.org/10.1063/1.1327260
98. Gaiduk AP, Staroverov VN (2009) How to tell when a model Kohn–Sham potential is not a functional derivative. J Chem Phys 131:044107. https://doi.org/10.1063/1.3176515
99. Baer R, Livshits E, Salzner U (2010) Tuned range-separated hybrids in density functional theory. Annu Rev Phys Chem 61:85–109. https://doi.org/10.1146/annurev.physchem.012809.103321
100. Tsuneda T, Hirao K (2014) Long-range correction for density functional theory. Wiley Interdiscip Rev Comput Mol Sci 4:375–390. https://doi.org/10.1002/wcms.1178
101. Herbert JM (2023) Density-functional theory for electronic excited states. In: García-Iriepa C, Marazzi M (eds) Theoretical and Computational Photochemistry: Fundamentals, Methods, Applications and Synergy with Experimental Approaches, chap 3. Elsevier, Amsterdam, pp 69–118. https://doi.org/10.1016/B978-0-323-91738-4.00005-1
102. Rohrdanz MA, Herbert JM (2008) Simultaneous benchmarking of ground- and excited-state properties with long-range-corrected density functional theory. J Chem Phys 129:034107. https://doi.org/10.1063/1.2954017

103. Rohrdanz MA, Martins KM, Herbert JM (2009) A long-range-corrected density functional that performs well for both ground-state properties and time-dependent density functional theory excitation energies, including charge-transfer excited states. J Chem Phys 130:054112. https://doi.org/10.1063/1.3073302
104. Alam B, Morrison AF, Herbert JM (2020) Charge separation and charge transfer in the low-lying excited states of pentacene. J Phys Chem C 124:24653–24666. https://doi.org/10.1021/acs.jpcc.0c07932
105. Modrzejewski M, Rajchel Ł, Chalasinski G, Szczesniak MM (2013) Density-dependent onset of the long-range exchange: A key to donor–acceptor properties. J Phys Chem A 117:11580–11586. https://doi.org/10.1021/jp4088404
106. Mandal A, Herbert JM (2025) Simplified tuning of long-range corrected time-dependent density functional theory. J Phys Chem Lett 16:2672–2680. https://doi.org/10.1021/acs.jpclett.5c00086
107. Karolewski A, Kronik L, Kümmel S (2013) Using optimally tuned range separated hybrid functionals in ground-state calculations: Consequences and caveats. J Chem Phys 138:204115. https://doi.org/10.1063/1.4807325
108. Phipps MJS, Fox T, Tautermann CS, Skylaris CK (2015) Energy decomposition analysis approaches and their evaluation on prototypical protein–drug interaction patterns. Chem Soc Rev 44:3177–3211. https://doi.org/10.1039/c4cs00375f
109. Herbert JM (2019) Fantasy versus reality in fragment-based quantum chemistry. J Chem Phys 151:170901. https://doi.org/10.1063/1.5126216
110. Řezáč, J, Huang Y, Hobza P, Beran GJO (2015) Benchmark calculations of three-body intermolecular interactions and the performance of low-cost electronic structure methods. J Chem Theory Comput 11:3065–3079. https://doi.org/10.1021/acs.jctc.5b00281
111. Alkan M, Xu P, Gordon MS (2019) Many-body dispersion in molecular clusters. J Phys Chem A 123:8406–8416. https://doi.org/10.1021/acs.jpca.9b05977
112. Yang J, Hu W, Usvyat D, Matthews D, Schütz M, Chan GKL (2014) Ab initio determination of the crystalline benzene lattice energy to sub-kilojoule/mole accuracy. Science 345:640–643. https://doi.org/10.1126/science.1254419
113. Kennedy MR, McDonald AR, DePrince III AE, Marshall MS, Podeszwa R, Sherrill CD (2014) Communication: Resolving the three-body contribution to the lattice energy of crystalline benzene: Benchmark results from coupled-cluster theory. J Chem Phys 140:121104. https://doi.org/10.1063/1.4869686
114. Borca CH, Glick ZL, Metcalf DP, Burns LA, Sherrill CD (2023) Benchmark coupled-cluster lattice energy of crystalline benzene and assessment of multi-level approximations in the many-body expansion. J Chem Phys 158:234102. https://doi.org/10.1063/5.0159410
115. Chen Y, Li H (2010) Intermolecular interaction in water hexamer. J Phys Chem A 114:11719–11724. https://doi.org/10.1021/jp104822e
116. Cisneros GA, Wikfeldt KT, Ojamäe L, Lu J, Xu Y, Torabifard H, Bartók AP, Csányi G, Molinero V, Paesani F (2016) Modeling molecular interactions in water: From pairwise to many-body potential energy functions. Chem Rev 116:7501–7528. https://doi.org/10.1021/acs.chemrev.5b00644
117. Jacobson LD, Williams CF, Herbert JM (2009) The static-exchange electron-water pseudopotential, in conjunction with a polarizable water model: A new Hamiltonian for hydrated-electron simulations. J Chem Phys 130:124115. https://doi.org/10.1063/1.3089425
118. Breneman CM, Wiberg KB (1990) Determining atom-centered monopoles from molecular electrostatic potentials. The need for high sampling density in formamide conformational analysis. J Comput Chem 11:361–373. https://doi.org/10.1002/jcc.540110311
119. Holden ZC, Richard RM, Herbert JM (2013) Periodic boundary conditions for QM/MM calculations: Ewald summation for extended Gaussian basis sets. J Chem Phys 139:244108. https://doi.org/10.1063/1.4850655

120. Marenich AV, Jerome SV, Cramer CJ, Truhlar DG (2012) Charge model 5: An extension of Hirshfeld population analysis for the accurate description of molecular interactions in gaseous and condensed phases. J Chem Theory Comput 8:527–541. https://doi.org/10.1021/ct200866d
121. Hirshfeld FL (1977) XVII. Spatial partitioning of charge density. Isr J Chem 16:198–201. https://doi.org/10.1002/ijch.197700033
122. Tkatchenko A, DiStasio Jr. RA, Car R, Scheffler M (2012) Accurate and efficient method for many-body van der Waals interactions. Phys Rev Lett 108:236402. https://doi.org/10.1103/PhysRevLett.108.236402
123. Tkatchenko A, Ambrosetti A, DiStasio Jr. RA (2013) Interatomic methods for the dispersion energy derived from the adiabatic connection fluctuation-dissipation theorem. J Chem Phys 138:074106. https://doi.org/10.1063/1.4789814
124. Ambrosetti A, Reilly AM, DiStasio Jr. RA, Tkatchenko A (2014) Long-range correlation energy calculated from coupled atomic response functions. J Chem Phys 140:18A508. https://doi.org/10.1063/1.4865104
125. Carter-Fenk K, Herbert JM (2023) Appraisal of dispersion damping functions for the effective fragment potential method. Mol Phys 121:e2055504. https://doi.org/10.1080/00268976.2022.2055504
126. Heßelmann A (2011) Comparison of intermolecular interaction energies from SAPT and DFT including empirical dispersion contributions. J Phys Chem A 115:11321–11330. https://doi.org/10.1021/jp205031e
127. Schriber JB, Sirianni DA, Smith DGA, Burns LA, Sitkoff D, Cheney DL, Sherrill CD (2021) Optimized damping parameters for empirical dispersion corrections to symmetry-adapted perturbation theory. J Chem Phys 154:234107. https://doi.org/10.1063/5.0049745
128. Wallace AM, Sherrill CD (2024) Optimization of damping functinos parameters for -D3 and -D4 dispersion models for Hartree–Fock based symmetry-adapted perturbation theory. J Chem Phys 161:114115. https://doi.org/10.1063/5.0219185
129. Grimme S, Antony J, Ehrlich S, Krieg H (2010) A consistent and accurate *ab initio* parameterization of density functional dispersion correction (DFT-D) for the 94 elements H–Pu. J Chem Phys 132:154104. https://doi.org/10.1063/1.3382344
130. Caldeweyher E, Bannwarth C, Grimme S (2017) Extension of the D3 dispersion coefficient model. J Chem Phys 147:034112. https://doi.org/10.1063/1.4993215
131. Grimme S (2006) Semiempirical GGA-type density functional constructed with a long-range dispersion correction. J Comput Chem 27:1787–1799. https://doi.org/10.1002/jcc.20495
132. Sedlak R, Janowski T, Pitoňák M, Řezáč J, Pulay P, Hobza P (2013) Accuracy of quantum chemical methods for large noncovalent complexes. J Chem Theory Comput 9:3364–3374. https://doi.org/10.1021/ct400036b
133. Goldey M, Head-Gordon M (2012) Attenuating away the errors in inter- and intramolecular interactions from second-order Møller–Plesset calculations in the small aug-cc-pVDZ basis set. J Phys Chem Lett 3:3592–3598. https://doi.org/10.1021/jz301694b
134. Dobson JF (2014) Beyond pairwise additivity in London dispersion interactions. Int J Quantum Chem 114:1157–1161. https://doi.org/10.1002/qua.24635
135. Tkatchenko A, Scheffler M (2009) Accurate molecular van der Waals interactions from ground-state electron density and free-atom reference data. Phys Rev Lett 102:073005. https://doi.org/10.1103/PhysRevLett.102.073005
136. von Lilienfeld OA, Tkatchenko A (2010) Two- and three-body interatomic dispersion energy contributions to binding in molecules and solids. J Chem Phys 132:234109. https://doi.org/10.1063/1.3432765
137. Proynov E, Liu F, Gan Z, Wang M, Kong J (2015) Density-functional approach to the three-body dispersion interaction based on the exchange dipole moment. J Chem Phys 143:084125. https://doi.org/10.1063/1.4929581
138. Al-Saidi WA, Voora VK, Jordan KD (2012) An assessment of the vdW-TS method for extended systems. J Chem Theory Comput 8:1503–1513. https://doi.org/10.1021/ct200618b

139. Chu X, Dalgarno A (2004) Linear response time-dependent density functional theory for van der Waals coefficients. J Chem Phys 121:4083–4088. https://doi.org/10.1063/1.1779576
140. De Proft F, Vivas-Reyes R, Peeters A, Van Alsenoy C, Geerlings P (2003) Hirshfeld partitioning of the electron density: Atomic dipoles and their relation with functional group properties. J Comput Chem 24:463–469. https://doi.org/10.1002/jcc.10241
141. Krishtal A, Senet P, Yan M, Van Alsenoy C (2006) A Hirshfeld partitioning of polarizabilities of water clusters. J Chem Phys 125:034312. https://doi.org/10.1063/1.2210937
142. Olasz A, Vanommeslaeghe K, Krishtal A, Vesszprémi T, Van Alsenoy C, Geerlings P (2007) The use of atomic intrinsic polarizabilities in the evaluation of the dispersion energy. J Chem Phys 127:224105. https://doi.org/10.1063/1.2805391
143. Brinck T, Murray JS, Politzer P (1993) Polarizability and volume. J Chem Phys 98:4305–4306. https://doi.org/10.1063/1.465038
144. Szabó P, Góger S, Charry J, Karimpour MR, Fedorov DV, Tkatchenko A (2022) Four-dimensional scaling of dipole polarizability in quantum systems. Phys Rev Lett 128:070602. https://doi.org/10.1103/PhysRevLett.128.070602
145. Longuet-Higgins HC (1965) Intermolecular forces. Discuss Faraday Soc 40:7–18. https://doi.org/10.1039/DF9654000007
146. Herbert JM (2021) Dielectric continuum methods for quantum chemistry. Wiley Interdiscip Rev Comput Mol Sci 11:e1519. https://doi.org/10.1002/wcms.1519
147. Ambrosetti A, Ferri N, DiStasio Jr. RA, Tkatchenko A (2016) Wavelike charge density fluctuations and van der Waals interactions at the nanoscale. Science 351:1171–1176. https://doi.org/10.1126/science.aae0509
148. Carter-Fenk K, Herbert JM (2020) Reinterpreting π-stacking. Phys Chem Chem Phys 22: 24870–24886. https://doi.org/10.1039/D0CP05039C
149. Bryenton KR, Johnson ER (2023) Many-body dispersion in model systems and the sensitivity of self-consistent screening. J Chem Phys 158:204110. https://doi.org/10.1063/5.0142465
150. Jurečka P, Šponer J, Černý J, Hobza P (2006) Benchmark database of accurate (MP2 and CCSD(T) complete basis set limit) interaction energies of small model complexes, DNA base pairs, and amino acid pairs. Phys Chem Chem Phys 8:1985–1993. https://doi.org/10.1039/B600027D
151. Flick JC, Kosenkov D, Hohenstein EG, Sherrill CD, Slipchenko LV (2012) Accurate prediction of noncovalent interaction energies with the effective fragment potential method: Comparison of energy components to symmetry-adapted perturbation theory for the S22 test set. J Chem Theory Comput 8:2835–2843. https://doi.org/10.1021/ct200673a
152. Grafova L, Pitonak M, Řezác J, Hobza P (2010) Comparative study of selected wave function and density functional methods for noncovalent interaction energy calculations using the extended S22 data set. J Chem Theory Comput 6:2365–2376. https://doi.org/10.1021/ct1002253
153. Zheng J, Xu X, Truhlar DG (2011) Minimally augmented Karlsruhe basis sets. Theor Chem Acc 128:295–305. https://doi.org/10.1007/s00214-010-0846-z
154. Ronca E, Belpassi L, Tarantelli F (2014) A quantitative view of charge transfer in the hydrogen bond: The water dimer case. ChemPhysChem 15:2682–2687. https://doi.org/10.1002/cphc.201402321
155. Stone AJ (2017) Natural bond orbitals and the nature of the hydrogen bond. J Phys Chem A 121:1531–1534. https://doi.org/10.1021/acs.jpca.6b12930
156. Mao Y, Ge Q, Horn PR, Head-Gordon M (2018) On the computational characterization of charge-transfer effects in noncovalently bound molecular complexes. J Chem Theory Comput 14:2401–2417. https://doi.org/10.1021/acs.jctc.7b01256
157. Taylor R (2024) Aerogen bond, halogen bond, chalcogen bond, pnictogen bond, tetrel bond, triel bond... why so many names? Crys Growth Des 24:4003–4012. https://doi.org/10.1021/acs.cgd.4c00303
158. Cavallo G, Metrangolo P, Milani R, Pilati T, Priimagi A, Resnati G, Terraneo G (2016) The halogen bond. Chem Rev 116:2478–2601. https://doi.org/10.1021/acs.chemrev.5b00484

159. Řezác J, de la Lande A (2017) On the role of charge transfer in halogen bonding. Phys Chem Chem Phys 19:791–803. https://doi.org/10.1039/C6CP07475H
160. Thirman J, Engelage E, Huber SM, Head-Gordon M (2018) Characterizing the interplay of Pauli repulsion, electrostatics, dispersion and charge transfer in halogen bonding with energy decomposition analysis. Phys Chem Chem Phys 20:905–915. https://doi.org/10.1039/C7CP06959F
161. Clark T, Heßelmann A (2018) The coulombic σ-hole model describes bonding in $CX_3I\cdots Y^-$ complexes completely. Phys Chem Chem Phys 20:22849–22855. https://doi.org/10.1039/C8CP03079K
162. Kellett CW, Kennepohl P, Berlinguette CP (2020) π covalency in the halogen bond. Nat. Commun. 11:3310. https://doi.org/10.1038/s41467-020-17122-7
163. Inscoe B, Rathnayake H, Mo Y (2021) Role of charge transfer in halogen bonding. J Phys Chem A 125:2944–2953. https://doi.org/10.1021/acs.jpca.1c01412
164. Murray JS, Politzer P (2021) Can counter-intuitive halogen bonding be Coulombic? ChemPhysChem 22:1201–1207. https://doi.org/10.1002/cphc.202100202
165. Mo Y, Danovich D, Shaik S (2022) The roles of charge transfer and polarization in non-covalent interactions: A perspective from ab initio valence bond methods. J Mol Model 28:274. https://doi.org/10.1007/s00894-022-05187-8
166. Feynman RP (1939) Forces in molecules. Phys Rev 56:340–343. https://doi.org/10.1103/PhysRev.56.340
167. Politzer P, Murray JS (2015) A unified view of halogen bonding, hydrogen bonding and other σ-hole interactions. In: Scheiner S (ed) Noncovalent forces. Challenges and Advances in Computational Chemistry and Physics, vol 19, chap 10. Springer International Publishing, Switzerland, pp 291–321. https://doi.org/10.1007/978-3-319-14163-3_10
168. Clark T, Politzer P, Murray JS (2015) Correct electrostatic treatment of noncovalent interactions: The importance of polarization. Wiley Interdiscip Rev Comput Mol Sci 5:169–177. https://doi.org/10.1002/wcms.1210
169. Clark T (2017) Polarization, donor–acceptor interactions, and covalent contributions in weak interactions: A clarification. J Mol Model 23:297. https://doi.org/10.1007/s00894-017-3473-y
170. Clark T, Murray JS, Politzer P (2018) A perspective on quantum mechanics and chemical concepts in describing noncovalent interactions. Phys Chem Chem Phys 20:30076–30082. https://doi.org/10.1039/c8cp06786d
171. Anderson PW (1972) More is different: Broken symmetry and the nature of the hierarchical structure of science. Science 177:393–396. https://doi.org/10.1126/science.177.4047.393
172. Thouless DJ (1960) Stability conditions and nuclear rotations in the Hartree-Fock theory. Nucl Phys 21:225–232. https://doi.org/10.1016/0029-5582(60)90048-1
173. Lao KU, Herbert JM (2016) Energy decomposition analysis with a stable charge-transfer term for interpreting intermolecular interactions. J Chem Theory Comput 12:2569–2582. https://doi.org/10.1021/acs.jctc.6b00155
174. Misquitta AJ (2013) Charge transfer from regularized symmetry-adapted perturbation theory. J Chem Theory Comput 9:5313–5326. https://doi.org/10.1021/ct400704a
175. Řezáč J, de la Lande A (2015) Robust, basis-set independent method for the evaluation of charge-transfer energy in noncovalent complexes. J Chem Theory Comput 11:528–537. https://doi.org/10.1021/ct501115m
176. Stone AJ, Misquitta AJ (2009) Charge-transfer in symmetry-adapted perturbation theory. Chem Phys Lett 473:201–205. https://doi.org/10.1016/j.cplett.2009.03.073
177. Kaduk B, Kowalczyk T, Van Voorhis T (2012) Constrained density functional theory. Chem Rev 112:321–370. https://doi.org/10.1021/cr200148b
178. Herbert JM, Carter-Fenk K (2021) Electrostatics, charge transfer, and the nature of the halide–water hydrogen bond. J Phys Chem A 125:1243–1256. https://doi.org/10.1021/acs.jpca.0c11356

179. Herbert JM, Paul SK (2021) Interaction energy analysis of monovalent inorganic anions in bulk water versus air/water interface. Molecules 26:6719. https://doi.org/10.3390/molecules26216719
180. Horn PR, Head-Gordon M (2015) Polarization contributions to intermolecular interactions revisited with fragment electric-field response functions. J Chem Phys 143:114111. https://doi.org/10.1063/1.4930534
181. Ochieng SA, Patkowski K (2023) Accurate three-body noncovalent interactions: The insights from energy decomposition. Phys Chem Chem Phys 25:28621–28637. https://doi.org/10.1039/D3CP03938B
182. van Meer R, Gritsenko OV, Baerends EJ (2014) Physical meaning of virtual Kohn–Sham orbitals and orbital energies: An ideal basis for the description of molecular excitations. J Chem Theory Comput 10:4432–4441. https://doi.org/10.1021/ct500727c
183. Rappoport D, Furche F (2010) Property-optimized Gaussian basis sets for molecular response calculations. J Chem Phys 133:134105. https://doi.org/10.1063/1.3484283
184. Sure R, Grimme S (2013) Corrected small basis set Hartree-Fock method for large systems. J Comput Chem 34:1672–1685. https://doi.org/10.1002/jcc.23317
185. Grimme S, Brandenburg JG, Bannwarth C, Hansen A (2015) Consistent structures and interactions by density functional theory with small atomic orbital basis sets. J Chem Phys 143:054107. https://doi.org/10.1063/1.4927476
186. Grimme S (2012) Supramolecular binding thermodynamics by dispersion-corrected density functional theory. Chem Eur J 18:9955–9964. https://doi.org/10.1002/chem.201200497
187. Sure R, Grimme S (2015) Comprehensive benchmark of association (free) energies of realistic host–guest complexes. J Chem Theory Comput 11:3785–3801. https://doi.org/10.1021/acs.jctc.5b00296
188. Mardirossian N, Head-Gordon M (2017) Thirty years of density functional theory in computational chemistry: An overview and extensive assessment of 200 density functionals. Mol Phys 115:2315–2372. https://doi.org/10.1080/00268976.2017.1333644
189. Lao KU, Schäffer R, Jansen G, Herbert JM (2015) Accurate description of intermolecular interactions involving ions using symmetry-adapted perturbation theory. J Chem Theory Comput 11:2473–2486. https://doi.org/10.1021/ct5010593
190. Gould T, Lebègue S, Àngyán JG, Bučko T (2016) A fractionally ionic approach to polarizability and van der Waals many-body dispersion calculations. J Chem Theory Comput 12: 5920–5930. https://doi.org/10.1021/acs.jctc.6b00925
191. Hermann J, Tkatchenko A (2020) Density functional model for van der Waals interactions: Unifying many-body atomic approaches with nonlocal functionals. Phys Rev Lett 124: 146401. https://doi.org/10.1103/PhysRevLett.124.146401
192. Hermann J, Stöhr M, Góger S, Chaudhuri S, Aradi B, Maurer RJ, Tkatchenko A (2023) libMBD: A general-purpose package for scalable quantum many-body dispersion calculations. J Chem Phys 159:174802. https://doi.org/10.1063/5.0170972
193. Bultinck P, Van Alsenoy C, Ayers PW, Carbó-Dorca R (2007) Critical analysis and extension of the Hirshfeld atoms in molecules. J Chem Phys 126:144111. https://doi.org/10.1063/1.2715563
194. Bučko T, Lebègue S, Hafner J, Ángyán JG (2013) Improved density dependent correction for the description of London dispersion forces. J Chem Theory Comput 9:4293–4299. https://doi.org/10.1021/ct400694h
195. Finzel K, Martín Pendás A, Francisco E (2015) Efficient algorithms for Hirshfeld-I charges. J Chem Phys 143:084115. https://doi.org/10.1063/1.4929469
196. Bučko T, Lebègue S, Ángyán JG, Hafner J (2014) Extending the applicability of the Tkatchenko-Scheffler dispersion correction via iterative Hirshfeld partitioning. J Chem Phys 141:034114. https://doi.org/10.1063/1.4890003
197. Hohenstein EG, Sherrill CD (2010) Density fitting and Cholesky decomposition approximations in symmetry-adapted perturbation theory: Implementation and application to probe the

nature of π-π interactions in linear acenes. J Chem Phys 132;184111. https://doi.org/10.1063/1.3426316
198. Hohenstein EG, Parrish RM, Sherrill CD, Turney JM, Schaefer III HF (2011) Large-scale symmetry-adapted perturbation theory computations via density fitting and Laplace transformation techniques: Investigating the fundamental forces of DNA-intercalator interactions. J Chem Phys 135:174107. https://doi.org/10.1063/1.3656681
199. Herbert JM, Head-Gordon M (2004) Curvy-steps approach to constraint-free extended-Lagrangian *ab initio* molecular dynamics, using atom-centered basis functions: Convergence toward Born-Oppenheimer trajectories. J Chem Phys 121:11542–11556. https://doi.org/10.1063/1.1814934
200. Gray M, Bowling PE, Herbert JM (2024) Comment on "Benchmarking basis sets for density functional theory thermochemistry calculations: Why unpolarized basis sets and the polarized 6-311G family should be avoided". J Phys Chem A 128:7739–7745. https://doi.org/10.1021/acs.jpca.4c00283
201. Paul SK, Herbert JM (2021) Probing interfacial effects on ionization energies: The surprising banality of anion–water hydrogen bonding at the air/water interface. J Am Chem Soc 143: 10189–10202. https://doi.org/10.1021/jacs.1c03131
202. Moran D, Simmonett AC, Leach FE, Allen WD, Schleyer PvR, Schaefer HF (2006) Popular theoretical methods predict benzene and arenes to be nonplanar. J Am Chem Soc 128:9342–9343. https://doi.org/10.1021/ja0630285
203. Grimme S, Steinmetz M, Korth M (2007) How to compute isomerization energies of organic molecules with quantum chemical methods. J Org Chem 72:2118–2126. https://doi.org/10.1021/jo062446p
204. Kruse H, Goerigk L, Grimme S (2012) Why the standard B3LYP/6-31G* model chemistry should not be used in DFT calculations of molecular thermochemistry: Understanding and correcting the problem. J Org Chem 77:10824–10834. https://doi.org/10.1021/jo302156p
205. Morgante P, Peverati R (2020) The devil in the details: A tutorial review on some undervalued aspects of density functional theory calculations. Int J Quantum Chem 120:e26332. https://doi.org/10.1002/qua.26332
206. Gavini V, Baroni S, Blum V, Bowler DR, Buccheri A, Chelikowsky JR, Das S, Dawson W, Delugas P, Dogan M, Draxl C, Galli G, Genovese L, Giannozzi P, Giantomassi M, Gonze X, Govoni M, Gygi F, Gulans A, Herbert JM, Kokott S, Kühne TD, Liou KH, Miyazaki T, Motamarri P, Nakata A, Pask JE, Plessl C, Ratcliff LE, Richard RM, Rossi M, Schade R, Scheffler M, Schütt O, Suryanarayana P, Torrent M, Truflandier L, Windus TL, Xu Q, Yu VWZ, Perez D (2023) Roadmap on electronic structure codes in the exascale era. Model Simul Mater Sci Eng 31:063301. https://doi.org/10.1088/1361-651X/acdf06
207. Stone AJ (2013) The Theory of Intermolecular Forces, 2nd edn. Oxford University Press, Oxford, UK. https://doi.org/10.1007/978-3-642-58183-0_4
208. Vigné-Maeder F, Claverie P (1988) The exact multicenter multipolar part of a molecular charge distribution and its simplified representations. J Chem Phys 88:4934–4948. https://doi.org/10.1063/1.454705
209. Stone AJ (1991) Classical electrostatics in molecular interactions. In: Maksić ZB (ed) Theoretical Treatment of Large Molecules and Their Interactions. Springer, Berlin, pp 103–131. https://doi.org/10.1007/978-3-642-58183-0_4
210. Kaplan IG (2006) Intermolecular Interactions: Physical Picture, Computational Methods and Model Potentials. Wiley, Chichester, UK. https://doi.org/10.1002/047086334X
211. Ryno SM, Risko C, Brédas JL (2016) Noncovalent interactions and impact of charge penetration effects in linear oligoacene dimers and single crystals. Chem Mater 28:3990–4000. https://doi.org/10.1021/acs.chemmater.6b01340
212. Podeszwa R, Bukowski R, Szalewicz K (2006) Potential energy surface for the benzene dimer and perturbational analysis of π–π interactions. J Phys Chem A 110:10345–10354. https://doi.org/10.1021/jp064095o

213. Sherrill CD (2013) Energy component analysis of π interactions. Acc Chem Res 46:1020–1028. https://doi.org/10.1021/ar3001124
214. Hohenstein EG, Duan J, Sherrill DC (2011) Origin of the surprising enhancement of electrostatic energies by electron-donating substituents in substituted sandwich benzene dimers. J Am Chem Soc 133:13244–13247. https://doi.org/10.1021/ja204294q
215. Carter-Fenk K, Herbert JM (2020) Electrostatics does not dictate the slip-stacked arrangement of aromatic π–π interactions. Chem Sci 11:6758–6765. https://doi.org/10.1039/D0SC02667K
216. Gray M, Herbert JM (2023) Origins of offset-stacking in porous frameworks. J Phys Chem C 127:2675–2686. https://doi.org/10.1021/acs.jpcc.2c08413
217. Schramm B, Gray M, Herbert JM (2025) Substituent and heteroatom effects on π–π interactions: Evidence that parallel-displaced π-stacking is not driven by quadrupolar electrostatics. J Am Chem Soc 147:3243–3260. https://doi.org/10.1021/jacs.4c13291
218. Franklin RE (1951) The structure of graphitic carbon. Acta Cryst 4:253–261. https://doi.org/10.1107/S0365110X51000842
219. Baldridge KK, Siegel JS (1997) Corannulene-based fullerene fragments $C_{20}H_{10}$–$C_{50}H_{10}$: When does a buckybowl become a buckytube? Theor Chem Acc 97:67–71. https://doi.org/10.1007/s002140050238
220. Scott LT, Bronstein HE, Preda DV, Ansems RBM, Bratcher MS, Hagen S (1999) Geodesic polyarenes with exposed concave surfaces. Pure Appl Chem 71:209–219. https://doi.org/10.1351/pac199971020209
221. Stuparu MC (2021) Corannulene: A curved polyarene building block for the construction of functional materials. Acc Chem Res 54:2858–2870. https://doi.org/10.1021/acs.accounts.1c00207
222. Hanson JC, Nordman CE (1976) The crystal and molecular structure of corannulene, $C_{20}H_{10}$. Acta Cryst B 32:1147–1153. https://doi.org/10.1107/S0567740876012430
223. Filatov AS, Scott LT, Petrukhina MA (2010) π–π interactions and solid state packing trends of polycyclic aromatic bowls in the indenocorannulene family: Predicting potentially useful bulk properties. Crys Growth Des 10:4607–4621. https://doi.org/10.1021/cg100898g
224. Chen R, Lu RQ, Shi PC, Cao XY (2016) Corannulene derivatives for organic electronics: From molecular engineering to applications. Chin Chem Lett 27:1175–1183. https://doi.org/10.1016/j.cclet.2016.06.033
225. Kennedy MR, Burns LA, Sherrill CD (2012) Buckyplates and buckybowls: Examining the effects of curvature on π–π interactions. J Phys Chem A 116:11920–11926. https://doi.org/10.1021/jp305700k
226. Lovas FJ, McMahon RJ, Grabow JU, Schnell M, Mack J, Scott LT, Kuczkowski RL (2005) Interstellar chemistry: A strategy for detecting polycyclic aromatic hydrocarbons in space. J Am Chem Soc 127:4345–4349. https://doi.org/10.1021/ja0426239
227. Kvashnin AG, Sorokin PB, Yakobson BI (2015) Flexoelectricity in carbon nanostructures: Nanotubes, fullerenes, and nanocones. J Phys Chem Lett 6:2740–2744. https://doi.org/10.1021/acs.jpclett.5b01041
228. Martin JW, Slavchov RI, Yapp EKY, Akroyd J, Mosbach S, Kraft M (2017) The polarizaton of polycyclic aromatic hydrocarbons curved by pentagon incorporation: The role of the flexoelectric dipole. J Phys Chem C 121:27154–17163. https://doi.org/10.1021/acs.jpcc.7b09044
229. Smith DW (1998) A simple molecular orbital treatment of the barrier to internal rotation in the ethane molecule. J Chem Educ 75:907–909. https://doi.org/10.1021/ed075p907
230. Zhou P, Huang J, Tian F (2012) Specific noncovalent interactions at protein–ligand interface: Implications for rational drug design. Curr Med Chem 19:226–238. https://doi.org/10.2174/092986712803414150
231. Sauers RR (1996) Steric attraction: The far side. J Chem Educ 73:114–16. https://doi.org/10.1021/ed073p114

232. Wolters LP, Koekkoek R, Bickelhaupt FM (2015) Role of steric attraction and bite-angle flexibility in metal-mediated C–H bond activation. ACS Catal 5:5766–5775. https://doi.org/10.1021/acscatal.5b01354
233. Gryn'ova G, Corminboeuf C (2018) Steric "attraction": Not by dispersion alone. Beilstein J Org Chem 14:1482–1490. https://doi.org/10.3762/bjoc.14.125
234. Holtrop F, Visscher KW, Jupp AR, Slootweg J (2020) Steric attraction: A force to be reckoned with. Adv Phys Org Chem 54:119–141. https://doi.org/10.1016/bs.apoc.2020.08.001
235. Hansen T, Sun X, Dalla Tiezza M, van Zeist WJ, van Stralen JNP, Geerke DP, Wolters LP, Poater J, Hamlin TA, Bickelhaupt FM (2022) C–X bond activation by palladium: Steric shielding versus steric attraction. Chem Eur J 28:e202201093. https://doi.org/10.1002/chem.202201093
236. Villot C, Ballesteros F, Wang D, Lao KU (2022) Coupled cluster benchmarking of large noncovalent complexes in L7 and S12L as well as the C_{60} dimer, DNA–ellipticine, and HIV–indinavir. J Phys Chem A 126:4326–4341. https://doi.org/10.1021/acs.jpca.2c01421
237. Ballesteros F, Dunivan S, Lao KU (2021) Coupled cluster benchmarks of large noncovalent complexes: The L7 dataset as well as DNA–ellipticine and buckycatcher–fullerene. J Chem Phys 154:154104. https://doi.org/10.1063/5.0042906
238. London F (1937) The general theory of intermolecular forces. Trans Faraday Soc 33:8–26. https://doi.org/10.1039/TF937330008B
239. Truhlar DG (2019) Dispersion forces: Neither fluctuating nor dispersing. J Chem Educ 96:1671–1675. https://doi.org/10.1021/acs.jchemed.8b01044
240. Sinnokrot MO, Sherrill CD (2004) Highly accurate coupled cluster potential energy curves for the benzene dimer: Sandwich, T-shaped, and parallel-displaced configurations. J Phys Chem A 108:10200–10207. https://doi.org/10.1021/jp0469517
241. Hohenstein EG, Sherrill DC (2009) Effects of heteroatoms on aromatic π–π interactions: Benzene–pyridine and pyridine dimer. J Phys Chem A 113:878–886. https://doi.org/10.1021/jp809062x
242. Geng Y, Takatani T, Hohenstein EG, Sherrill CD (2010) Accurately characterizing π–π interaction energies of indole–benzene complexes. J Phys Chem A 114:3576–3582. https://doi.org/10.1021/jp9099495
243. Parrish RM, Sherrill CD (2014) Quantum-mechanical evaluation of the π–π versus substituent–π interactions in π stacking: Direct evidence for the Wheeler–Houk picture. J Am Chem Soc 136:17386–17389. https://doi.org/10.1021/ja5101245
244. Wheeler SE, Houk KN (2008) Substituent effects in the benzene dimer are due to direct interactions of the substituents with the unsubstituted benzene. J Am Chem Soc 130:10854–10855. https://doi.org/10.1021/ja802849j
245. Wheeler SE, Houk KN (2009) Through-space effects of substituents dominate molecular electrostatic potentials of substituted arenes. J Chem Theory Comput 5:2301–2312. https://doi.org/10.1021/ct900344g
246. Wheeler SE, Houk KN (2009) Origin of substituent effects in edge-to-face aryl–aryl interactions. Mol Phys 107:749–760. https://doi.org/10.1080/00268970802537614
247. Wheeler SE (2011) Local nature of substituent effects in stacking interactions. J Am Chem Soc 133:10262–10274. https://doi.org/10.1021/ja202932e
248. Raju RK, Bloom JWG, Wheeler SE (2013) Broad transferability of substituent effects in π-stacking interactions provides new insights into their origin. J Chem Theory Comput 9:3479–3490. https://doi.org/10.1021/ct400481r
249. Wheeler SE, Bloom JWG (2014) Toward a more complete understanding of noncovalent interactions involving aromatic rings. J Phys Chem A 118:6133–6147. https://doi.org/10.1021/jp504415p
250. Bootsma AN, Doney AC, Wheeler SE (2019) Predicting the strength of stacking interactions between heterocycles and aromatic amino acid side chains. J Am Chem Soc 141:11027–11035. https://doi.org/10.1021/jacs.9b00936

251. Sinnokrot MO, Valeev EF, Sherrill CD (2002) Estimates of the ab initio limit for π-π interactions: The benzene dimer. J Am Chem Soc 124:10887–10893. https://doi.org/10.1021/ja025896h
252. Bludský O, Rubeš M, Soldán P, Nachtigall P (2008) Investigation of the benzene-dimer potential energy surface: DFT/CCSD(T) correction scheme. J Chem Phys 128:114102. https://doi.org/10.1063/1.2890968
253. Hunter CA, Lawson KR, Perkins J, Urch CJ (2001) Aromatic interactions. J Chem Soc Perkin Trans 2:651–669. https://doi.org/10.1039/b008495f
254. Fagnani DE, Sotuyo A, Castellano RK (2017) π–π interactions. In: Comprehensive Supramolecular Chemistry II, vol 1, chap 6. Elsevier, Oxford, pp 121–148. https://doi.org/10.1016/B978-0-12-409547-2.12485-0
255. Molčanov K, Kojić-Prodić B (2019) Towards understanding π-stacking interactions between non-aromatic rings. IUCrJ 6:156–166. https://doi.org/10.1107/S2052252519000186
256. Battaglia MR, Buckingham AD, Williams JH (1981) The electric quadrupole moments of benzene and hexafluorobenzene. Chem Phys Lett 78:421–423. https://doi.org/10.1016/0009-2614(81)85228-1
257. Hunter CA, Singh J, Thornton JM (1991) π–π interactions: The geometry and energetics of phenylalanine-phenylalanine interactions in proteins. J Mol Biol 218:837–846. https://doi.org/10.1016/0022-2836(91)90271-7
258. Carter-Fenk K, Liu M, Pujal L, Loipersberger M, Tsanai M, Vernon RM, Forman-Kay JD, Head-Gordon M, Heidar-Zadeh F, Head-Gordon T (2023) The energetic origins of pi–pi contacts in proteins. J Am Chem Soc 145:24836–24851. https://doi.org/10.1021/jacs.3c09198
259. Gung BW, Amicangelo JC (2006) Substituent effects in C_6F_6–C_6H_5X stacking interactions. J Org Chem 71:9261–9270. https://doi.org/10.1021/jo061235h
260. Tsuzuki S, Uchimaru T, Mikami M (2006) Intermolecular interaction between hexafluorobenzene and benzene: Ab initio calculations including CCSD(T) level electron correlation correction. J Phys Chem A 110:2027–2033. https://doi.org/10.1021/jp054461o
261. Wang W, Zhang Y, Wang YB (2017) Highly accurate benchmark calculations of the interaction energies in the complexes $C_6H_6 \cdots C_6X_6$ (X = F, Cl, Br, and I). Int J Quantum Chem 117:e25345. https://doi.org/10.1002/qua.25345
262. Gray CG, Gubbins KE (1984) Theory of Molecular Fluids, vol I. Oxford University Press, New York. https://doi.org/10.1093/oso/9780198556022.001.0001
263. Hunter CA, Sanders JKM (1990) The nature of π–π interactions. J Am Chem Soc 112:5525–5534. https://doi.org/10.1021/ja00170a016
264. Vinter JG (1994) Extended electron distribution applied to the molecular mechanics of some intermolecular interactions. J Comput Aided Mol Des 8:653–668. https://doi.org/10.1007/BF00124013
265. Carter-Fenk K, Lao KU, Herbert JM (2025) Correction to "Predicting and understanding noncovalent interactions using novel forms of symmetry-adapted perturbation theory". Acc Chem Res 58:1051–1053. https://doi.org/10.1021/acs.accounts.5c00107
266. Côté AP, Benin AI, Ockwig NW, O'Keeffe M, Matzger AJ, Yaghi OM (2005) Porous, crystalline, covalent organic frameworks. Science 310:1166–1170. https://doi.org/10.1126/science.1120411
267. Du Y, Calabro D, Wooler B, Li Q, Cundy S, Kamakoti P, Colmyer D, Mao K, Ravikovitch P (2014) Kinetic and mechanistic study of COF-1 phase change from a staggered to eclipsed model upon partial removal of mesitylene. J Phys Chem C 118:399–407. https://doi.org/10.1021/jp4097293
268. Gao J, Jiang D (2016) Covalent organic frameworks with spatially confined guest molecules in nanochannels and their impacts on crystalline structures. Chem Commun 52:1498–1500. https://doi.org/10.1039/C5CC09225F
269. Wonanke ADD, Addicoat MA (2022) Effect of unwanted guest molecules on the stacking configuration of covalent organic frameworks: A periodic energy decomposition analysis. Phys Chem Chem Phys 24:15494–15501. https://doi.org/10.1039/D2CP00017B

270. Lohse MS, Bein T (2018) Covalent organic frameworks: Structures, synthesis, and applications. Adv Funct Mater 28:1705553. https://doi.org/10.1002/adfm.201705553
271. Song Y, Sun Q, Aguila B, Ma S (2019) Opportunities of covalent organic frameworks for advanced applications. Adv Sci 6:1801410. https://doi.org/10.1002/advs.201801410
272. Birowska M, Milowska K, Majewski JA (2011) Van der Waals density functionals for graphene layers and graphite. Acta Phys Pol A 120:845–848. https://doi.org/10.12693/APhysPolA.120.845
273. Rozhkov AV, Sboychakov AO, Rakhmanov AL, Nori F (2016) Electronic properties of graphene-based bilayer systems. Phys Rep 648:1–104. https://doi.org/10.1016/j.physrep.2016.07.003
274. Lukose B, Kuc A, Frenzel J, Heine T (2010) On the reticular construction concept of covalent organic frameworks. Beilstein J Nanotechnol 1:60–70. https://doi.org/10.3762/bjnano.1.8
275. Spitler EL, Koo BT, Novotney JL, Colson JW, Uribe-Romo FJ, Gutierrez GD, Clancy P, Dichtel WR (2011) A 2D covalent organic framework with 4.7-nm pores and insight into its interlayer stacking. J Am Chem Soc 133:19416–19421. https://doi.org/10.1021/ja206242v
276. Lukose B, Kuc A, Heine T (2011) The structure of layered covalent-organic frameworks. Chem Eur J 17:2388–2392. https://doi.org/10.1002/chem.201001290
277. Koo BT, Dichtel WR, Clancy P (2012) A classification scheme for the stacking of two-dimensional boronate ester-linked covalent organic frameworks. J Mater Chem 22: 17460–17469. https://doi.org/10.1039/C2JM32009F
278. Haase F, Gottschling K, Stegbauer L, Germann LS, Gutzler R, Duppel V, Vyas VS, Kern K, Dinnebier RE, Lotsch BV (2017) Tuning the stacking behaviour of a 2D covalent organic framework through non-covalent interactions. Mater Chem Front 1:1354–1361. https://doi.org/10.1039/c6qm00378h
279. Wu X, Han X, Liu Y, Cui Y (2018) Control interlayer stacking and chemical stability of two-dimensional covalent organic frameworks via steric tuning. J Am Chem Soc 140:16124–16133. https://doi.org/10.1021/jacs.8b08452
280. Geng K, He T, Liu R, Dalapati S, Tan KT, Li Z, Tao S, Gong Y, Jiang Q, Jiang D (2020) Covalent organic frameworks: Design, synthesis, and functions. Chem Rev 120:8814–8933. https://doi.org/10.1021/acs.chemrev.9b00550
281. Grimme S (2008) Do special noncovalent π–π stacking interactions really exist? Angew Chem Int Ed Engl 47:3430–3434. https://doi.org/10.1002/anie.200705157
282. Martinez CR, Iverson BL (2012) Rethinking the term "pi-stacking". Chem Sci 3:2191–2201. https://doi.org/10.1039/C2SC20045G
283. Ninković DB, Blagojević Filipović JP, Hall MB, Brothers EN, Zarić SD (2020) What is special about aromatic–aromatic interactions? Significant attraction at large horizontal displacement. ACS Cent Sci 6:420–425. https://doi.org/10.1021/acscentsci.0c00005
284. Janowski T, Pulay P (2012) A benchmark comparison of σ/σ and π/π dispersion: The dimers of naphthalene and decalin, and coronene and perhydrocoronene. J Am Chem Soc 134:17520–17525. https://doi.org/10.1021/ja303676q
285. Fokin AA, Gerbig D, Schreiner PR (2011) σ/σ- and π/π-interactions are equally important: Multilayered graphanes. J Am Chem Soc 133:20036–20039. https://doi.org/10.1021/ja206992j
286. Boto RA, Contreras-García J, Tierny J, Piquemal JP (2016) Interpretation of the reduced density gradient. Mol Phys 114:1406–1414. https://doi.org/10.1080/00268976.2015.1123777
287. Johnson ER, Keinan S, Mori-Sánchez P, Contreras-García J, Cohen AJ, Yang W (2010) Revealing noncovalent interactions. J Am Chem Soc 132:6498–6506. https://doi.org/10.1021/ja100936w
288. Contreras-García J, Johnson ER, Keinan S, Chaudret R, Piquemal JP, Beratan DN, Yang W (2011) NCIPLOT: A program for plotting noncovalent interaction regions. J Chem Theory Comput 7:625–632. https://doi.org/10.1021/ct100641a

289. Otero-de-la-Roza A, Johnson ER, Contreras-García J (2012) Revealing non-covalent interactions in solids: NCI plots revisited. Phys Chem Chem Phys 14:12165–12172. https://doi.org/10.1039/C2CP41395G
290. Contreras-García J, Calatayud M, Piquemal JP, Recio JM (2012) Ionic interactions: Comparative topological approach. Comput Theor Chem 998:193–201. https://doi.org/10.1016/j.comptc.2012.07.043
291. Boto RA, Peccati F, Laplaza R, Quan C, Carbone A, Piquemal JP, Maday Y, Contreras-García J (2020) NCIPLOT4: Fast, robust, and quantitative analysis of noncovalent interactions. J Chem Theory Comput 16:4150–4158. https://doi.org/10.1021/acs.jctc.0c00063
292. Corminboeuf C, von Ragué Schleyer P, Warner P (2007) Are antiaromatic rings stacked face-to-face aromatic? Org Lett 9:3263–3266. https://doi.org/10.1021/ol071183y
293. Björk J, Hanke F, Palma CA, Samori P, Cecchini M, Persson M (2010) Adsorption of aromatic and anti-aromatic systems on graphene through π–π stacking. J Phys Chem Lett 1:3407–3412. https://doi.org/10.1021/jz101360k
294. Bloom JWG, Wheeler SE (2011) Taking the aromaticity out of aromatic interactions. Angew Chem Int Ed Engl 50:7847–7849. https://doi.org/10.1002/anie.201102982
295. Vernon RM, Chong PA, Tsang B, Kim TH, Bah A, Farber P, Lin H, Forman-Kay JD (2018) Pi-pi contacts are an overlooked protein feature relevant to phase separation. eLife 7:e31486. https://doi.org/10.7554/eLife.31486
296. Ayotte P, Weddle GH, Kim J, Johnson MA (1998) Vibrational spectroscopy of the ionic hydrogen bond: Fermi resonances and ion–molecule stretching frequencies in the binary $X^- \cdot H_2O$ (X = Cl, Br, I) complexes via argon predissociation spectroscopy. J Am Chem Soc 120: 12361–12362. https://doi.org/10.1021/ja981979f
297. Robertson WH, Johnson MA (2003) Molecular aspects of halide ion hydration: The cluster approach. Annu Rev Phys Chem 54:173–213. https://doi.org/10.1146/annurev.physchem.54.011002.103801
298. Roscioli JR, Diken EG, Johnson MA, Horvath S, McCoy AB (2006) Prying apart a water molecule with anionic H-bonding: A comparative spectroscopic study of the $X^- \cdot H_2O$ (X = OH, O, F, Cl, and Br) binary complexes in the 600–3800 cm^{-1} region. J Phys Chem A 110: 4943–4952. https://doi.org/10.1021/jp056022v
299. Thompson WH, Hynes JT (2000) Frequency shifts in the hydrogen-bonded OH stretch in halide–water clusters. The importance of charge transfer. J Am Chem Soc 122:6278–6286. https://doi.org/10.1021/ja993058q
300. Herbert JM, Head-Gordon M (2006) Charge penetration and the origin of large O–H vibrational red-shifts in hydrated-electron clusters, $(H_2O)_n^-$. J Am Chem Soc 128:13932–13939. https://doi.org/10.1021/ja064949i
301. Sharpe AG (1990) The solvation of halide ions and its chemical significance. J Chem Educ 67: 309–315. https://doi.org/10.1021/ed067p309
302. Antalek M, Pace E, Hedman B, Hodgson KO, Chillemi G, Benfatto M, Sarangi R, Frank P (2016) Solvation structure of the halides from x-ray absorption spectroscopy. J Chem Phys 145:044318. https://doi.org/10.1063/1.4959589
303. Brown TL, LeMay, Jr. HE, Bursten BE, Murphy CJ, Woodward PM, Stoltzfus MW: Chemistry: The Central Science, 13th edn. Pearson, Boston (2015)
304. Arunan E, Desiraju GR, Klein RA, Sadlej J, Scheiner S, Alkorta I, Clary DC, Crabtree RH, Dannenberg JJ, Hobza P, Kjaergaard HG, Legon AC, Mennucci B, Nesbitt DJ (2011) Defining the hydrogen bond: An account (IUPAC technical report). Pure Appl Chem 83:1619–1637. https://doi.org/10.1351/PAC-REP-10-01-01
305. Arunan E, Desiraju GR, Klein RA, Sadlej J, Scheiner S, Alkorta I, Clary DC, Crabtree RH, Dannenberg JJ, Hobza P, Kjaergaard HG, Legon AC, Mennucci B, Nesbitt DJ (2011) Definition of the hydrogen bond (IUPAC recommendations 2011). Pure Appl Chem 83:1637–1641. https://doi.org/10.1351/PAC-REC-10-01-02

306. Petersen PB, Saykally RJ (2004) Confirmation of enhanced anion concentration at the liquid water surface. Chem Phys Lett 397:51–55. https://doi.org/10.1016/j.cplett.2004.08.049
307. Petersen PB, Saykally RJ, Mucha M, Jungwirth P (2005) Enhanced concentration of polarizable anions at the liquid water surface: SHG spectroscopy and MD simulations of sodium thiocyanide. J Phys Chem B 109:10915–10921. https://doi.org/10.1021/jp050864c
308. Mucha M, Frigato T, Levering LM, Allen HC, Tobias DJ, Dang LX, Jungwirth P (2005) Unified molecular picture of the surfaces of aqueous acid, base, and salt solutions. J Phys Chem B 109:7617–7623. https://doi.org/10.1021/jp0445730
309. Ghosal S, Hemminger JC, Bluhm H, Mun BS, Hebenstreit ELD, Ketteler G, Olgetree DF, Requejo FG, Salmeron M (2005) Electron spectroscopy of aqueous solution interfaces reveals surface enhancement of halides. Science 307:563–566. https://doi.org/10.1126/science.1106525
310. Petersen PB, Saykally RJ (2006) On the nature of ions at the liquid water surface. Annu Rev Phys Chem 57:333–364. https://doi.org/10.1146/annurev.physchem.57.032905.104609
311. Otten DE, Shaffer PR, Geissler PL, Saykally RJ (2012) Elucidating the mechanism of selective ion adsorption to the liquid water surface. Proc Natl Acad Sci USA 109:701–705. https://doi.org/10.1073/pnas.1116169109

Struct Bond (2026) 190: 223–262
https://doi.org/10.1007/430_2025_99

Published online: 15 July 2025

A Survey of Contemporary Applications of Quantum Chemical Topology

Daniel Barrena-Espés, Fernando Jiménez-Grávalos, Julen Munárriz, and Ángel Martín Pendás

Contents

Abstract This chapter explores contemporary applications of quantum chemical topology, a theoretical framework that enables the real-space analysis of electronic structures from a chemically meaningful perspective. Among the partitioning methods discussed, the quantum theory of atoms in molecules (QTAIM) defines atomic basins based on electron density gradients, while the electron localization function (ELF) identifies Lewis entities such as atomic cores, bonds, and lone pairs.

D. Barrena-Espés and Á. Martín Pendás (✉)
Departamento de Química Física y Analítica, Universidad de Oviedo, Oviedo, Spain
e-mail: barrenadaniel@uniovi.es; ampendas@uniovi.es

F. Jiménez-Grávalos
Departamento de Química, Universidad de La Rioja, Logroño, Spain
e-mail: fernando.jimenez@unirioja.es

J. Munárriz (✉)
Departamento de Química Física and Instituto de Biocomputación y Física de Sistemas Complejos, Universidad de Zaragoza, Zaragoza, Spain
e-mail: julen@unizar.es

In addition, the QTAIM-ELF intersection (QEI) is introduced, which further refines these partitions by attributing Lewis entities to atomic contributions, providing a highly detailed view of molecular systems. These partitions are integrated with two key interpretative frameworks: electron distribution functions (EDFs) and the interacting quantum atoms (IQA) energy decomposition scheme. EDFs offer a statistical perspective on chemical bonding by quantifying the probability of electron distributions within predefined real-space regions (derived from QTAIM, ELF, or QEI). Meanwhile, IQA decomposes molecular energy into contributions from individual atomic regions and their interactions, though in this chapter, it is applied exclusively with QTAIM. Finally, we demonstrate the utility of these tools through applications to chemically relevant systems, including pyramidanes, halogen bonds, and the analysis of bond polarity and electron distribution in various molecules.

Keywords Chemical bonding · QTAIM · Quantum chemical topology · Real-space analyses

1 Introduction

Theoretical chemistry has successfully permeated all chemical disciplines. Thanks to the simultaneous advance in the efficiency of methods, algorithms, and computational hardware, the electronic structure of gas-phase molecules and condensed phases can now be simulated with relative ease and reasonable accuracy. On top of this scenario, the advent of accelerated techniques based on machine learning or quantum computing promises a new paradise for chemical theory. However, since Hilbert space is not aligned with conventional chemical concepts, the interpretation of any electronic structure calculation remains a challenge, typically resting on ill-defined assumptions and approximations. This has led to a range of uncoupled interpretative languages associated with the different computational paradigms developed so far, including the valence bond (VB) method inherited from Heitler, London, and Pauling [1], or the molecular orbital (MO) theory of Mulliken and coworkers [2–4]. Although often considered mutually exclusive [5], these methods must necessarily converge to the same description for any system as their wavefunctions approach completeness [6–8]. However, although this must happen with the energy, it does not necessarily occur with the interpretations themselves, which may still diverge in this limit [9, 10].

It is our basic tenet that the chemical insight extracted from a system described by a certain wavefunction, Ψ, regardless of the procedure that has led to its construction, must be unequivocal and stable. This requirement imposes severe restrictions on the type of descriptors that we are allowed to use to build chemical interpretations. A possible solution lies in using objects, be them scalar or vector in nature, that remain invariant under orbital transformations that leave the wavefunction intact [11]. Moreover, and although the following is not truly necessary, we expect these objects to live in the physical 3D space in which chemists typically think, and not in, e.g., momentum space.

Over the years, quite a few real-space techniques have been developed that use the so-called reduced density matrices. This makes them inherently invariant under orbital transformations. After several decades, there is no doubt that they can offer appealing chemical interpretations, and after an influential work by Paul Popelier [12], they are now gathered under the so-called quantum chemical topology (QCT) umbrella. Although abandoning Hilbert space has many advantages, it has also a clear drawback, since the atoms that somehow persist when we use atom-centered basis sets have now faded. Provided that no chemical interpretation is possible without the atomic or functional group concept, QCT has devised clever methods to partition the space into discrete entities—such as atoms, electron pairs, or charge accumulation regions—based on the mathematical properties of the quantities that it employs.

In addition to the aforementioned diversity of entities, there is another necessary element in a chemical description that is lacking in any pure spatial partitioning. Chemistry being the science of atoms, electron pairs, or whatever entitites we decide to use, they must *interact* with each other. Only in this way will we have access to bonding or reactivity. Once this idea is accepted, we can build pair descriptors on top of a given partitioning. These are typically energetic, although they can also be based on electron counting, and lead us directly to the field of intermolecular interactions [13]. Pair descriptors can immediately be used to classify interactions, starting from a stage in which two distinct entities coexist without interacting until they reach a given descriptor threshold. QCT techniques can thus be equally applied to both strong and weak (e.g., non-covalent) interactions, something that cannot be easily done with other methodologies [14].

2 Real-Space Partitions

The utility of QCT lies in its ability to efficiently provide chemical information. As stated previously, several scalar or vector fields have been employed to partition the R^3 space over the years, leading to decompositions associated with chemically meaningful objects such as atoms, functional groups, or electron pairs.

A general approach to deal with spatial partitionings starts by considering a set of weight functions, $\omega_{\mathrm{A}}(\mathbf{r})$, that we associate with each of the chemical objects, A, that the partition leads to. They can be understood as the share that each of these objects has at any given point in space, and thus must fulfill the following completeness relation:

$$\sum_{\mathrm{A}} \omega_{\mathrm{A}}(\mathbf{r}) = 1, \tag{1}$$

at every **r** point of R^3 [15]. If the weights are continuous functions, we talk about a fuzzy decomposition, and the different objects interpenetrate each other. Notice that we expect ω_A to be large in the vicinity of A and decay to zero as we move away from it. In contrast, non-fuzzy or exhaustive partitions are defined by the rule that each point in space only belongs to a single chemical object. Consequently, for these cases, Eq. 1 simplifies to $\omega_{\mathrm{A}}(\mathbf{r}) = 1$ for a given object A, while $\omega_{\mathrm{B}}(\mathbf{r}) = 0$ for all the others at each **r**.

Several fuzzy partitionings have been proposed. For instance, Becke introduced an atomic partition based on geometrical criteria that provides some controllable fuzziness onto the otherwise exhaustive division provided by the Voronoi decomposition [16]. The size of atoms can be further tuned by using different types of atomic radii, giving rise to several possibilities known as fuzzy Voronoi polyhedra [11, 17]. Other possibilities include the Hirshfeld or stockholder partition [18]. In its simplest version, the weights are obtained from the share of the total promolecular electron density at a given point with respect to a reference atomic value, typically taken as that of the isolated in vacuo electron density ρ^0_{A}.

The remainder of this discussion will focus on non-fuzzy partitions. It is clear that the simplest possibility is the use of Voronoi cells, exclusively based on geometrical criteria. Each point in space is associated with its closest atomic nucleus. Since different atoms should display different sizes, the planes that separate the different Voronoi cells can be modified to reflect an appropriate atomic size [19]. However, the simplest Voronoi partition is devoid of chemical meaning [20], and, indeed, other approaches have surpassed it in terms of completeness, applicability, and success [20–23]. We will therefore not consider it further since, moreover, by far the most successful procedure to build an atomic spatial partition uses the electron density, $\rho(\mathbf{r})$. Its physical meaning is rooted in the Born probability density [24], which is a Dirac observable,

$$\rho(\mathbf{x}) = \langle \psi | \sum_{i=1}^{N} \delta(\mathbf{x} - \mathbf{x}_i) | \psi \rangle . \tag{2}$$

In this expression, the sum runs over the N electrons of the system. If we are not interested in spin, we can obtain a spinless density by integrating out the spin component of the reference electron, $d\mathbf{x}$, and arrive at

$$\rho(\mathbf{r}) = N \int \psi(\mathbf{x}, \mathbf{x}_2, \ldots, \mathbf{x}_N) \psi^*(\mathbf{x}, \mathbf{x}_2, \ldots, \mathbf{x}_N) ds d\mathbf{x}_2 \ldots d\mathbf{x}_N . \tag{3}$$

As it is well known from density matrix theory, the electron density is just the first of a set of important objects, the n-th-order densities. The next is the pair density, which provides the probability density of finding pairs of electrons. If we are again not interested in spin, we can easily generalize Eq. 3, and

$$\rho_2(\mathbf{r}_i, \mathbf{r}_j) = N(N-1) \int \psi(\mathbf{x}_i, \mathbf{x}_j, \ldots, \mathbf{x}_N)\psi^*(\mathbf{x}_i, \mathbf{x}_j, \ldots, \mathbf{x}_N) d\mathbf{s}_i d\mathbf{s}_j d\mathbf{x}_2 \ldots d\mathbf{x}_N. \quad (4)$$

Although we will not use them, this magnitude, as the electron density, can be decomposed into the different spin distributions. In this case, we would have $\alpha\alpha$, $\beta\beta$, $\alpha\beta$, and $\beta\alpha$ terms. Notice that the pair density that we will use is normalized to the number of ordered pairs, $N(N-1)$, although there are other possibilities.

When the coordinates of the non-integrate electrons in ψ and ψ^* are different, instead of the nth-order densities (nDs), also known as reduced densities (nRDs), we come to the density matrices themselves (nDMs or nRDMs). This way, the spinless 1RDM would be given by

$$\rho_1(\mathbf{r}; \mathbf{r}') = N \iint \psi(\mathbf{x}, \ldots, \mathbf{x}_N)\psi^*(\mathbf{x}, \ldots, \mathbf{x}_N) d\mathbf{s} d\mathbf{s}' d\mathbf{x}_3 \ldots d\mathbf{x}_N. \quad (5)$$

The general (spin-full) nRDM is, in our normalization convention, defined as

$$\rho_n(\mathbf{x}_1, \ldots, \mathbf{x}_n; \mathbf{x}_1', \ldots, \mathbf{x}_n') = \frac{N!}{(N-n)!} \int \psi(\mathbf{x}_1, \ldots, \mathbf{x}_N)\psi^*(\mathbf{x}_1', \ldots, \mathbf{x}_N') d\mathbf{x}_{1+n} \ldots d\mathbf{x}_N. \quad (6)$$

It can be shown that the expectation value of any n-electron operator can be obtained from the information contained in the nRDM directly, without the need of the full wavefunction. We also note, in passing, that the nRDs are just the diagonal elements of the nRDMs. For instance, $\rho_1(\mathbf{r}; \mathbf{r}) = \rho(\mathbf{r})$ if we take the 1-RDM, or $\rho_2(\mathbf{r}_1, \mathbf{r}_2; \mathbf{r}_1, \mathbf{r}_2) = \rho_2(\mathbf{r}_1, \mathbf{r}_2)$.

The idea of using the electron density as a topological tool to obtain an exhaustive partition of the space is the central tenet of the quantum theory of atoms in molecules (QTAIM) developed by Richard F. W. Bader [25]. Since Kato's cusp theorem [26] shows that the electron density has necessarily a cusp (which can be assimilated to a maximum in topology) at the nuclear positions, the attraction basins of the electron density scalar field will typically divide the space in as many disjoint regions as atoms. Using these insights, it thus becomes possible to find partitions of space for any chemically relevant scalar field. For instance, the electron localization function (ELF), developed by Becke and Edgecombe [27], has maxima outside nuclei related to localized electron pairs and has proven to be an effective means of relating chemical thinking with orbital invariant reasoning [21, 28, 29]. Within this approach, which will be reviewed in the following subsections, the different regions in which space is divided are associated with electron pairs—such as core regions, lone pairs, and chemical bonds [21]. It is worth noting that there are numerous additional scalar fields, such as the electron localizability indicator (ELI) [30], or the localized orbital locator (LOL) [31], which will not be covered here but are worth mentioning, and the revision of which is recommended to interested readers.

2.1 Atoms in Molecules

The topological analysis of a given scalar field starts by building its gradient and studying the points where it vanishes, the so-called critical points (CPs). The number and type of these CPs allow for a thorough characterization of the properties of the field, since all its gradient lines will start or end at these specific points and never intersect with each other outside CPs. Grossly speaking, when a 3D bundle of gradient lines starts or ends up at a given CP, the latter will have a 3D basin associated with it. This is the case for maxima or minima, so that any scalar can induce two complementary partitions, one based on its maxima and the other on its minima [12].

Let us particularize to the electron density, $\rho(\mathbf{r})$, and build its gradient,

$$\nabla\rho(\mathbf{r}) = \frac{\partial\rho(\mathbf{r})}{\partial x}\boldsymbol{i} + \frac{\partial\rho(\mathbf{r})}{\partial y}\boldsymbol{j} + \frac{\partial\rho(\mathbf{r})}{\partial z}\boldsymbol{k}. \tag{7}$$

A CP $\mathbf{r}_c$ satisfies $\nabla\rho(\mathbf{r}_c) = \mathbf{0}$ and can be classified by means of its Hessian matrix,

$$H = \begin{pmatrix} \frac{\partial^2\rho(\mathbf{r})}{\partial x^2} & \frac{\partial^2\rho(\mathbf{r})}{\partial x\partial y} & \frac{\partial^2\rho(\mathbf{r})}{\partial x\partial z} \\ \frac{\partial^2\rho(\mathbf{r})}{\partial y\partial x} & \frac{\partial^2\rho(\mathbf{r})}{\partial y^2} & \frac{\partial^2\rho(\mathbf{r})}{\partial y\partial z} \\ \frac{\partial^2\rho(\mathbf{r})}{\partial z\partial x} & \frac{\partial^2\rho(\mathbf{r})}{\partial z\partial y} & \frac{\partial^2\rho(\mathbf{r})}{\partial z^2} \end{pmatrix}, \tag{8}$$

which is the real, symmetric matrix of its second derivatives. The local behavior of the field around a CP is contained in H. If we rotate the reference system such that the matrix is diagonal, i.e., if we diagonalize the matrix, then the three diagonal curvatures or eigenvalues λ_i of H can be either positive, negative, or zero. Excluding the latter case, which is typically unstable and of no chemical interest at general molecular geometries, we have thus several possibilities according to the number of positive and negative eigenvalues, namely, the number of directions in which the field behaves as a minimum or a maximum, respectively. If we count the number of positive (p) and negative (n) eigenvalues, so that for the nondegenerate case here considered the rank of the CP $r = p + n = 3$, the so-called signature, $\sigma = p - n$, can display four different values: -3, -1, $+1$, $+3$. Four distinct types of CPs thus appear that are typically labeled as (r, σ) CPs. In the case where all the eigenvalues are negative, a $(3, -3)$ CP, we have a maximum in all three directions. Since this is what must happen at nuclei (disregarding the fact that the density is not differentiable at cusps), we call these CPs nuclear critical points, NCPs. The bunch of gradient lines that converge onto a given atom A form its 3D attraction basin, Ω_A. Practically speaking, computed electron densities never satisfy the cusp condition, particularly when Gaussian basis sets are used. Moreover, in the latter case, there is no cusp at all, and the density is differentiable at all points. Thus nuclei are marked by true $(3, -3)$ CPs, which may lie slightly offsite, as it usually happens with H atoms. In very rare cases, nonnuclear maxima may also appear. They have been much debated in the literature and may have genuine physical meaning as in the case of localized electrides [32–34], but we will not consider them here.

The next type of CP that we may encounter in a molecular system is a $(3, -1)$ point or first-order saddle, which displays two maxima and one minimum directions. These are commonly known as bond critical points (BCPs) and usually appear between particular pairs (A, B) of $(3, -3)$ 3D attractors (which correspond to atomic nuclei). In this case, the two gradient lines that emerge from the BCP along its positive λ eigenvector flow toward these two different attractors, A and B, respectively, drawing the so-called atomic interaction line or bond path. Traditionally, they are associated with chemical bonds, although this has been much debated. Notice that the set of gradient lines that end up at a BCP form a two-dimensional region that separates the attraction basins of A and B. This is called the $A - B$ interatomic surface and exhibits the so-called local zero-flux condition: $\nabla\rho(\mathbf{r})\cdot\mathbf{n}(\mathbf{r}) = 0$, at all the points, r belonging to the surface. **n** is the exterior normal to the surface at each point.

The opposite case would be given by CPs displaying a single positive eigenvalue, $(3, +1)$: these are called ring critical points (RCP) and typically arise close to the center of rings formed by several BCPs. The set of gradient lines starting in the plane formed by their two positive eigenvectors can be used to obtain a two-dimensional region that defines the ring. Finally, the last type of nondegenerate CP we may find is that in which its three λ's are positive. This is a minimum, $(3, +3)$, or cage critical point (CCP), that tends to appear whenever a set of RCPs encloses a 3D region. This cage-like domain can be recovered by building the 3D bundle of gradient lines that end at the CCP.

As an example, all the previous cases of CPs are shown in Fig. 1 for the $Si[C_4H_4]$ pyramidane-like structure [35, 36]. Besides the nuclei, we find the BCPs associated with each chemical bond. It is noteworthy that the BCPs between atoms with the same Z_A are typically situated at the midpoint of the bond path, as long as their atomic environment is equivalent. On the contrary, when the identity of the nuclei and/or their chemical environment differs significantly, the BCP position is considerably displaced, giving information about the local size of each atom along the interatomic direction. The former case can be illustrated by the C–C bonds in the base of the pyramidane, while the latter is exemplified by the C–H and Si–C bonds. Notice that the topological charge of the Si atom is rather large, about $+1$ au, so that its effective size is smaller than that of the C atom. Regarding the rest of the CP typologies, one RCP is found on each face of the pyramidane structure (depicted in red), and a single CCP is located within the cage formed by the pyramidane-like structure (shown in blue).

The number and type of critical points within a system are determined by the Poincaré-Hopf or Morse relationships, which establish different rules depending on mathematical features of the space in which we embed the ρ scalar field. In the case

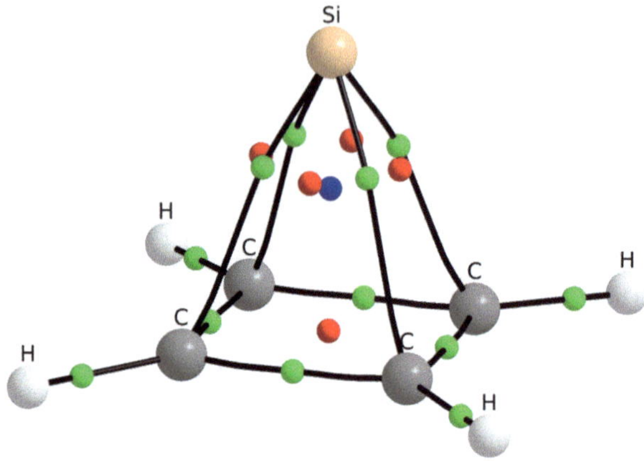

Fig. 1 Critical points of the $\rho(\mathbf{r})$ field for the Si-substituted [4]-pyramidane, $Si[C_4H_4]$. Structure obtained at B3LYP-D3BJ/def2-TZVP level. CPs are represented by colored spheres: green, BCPs; red, RCPs; blue, CCP. NCPs are aligned with the nuclear coordinates of all atoms. The topological analysis has been performed with AIMAll package [37]

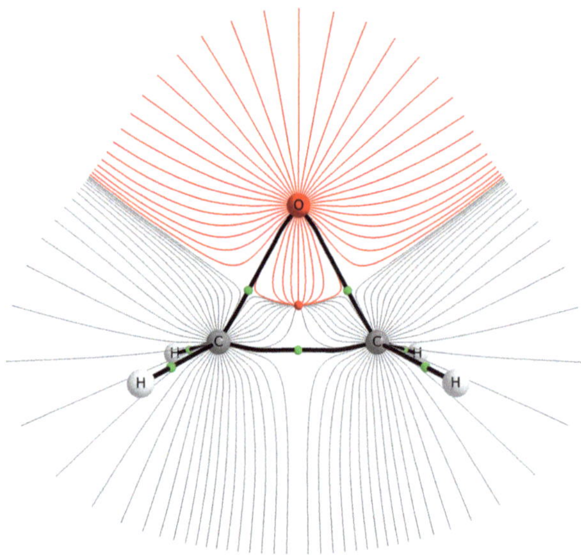

Fig. 2 Gradient lines of $\rho(\mathbf{r})$ for the oxirane molecule in the C-O-C plane. Structure obtained at B3LYP-D3BJ/def2-TZVP level. Green spheres: BCPs; red spheres: RCPs. NCPs match nuclear coordinates. Cutoff for $\rho(\mathbf{r}) = 0.001$ a.u. The figure has been reproduced with AIMAll [37]

of isolated molecules in R^3, $n^{\text{NCP}} - n^{\text{BCP}} + n^{\text{RCP}} - n^{\text{CCP}} = 1$. In a periodic system, however, the condition changes to $n^{\text{NCP}} - n^{\text{BCP}} + n^{\text{RCP}} - n^{\text{CCP}} = 0$.

An example of the behavior of the gradient field $\nabla\rho$ is provided in Fig. 2 for the oxirane molecule. As previously mentioned, the gradient lines end up at one of the possible NCPs and divide exhaustively the space into atomic basins that fill the whole R^3. As illustrated, these gradient lines never cross the boundary of the interatomic surfaces where the BCPs are situated.

The ability of the quantum theory of atoms in molecules (QTAIM) to partition R^3 into space-filling atomic regions allows also for a decomposition of the expectation value of any quantum mechanical operator into basin (atomic) contributions. This is immediately understood as soon as one recalls that in the Schrödinger representation these expectation values are obtained as integrals where the electron coordinates of all the electrons run over the whole space. If the space is now divided into a set of atomic-like regions, each integral will be the sum of atomic terms. For instance, given a spinless, symmetric, N-electron operator, $\hat{O} = \sum_{i_1 < \ldots < i_N} \hat{o}(\mathbf{r}_{i_1}, \ldots, \mathbf{r}_{i_N})$, its expectation value would be given by

$$\langle O \rangle = \frac{1}{n!} \int \hat{o} \rho_n(\mathbf{x}_1, \ldots, \mathbf{x}_n; \mathbf{x}'_1, \ldots, \mathbf{x}'_n) d\mathbf{x}_1 \ldots d\mathbf{x}_n, \tag{9}$$

where ρ_n is the nRDM. If we now partition every electron integral as a sum over the m nuclear basins A in which the space has been divided, we arrive at

$$\langle O \rangle = \frac{1}{n!} \sum_{\mathrm{A}_1} \cdots \sum_{\mathrm{A}_n} O_{\mathrm{A}_1, \ldots, \mathrm{A}_n}. \tag{10}$$

For a one-particle operator, like the kinetic energy or the electron-nucleus attraction, the expectation value $\langle O \rangle = \sum_A O_A$ becomes a sum of atomic components, the so-called atomic properties. Similarly, a two-electron operator would be reconstructed as $\langle O \rangle = (1/2) \sum_{A,B} O_{A,B}$, a sum of pairwise additive intra- $(B = A)$ and interatomic $(B \neq A)$ contributions. If, for instance, we take the one-electron density operator in Eq. 2, we can partition the total number of electrons N into basin contributions:

$$\int_{\Omega_\mathrm{A}} \rho(\mathbf{r}) d\mathbf{r} = N_\mathrm{A}. \tag{11}$$

Obviously, N_A is nothing but the electron population in basin A, so that partial atomic charges are straightforwardly obtained as the difference between the atomic number of the nucleus, Z_A, and its electron population, $q_\mathrm{A} = Z_\mathrm{A} - N_\mathrm{A}$. The concept of the density operator can be naturally extended to represent the density of electron pairs, $\sum_{i \neq j}^{N} \delta(\mathbf{x} - \mathbf{x}_i) \delta(\mathbf{x} - \mathbf{x}_j)$. This extension enables the calculation of properties such as the variance of the electron population within a specific atomic domain, which quantifies the uncertainty in determining the number of electrons in that region.

$$\sigma^2_{N_\mathrm{A}} = \langle N^2_\mathrm{A} \rangle - N^2_\mathrm{A}. \tag{12}$$

The difference between the average electron count in an atomic domain and its variance is called the localization index (LI), a measure of the number of localized electrons in the basin. Only if there is no fluctuation in the electron count does the LI coincide with N_A:

$$\lambda_{\mathrm{A}} = N_{\mathrm{A}} - \sigma^2_{N_{\mathrm{A}}}. \tag{13}$$

Similarly, one can obtain the covariance between the electron populations in two different basins A, B, $\mathrm{cov}(N_{\mathrm{A}}, N_{\mathrm{B}})$, and from it obtain the total variance of the full electron population,

$$\sigma^2_N = \sum_{\mathrm{A}}^{M} \sigma^2_{N_{\mathrm{A}}} + 2\sum_{\mathrm{A}}^{M}\sum_{\mathrm{B}>\mathrm{A}}^{M} \mathrm{cov}(N_{\mathrm{A}}, N_{\mathrm{B}}) = 0. \tag{14}$$

This implies that both terms in the right-hand side of the equation must be equal with opposite sign, connecting the uncertainty of the electron count in a basin and the rest of the domains in the system. Minus twice the covariance between two basins is known as the delocalization index (DI) between them,

$$\delta_{\mathrm{A,B}} = -2\mathrm{cov}(N_{\mathrm{A}}, N_{\mathrm{B}}). \tag{15}$$

Localized and delocalized electron counts, as measured by LIs and DIs, reconstruct the total number of electrons,

$$N = \sum_{\mathrm{A}}^{M} \lambda_{\mathrm{A}} + \sum_{\mathrm{A}}^{M}\sum_{\mathrm{B}>\mathrm{A}}^{M} \delta_{\mathrm{A,B}}. \tag{16}$$

The meaning of all these quantities can also be grasped from other lines of reasoning. The integration of the pair density operator defined above over two distinct atomic regions (A and B) yields the total number of electron pairs between them (N_{AB}). If we assume that the electron populations in the two areas are independent (N_{A} and N_{B}), then their product will be equal to the predicted number of pairs. Consequently, the discrepancy between N_{AB} and the previous product is measuring how many electrons are being shared between the two regions, a concept obviously associated with that of the bond order [38, 39]. Thus, the covariance of the electron populations is one possible direct measure of bond orders in real space. It can be shown that the DI collapses over the Mayer bond order [40] under suitable approximations. Other real-space alternatives to the measure of the bond order have also been suggested [41, 42] but will not be covered in this contribution.

Notice that since the Coulomb Hamiltonian operator does only contain one- and two-electron components, the QTAIM allows for a partition of the total energy of any molecular system into one- and two-basin contributions. Due to the importance of this energy decomposition, it will be treated separately in this chapter.

2.2 The Electron Localization Function

Electron density plays a fundamental role in computational chemistry thanks to the foundational theorems of Hohenberg and Kohn that lie behind density functional theory (DFT) [43]. Consequently, related fields, such as its Laplacian, have also been investigated from the point of view of QCT [23, 44]. We will shift now to examine some other chemical bonding descriptors that are, in principle, based on further-order reduced density matrices. Among them, we must immediately focus our attention on the electron localization function (ELF) of Becke and Edgecombe [27] that builds an electron localization measure from the same-spin electron pair density.

The concept of electron localization and the very existence of Lewis pairs is deeply connected to Pauli's exclusion principle and the Fermi hole. Succinctly, around an electron, a region with depleted probability of finding another electron with the same spin is formed. This is the so-called Fermi hole [45], and, thanks to the conservation of the particle number, it also implies an increased probability of finding an opposite spin electron in its surroundings, the Lewis pair. The larger the hole, the larger the same-spin exclusion area and the more localized the related pair. Thus, a measure of the size of the Fermi hole provides a way to find regions of localization of (Lewis) electron pairs.

This size can be quantified if we construct the same-spin conditional pair density, $\rho_2^{\sigma\sigma}(\mathbf{r}_1|\mathbf{r}_2)$, the conditional probability of finding a same-spin electron at point $\mathbf{r}_1$ when another is known to be at $\mathbf{r}_2$,

$$\rho_2^{\sigma\sigma}(\mathbf{r}_1|\mathbf{r}_2) = \frac{\rho_2^{\sigma\sigma}(\mathbf{r}_1, \mathbf{r}_2)}{\rho^{\sigma}(\mathbf{r}_2)}. \tag{17}$$

In this expression, $\rho_2^{\sigma\sigma}(\mathbf{r}_1, \mathbf{r}_2)$ is simply the α, α or β, β component of the 2RD defined before. It is clear that $\rho_2^{\sigma\sigma}(\mathbf{r}|\mathbf{r}) = 0$ due to Pauli exclusion, and this implies that the same-spin conditional density must grow parabolically from zero as we move away from the reference electron. Becke and Edgecombe proposed to use the spherically averaged second derivative (curvature) of this function as a measure of the size of the Fermi exclusion region. Since the pair density for a correlated wavefunction is a complex mathematical object, Becke and Edgecombe examined the single determinant case. If the reference electron lies at $\mathbf{r}$, they showed that the spherically averaged conditional same-spin pair density evolved with the distance $\mathbf{s}$ to this reference electron as

$$\langle \rho_2^{\sigma\sigma}(\mathbf{r}+\mathbf{s}|\mathbf{r}) \rangle_{av} = \left(1 + \frac{(\mathbf{s}\nabla)^2}{3!} + \dots \right) \rho_2^{\sigma\sigma}(\mathbf{r}+\mathbf{s}|\mathbf{r})|_{\mathbf{s}=0} = \left(\frac{\mathbf{s}^2}{3}\right) \left[\tau_\sigma - \frac{(\nabla\rho^{\sigma}(\mathbf{r}))^2}{4\rho^{\sigma}(\mathbf{r})}\right] + \dots. \tag{18}$$

In this expression, τ_σ is the positive-definite same-spin kinetic energy density for a single determinant wavefunction built from orbitals χ_i, $\tau_\sigma = \sum_i^\sigma |\nabla\chi_i^\sigma|^2$), and ρ^σ is the σ component of the electron density. The term in brackets is obviously a measure of the spherically averaged curvature of the Fermi hole itself [27] and is known as the ELF kernel,

$$D_\sigma(\mathbf{r}) = \tau_\sigma - \frac{1}{4}\frac{(\nabla\rho^\sigma(\mathbf{r}))^2}{\rho^\sigma(\mathbf{r})}. \tag{19}$$

Notice that the kernel is not an intensive quantity, since it depends on the number of electrons of the system, and this precludes an easy comparison of its values across grossly different chemical systems. For that reason, Becke and Edgecombe proposed to use its ratio with respect to a reference system, a homogeneous electron gas (HEG) of the same-spin density, whose kernel $D_\sigma^0(\mathbf{r})$ is, in the case of a single determinant function, $D_\sigma^0 = \frac{3}{5}(6\pi^2)^{2/3}(\rho^\sigma(\mathbf{r}))^{5/3} = c_F(\rho^\sigma(\mathbf{r}))^{5/3}$, where c_F is known as the Fermi constant. With this,

$$\chi^{\mathrm{ELF}}(\mathbf{r}) = \frac{D_\sigma(\mathbf{r})}{D_\sigma^0(\mathbf{r})}. \tag{20}$$

Since large curvatures imply small localizations and vice versa, a final Lorentzian map was made to define a function lying in the $[0, 1]$ interval that approaches 1 in the case of complete localization (zero curvature):

$$\eta(\mathbf{r}) = \left(1 + (\chi^{\mathrm{ELF}}(\mathbf{r}))^2\right)^{-1}. \tag{21}$$

Unfortunately, this definition provides $\eta = 1/2$ for the HEG itself, a fully delocalized system, so that the meaning of ELF values smaller than 0.5 has to be carefully interpreted.

Soon after the formulation of ELF in terms of the curvature of the Fermi hole was presented, Savin recognized that the ELF kernel could also be interpreted in terms of the difference between the positive-definite kinetic energy density τ and that of a bosonic system characterized by the same electron density ρ. The latter is usually known as the von Weizsäcker kinetic energy density, τ_w,

$$D(\mathbf{r}) = \frac{1}{2}\sum_i^N (\nabla\chi_i(\mathbf{r}))^2 - \frac{(\nabla\rho(\mathbf{r}))^2}{8\rho(\mathbf{r})} = \tau - \tau_w. \tag{22}$$

We notice that the $\tau - \tau_w$ difference can be understood as the excess kinetic energy of a system due to its fermionic nature, the so-called Pauli kinetic energy. Savin's construction allowed the use of ELF through density functional theory directly by building τ from the Kohn-Sham orbitals.

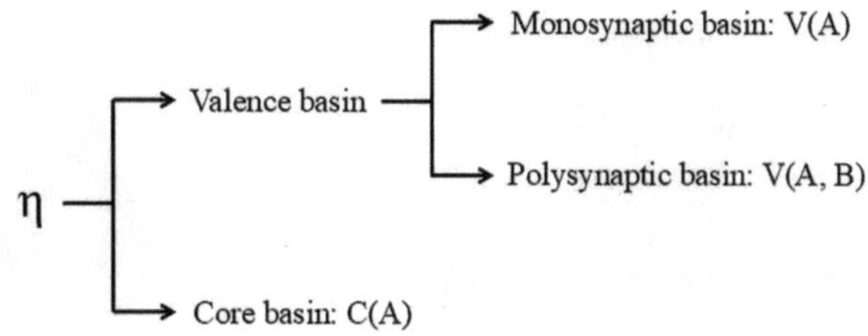

Fig. 3 ELF basin hierarchy attending to their nature and synapticity

The topological analysis of the ELF has led to many very fruitful images. As expected from its curvature definition, the $\eta(\mathbf{r})$ field has $(3, -3)$ attractors that coincide with electron pair localization regions. Typically, one can distinguish core, bonding, and lone pair attractors [21]. These are typically classified according to the number of nuclear attractors with which they interact, the so-called synaptic order. Thus, core domains are monosynaptic, while valence ones are typically monosynaptic in the case of lone pairs and disynaptic for bonding basins (see Fig. 3). Polysynaptic basins have also been found and have been successfully associated with multicenter bonds. In this way, the topological analysis of the ELF field recovers the Lewis-Langmuir picture in a rather straightforward manner. It has thus become an essential ingredient of real-space analyses.

As shown with the QTAIM partition, one can decompose expectation values as sums of ELF attractor contributions. This has been done mainly with the electron population and its variance. It has been found that core basins and lone pairs harbor close to two electrons, in agreement with expectations, and that disynaptic valence basins associated with 2c,2e bonds also display populations reasonably close to two. Figure 4 shows the ELF in two simple, albeit illustrative, examples: ammonia and diborane. In NH_3, the Lewis picture is readily recovered. Three disynaptic V(N,H) basins, represented in blue, are detected with a population close to two electrons, suggesting (2c,2e) bonds. Notice that due to the one-electron nature of isolated H atoms, V(A,H) basins (A being a heteroatom) typically include the H nucleus itself. In orange, a valence monosynaptic basin, associated with the nitrogen's lone pair, is depicted, with a population again close to two electrons. Finally, there is a basin that includes the core electrons of N. In the diborane molecule, two trisynaptic B–H–B basins are observed (depicted in green), matching the expected two (3c,2e) bonds. It should be noticed that the valence basins found in the case of multiple bonds (like in dinitrogen) tend to harbor a smaller electron population than that expected, e.g., less than six electrons in a triple bond.

2.3 *QTAIM-ELF Intersections*

It is clear that one can play the topological game here described with any scalar field, each possibility displaying its own advantages and disadvantages. For the sake of presenting these techniques, atomic and electron-pair partitions suffice to explore

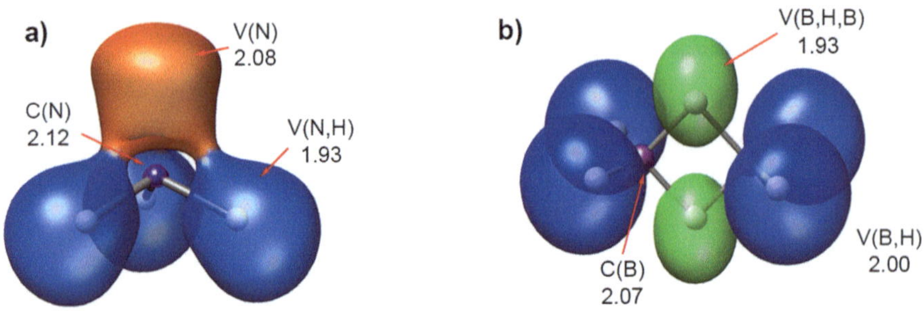

Fig. 4 ELF representation (isovalues of 0.8) for (**a**) ammonia and (**b**) diborane molecules. In purple: core basins; green: trisynaptic valence basins; blue: disynaptic basins involving H atoms; orange: monosynaptic basin. Electron populations are indicated next to their labels. ELF computed at Hartree-Fock level from the geometries optimized at CCSD/aug-cc-pVDZ. Images rendered with Chimera suite [46]

most of the basic ideas in chemistry and save that none of these decompositions allows the electrons of a given pair to be associated with a given atom. This is important in the case of heteroatomic links, in order to build some sense of ionicity, and when we try to build a notion of electronegativity from purely spatial analyses. A solution to this conundrum is to combine the partitions from several fields—effectively uniting different methods for defining real-space regions within a single system. It is interesting to examine what new insights emerge when we combine the QTAIM atomic partitioning with the ELF framework, for instance through the so-called QTAIM∩ELF intersection (QEI) [47, 48].

The QEI appears after superimposing the QTAIM and ELF partitions for a given system. By doing this, each of the QTAIM or ELF domains may or may not become divided into several subregions. For instance, we expect that an ELF atomic core domain for atom A will be fully contained in its appropriate QTAIM basin, Ω_A, while a disynaptic V(A,B) region will likely be divided into two new smaller basins, one belonging to the Ω_A and the other to Ω_B. Similarly, we expect polysynaptic basins to be split into as many components as their synaptic order. Figure 5 provides a pictorial example of a diatomic A–B in which the QTAIM interatomic surface is a plane perpendicular to the internuclear axis closer to the B atom. The V(A,B) basin is thus divided into two components $B^{A}_{A\text{-}B}$ and $B^{B}_{A\text{-}B}$ such that $V(A,B) \equiv B^{A}_{A\text{-}B} \cup B^{B}_{A\text{-}B}$.

The QEI partition was initially introduced as a measure of bond polarity, since one of the two B domains in which a bond is divided in such cases will harbor more electrons than the other [47]. This asymmetry, inherently linked to the notion of electronegativity, has long been the focus of extensive efforts aimed at achieving a unified and comprehensive description of bonding [49–51]. The QEI partition offers a notable alternative to the already many electronegativity scales available in the literature. It has since then been used to deal with the anomeric effect and hyperconjugation from a real-space point of view [52] and to provide an exquisitely

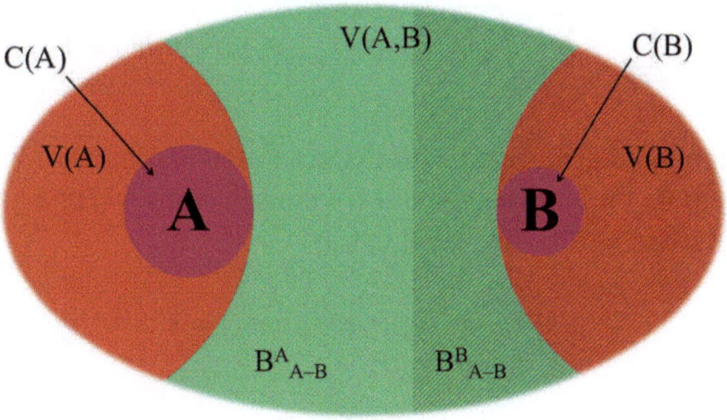

Fig. 5 QTAIM∩ELF schematic framework in a general diatomic A–B system. ELF basins by colors: in green, disynaptic V(A,B) basin; in orange, monosynaptic V(A) and V(B) basins; in purple, core basins C(A) and C(B). Non-shaded and shaded regions correspond to Ω_A and Ω_B basins, respectively. Note that the only basin that is split in this process is the disynaptic basin between A and B

detailed statistical analysis of the electron distribution in molecules [48] (see next section).

3 Electron Distribution Functions

Before turning to the spatial decomposition of the molecular energy, one can take population analyses to the limit by studying not only the average number of electrons or its variance in a given basin, but the full statistical distribution of the electron number in the set of domains we have decided to divide the space into. We thus associate each of the N (indistinguishable) electrons of the system to one of the m possible basins. To do this, we need the full nRD. For a general case where the integral of a function $\mathcal{F}(\mathbf{r}_1, \ldots, \mathbf{r}_N)$ is partitioned this way,

$$\begin{aligned}\mathcal{F}(\mathbf{r}_1, \ldots, \mathbf{r}_N) &= \sum_{B_1}^{m} \cdots \sum_{B_N}^{m} \mathcal{F}_{B_1 \ldots B_N}(\mathbf{r}_1, \ldots, \mathbf{r}_N) \\ &= \sum_{B_1}^{m} \cdots \sum_{B_N}^{m} \omega_{B_1}(\mathbf{r}_1) \ldots \omega_{B_N}(\mathbf{r}_N) \mathcal{F}(\mathbf{r}_1, \ldots, \mathbf{r}_N),\end{aligned} \tag{23}$$

we obtain m^N number of contributions. Now imagine that $\mathcal{F}$ is the full Born probability, $\mathcal{F} = |\Psi(\mathbf{r}_1, \ldots, \mathbf{r}_N)|^2$. Then,

$$\sum_{B_1}^{m} \cdots \sum_{B_N}^{m} \int \mathcal{F}_{B_1 \ldots B_N}(\mathbf{r}_1, \ldots, \mathbf{r}_N) d\mathbf{r}_1 \ldots d\mathbf{r}_N = 1. \tag{24}$$

Since the integrand is symmetric with respect to the exchange of any two coordinates, we can group these terms into a set of contributions where a given number of electrons lie in basin $\Omega_1, \cdots, \Omega_m$. In this way,

$$\sum_{n_1, n_2, \ldots, n_m} p(n_1, n_2, \ldots, n_m) = 1. \tag{25}$$

Here, the collection $n_1, n_2, \ldots, n_m$ comprises the set of electron counts such that $0 \leq n_j \leq N$ and altogether add to $\sum_j^m n_j = N$. The p's are the probabilities of finding a given partition of the N electrons in the m basins [53]. When we use a spatially exhaustive decomposition, one comes to

$$p(n_1, \ldots, n_m) = \frac{N!}{n_1! \ldots n_m!} \int_D |\Psi(\mathbf{r}_1, \ldots, \mathbf{r}_N)|^2 d\mathbf{r}_1 \ldots d\mathbf{r}_N, \tag{26}$$

where we integrate over a domain D in such a way that the first n_1 electrons lie in Ω_1, the next n_2 electrons in Ω_2, and so on. This gives the probability of finding an n_1 integer number of electrons in Ω_1, n_2 in Ω_2, etc. The partition $S = (n_1, \ldots, n_m) \equiv S(n_j)$ describes a real-space resonance structure (RSRS) [54]. The complete set of probabilities provides the full electron distribution function (EDF) [55–59].

The number of RSRSs grows combinatorially with the number of electrons and basins and soon becomes unmanageable. EDFs are accessible—in terms of computational cost—up to $m \approx 5$–6 as long as the system is not too large ($N \lesssim 30-40$) [60]. In the case of considering just two domains, i.e., a partition of space into a region Ω and its complement Ω' such that $R^3 = \Omega \cup \Omega'$, where $\Omega' = R^3 - \Omega$, the set (n_j) becomes the doublet $(n, N-n)$, and Eq. 26 simplifies to

$$p^{\Omega}(n) = \frac{N!}{n!(N-n)!} \int_{\Omega} d\mathbf{r}_1 \ldots d\mathbf{r}_n \int_{\Omega'} |\Psi(\mathbf{r}_1, \ldots, \mathbf{r}_N)| d\mathbf{r}_{n+1} \ldots d\mathbf{r}_N. \tag{27}$$

Eric Cancès has shown that a very fast recursive procedure to obtain the $p(n)$ probabilities exists in the case of single determinants [60] that only requires the so-called domain overlap matrix (DOM), the matrix of orbital overlaps restricted to a given region,

$$S_{ij}^{\Omega} = \int_{\Omega} \chi_i^*(\mathbf{r}) \chi_j(\mathbf{r}) d\mathbf{r}. \tag{28}$$

Actually, the S_{ij}^{Ω} matrix lies at the root of any other algorithm that has been devised to access the general EDF or even $p(n)$ in the case of correlated wavefunctions [57–

59]. When the number of electrons becomes large, several other approximations that improve the efficiency of EDF calculations can be invoked. The probability of finding n electrons in an atomic region is non-zero from $n = 0$ to $n = N$, the total number of electrons. However, for a large Z atom, only the valence population fluctuates, and the probability of finding a very large or very small number of electrons in its basin is negligible, so all those probabilities can be ignored. This is known as the core approximation, wherein only a few valence *active* electrons are considered.

Once the EDF is known, all statistical measures are available from it, including LIs, DIs, and all sorts of many-domain fluctuation measures that can be used to define multicenter bond indices [59, 61]. For instance, the average electron population of a basin can be obtained from the $p(n)$ probabilities,

$$\langle n_{\Omega} \rangle = \sum^{N} n \times p^{\Omega}(n), \tag{29}$$

and this $\langle n_{\Omega} \rangle$ value is exactly the same as that in Eq. 11. Notice that the amount of information contained in the set of $p(n)$ values is much larger than that in a given average, and this can be leveraged for chemical purposes. The statistical analysis coming from EDFs can be extended to the study of the electron delocalization [62]. Let us assume that A and B are two domains whose average electron population corresponds to $\langle n_{\mathrm{A}} \rangle$ and $\langle n_{\mathrm{B}} \rangle$. Then the N_{AB} pair population [63] that can be obtained from the integration of the pair density over the two basins can also be reconstructed with two-domain probabilities $p(n_A, n_B)$,

$$\langle n_{\mathrm{A}} n_{\mathrm{B}} \rangle = \sum_{S} n_{\mathrm{A}} n_{\mathrm{B}} p(n_{\mathrm{A}}, n_{\mathrm{B}}) = \int \rho_2^{\mathrm{AB}}(\mathbf{r}_1, \mathbf{r}_2) d\mathbf{r}_1 d\mathbf{r}_2. \tag{30}$$

From this expression, it is straightforward to obtain the DI (minus twice the covariance) as

$$\delta^{\mathrm{AB}} = -2\sum_{S} (n_{\mathrm{A}} - \langle n_{\mathrm{A}} \rangle)(n_{\mathrm{B}} - \langle n_{\mathrm{B}} \rangle) p(n_{\mathrm{A}}, n_{\mathrm{B}}), \tag{31}$$

which shows in a crystal clear manner that whenever a single RSRS contributes to the EDF the DI vanishes.

4 Energy Decompositions

At this point, it should be clear that the partition of expectation values into real-space domain contributions is a particularly fruitful strategy to access chemical bonding descriptors and that if this is done with the n-particle density operators, we obtain all kinds of electron counting insights. The theory of chemical bonding, however, has another, probably more important dimension related to the energy. Chemical bonds stabilize atomic aggregates, and no full understanding of the electronic structure of a system exists without considering its energetic dimension. However, finding the contribution of each interaction or bond to the stability of a molecular system is not a well-posed problem in polyatomics. As soon as we shift from a diatomic, where its dissociation energy E_b or its force constant k provides clear measures of bond strength, there is no unique way to disentangle the total binding energy into clearly defined contributions or to avoid the mixture of nuclear movements in normal modes to provide local force constants. Chemists expect to identify the contributions of individual fragments, bonds, or structural units to a molecule's stabilization [14]. This need has spurred the development of various energy decomposition analyses (EDAs), and, surely, those coming from real-space partitions stand on their own.

Contemporary computational chemistry offers an abundance of EDA methodologies, each varying in suitability depending on the phenomena being studied. While some methods excel in certain contexts, they may lack accuracy or reliability in others. For instance, symmetry-adapted perturbation theory (SAPT) [64], despite widespread deserved acclaim, struggles to effectively describe interactions between fragments at short ranges. Other prominent energy decomposition analyses rooted in early efforts to describe intermolecular interactions stem from the foundational work of Morokuma [65], Ziegler and Rauk [66]. These methods are particularly advantageous in the context of DFT studies [67, 68]. However, they come with inherent limitations, primarily due to their dependence on how interacting fragments are defined—a factor that significantly influences the resulting conclusions. Building on earlier methods, Martin Head-Gordon has employed fragments defined in the orbital space [69], typically based on atoms. As with its predecessors, its effectiveness relies on the specific criteria used to define atoms. In contrast, the interacting quantum atom (IQA) method adopts an orbital invariant real-space atomic description [11, 70, 71]. While it avoids the reliance on arbitrary references, it remains influenced by the specific partition applied to define the atoms in real space.

In essence, the IQA methodology proposes the decomposition of the expectation value of the Hamiltonian into atomic domains following the general principles summarized in the previous sections. Since $\hat{H}$ is a sum of one- and two-electron operators, the IQA energy becomes a sum of one- and two-body (domain) energy contributions [71] that we typically write as

$$E = \sum_{\mathrm{A}} E^{\mathrm{A}}_{self} + \sum_{\mathrm{A}>\mathrm{B}} E^{\mathrm{AB}}_{inter}. \tag{32}$$

In the above expression, we have gathered all terms depending on only one basin (including two-electron ones in which the two electrons are integrated over the same domain) in the so-called self-, net-, or intra-atomic energy term, E^{A}_{self}. If the basin is atomic and holds nucleus A, E^{A}_{self} is the sum of the kinetic energy of the electrons in A, the nucleus-electron attraction of those electrons to the nucleus A, and their mutual interelectron repulsion,

$$E_{self}^{A} = T^{A} + V_{ee}^{A} + V_{en}^{A}. \tag{33}$$

When a molecule dissociates, the self-energy of all of its atoms tends to that of the atoms in vacuo, E_0^A. E_{self}^A is a large quantity, of the same size as an atomic energy. Since the expectation value of the atomic Hamiltonian is minimized by the in vacuo atomic energy, in the absence of large charge transfers, the atomic self-energy is less negative than E_0^A, so that

$$E_{def}^{A} = E_{self}^{A} - E_0^{A}, \tag{34}$$

defined as the deformation energy of atom A, is typically a positive quantity [72, 73].

The two-basin terms in Eq. 32 are called interatomic interaction energies and gather all terms depending on a given pair of atomic domains A and B. These are the nuclear repulsion between A and B, the attraction between electrons in A and nucleus B, and between electrons in B and nucleus A, as well as the repulsion between the electrons in A and those in B,

$$E_{inter}^{AB} = V_{nn}^{AB} + V_{en}^{AB} + V_{ne}^{AB} + V_{ee}^{AB}. \tag{35}$$

The electron-electron repulsion term in Eq. 35 is obtained by integrating the pair density for electrons in two different domains,

$$V_{ee}^{AB} = \int_{\Omega_A}\int_{\Omega_B} \frac{\rho_2(\mathbf{r}_1, \mathbf{r}_2)}{|\mathbf{r}_1 - \mathbf{r}_2|} d\mathbf{r}_1 d\mathbf{r}_2, \tag{36}$$

and can be further divided into a term accounting for the repulsion of electrons where their densities are those of independent sets of particles, and a quantum mechanical correction due to exchange and correlation (*xc*),

$$V_{ee}^{AB} = \int_{\Omega_A}\int_{\Omega_B} \frac{\rho(\mathbf{r}_1)\rho(\mathbf{r}_2)}{|\mathbf{r}_1 - \mathbf{r}_2|} d\mathbf{r}_1 d\mathbf{r}_2 - \int_{\Omega_A}\int_{\Omega_B} \frac{\rho_2^{xc}(\mathbf{r}_1, \mathbf{r}_2)}{|\mathbf{r}_1 - \mathbf{r}_2|} d\mathbf{r}_1 d\mathbf{r}_2. \tag{37}$$

The first term is nothing but the purely Coulombic interaction of electrons described by the first-order density and is known as the classical electron repulsion, $V_{ee,cl}^{AB}$. The second is the exchange-correlation energy, $V_{ee,xc}^{AB}$. In this way, $V_{ee}^{AB} = V_{ee,cl}^{AB} + V_{ee,xc}^{AB}$. Joining all the terms that depend on the 1RDM into a classical interaction energy, in Eq. 35,

$$E_{inter}^{AB} = V_{nn}^{AB} + V_{en}^{AB} + V_{ne}^{AB} + V_{ee,cl}^{AB} + V_{ee,xc}^{AB} = V_{cl}^{AB} + V_{xc}^{AB}, \tag{38}$$

where $V_{xc}^{AB} = V_{ee,xc}^{AB}$, and the rest are collected into V_{cl}^{AB}. It has been clearly shown that the classical interaction energy can be associated with the electrostatic or ionic contribution to a bond and that the exchange-correlation term is a measure of the covalent contribution to the interaction energy [74].

One of the major advantages offered by IQA is the possibility of gathering sets of atoms into groups $\mathcal{G}, \Omega_{\mathcal{G}} = \bigcup_{A \in \mathcal{G}} \Omega_A$ such that the different groups are disjoint and contain all the atoms in the system, e.g., dividing a molecule into functional groups or a molecular aggregate into its different molecular units. This way, the total energy can be written in exactly the same IQA manner, now at the *group* level,

$$E = \sum_{\mathcal{G}} E_{self}^{\mathcal{G}} + \frac{1}{2} \sum_{\mathcal{G} \neq \mathcal{H}} E_{inter}^{\mathcal{G}\mathcal{H}}, \tag{39}$$

where $\mathcal{G}$ and $\mathcal{H}$ are two different fragments collecting M and M' atoms. A fragment's self-energy is the sum of all the self-energies of its atoms plus their mutual interaction energies,

$$E_{self}^{\mathcal{G}} = \sum_{A \in \mathcal{G}} E_{self}^{A} + \frac{1}{2} \sum_{A,B \in \mathcal{G}} \sum_{A \neq B} E_{inter}^{AB}, \tag{40}$$

while the interfragment interaction energies are simply obtained by adding all the interatomic interaction energies between atoms belonging to the two different groups,

$$E_{inter}^{\mathcal{G}\mathcal{H}} = \sum_{A \in \mathcal{G}, B \in \mathcal{H}} E_{inter}^{AB}. \tag{41}$$

This approach allows us to consider, on the same footing, strong intramolecular interactions between specific atoms or specific functional groups within a molecule and intermolecular interactions, covalent or non-covalent. In this sense, IQA offers a full window to deal with the energetics of chemical bonding at all possible graining levels. Moreover, since it is independent of external references, it is free from many of the objections that other EDAs have received in the last years [75].

5 Practical Applications

The true utility of QCT lies in its application to specific molecular systems, where theoretical constructs are translated into tangible chemical phenomena. To illustrate its versatility, the following subsections explore practical examples ranging from the analysis of electron distribution in different scenarios to the assessment of the energetics governing interatomic and interfragment interactions. By moving from theory to application, the upcoming discussion highlights the QCT's ability to bridge quantum mechanics and chemical intuition, thus reinforcing its relevance in contemporary research. Electronic structure calculations were performed with the Gaussian16 [76] and ORCA 5.0 [77] suites. Topological analyses were done with AIMAll [37], IQA decompositions with PROMOLDEN [78], EDFs found with the EDF code [79], and ELF and QEI decompositions with a locally modified version of TOPMOD [80].

5.1 EDFs Combined with the QTAIM-ELF Partition

The combination of two prominent methodologies, such as QTAIM and ELF, paves the way to obtaining a deep, fine-grained view of the electron arrangement in molecules, the *electron in the molecule*, whose implications extend to bonding and molecular properties. This objective is included in the fundamental goal of QCT: the rigorous characterization of the electron distribution in real space. In pursuit of this, a series of hydrides with representative elements (groups 1, 2, and 13–17 of the periodic table) belonging to rows 2–4 (represented as AH_n, where A is the heteroatom) have been studied [48], as shown in Fig. 6. Raw data of Figs. 6, 7, 8, 9, 10, 11 are adapted with permision from Barrena-Espés et al. [48].

Different perspectives can be selected at the outset by choosing the partition and the method. If we are interested in studying the consequences of changing the electronegativity of the central (A) atom, we can choose the EDF representation in the QTAIM partitioning and group together all non-A atoms, resulting in a simple model with two groups: the heteroatom A and the rest of the system (see Fig. 6). This framework has been widely used as a fruitful way to identify patterns that depend on electronegativity differences [48, 53].

Figure 6 shows that the systems under study can be easily categorized into three main groups based on (i) the position of the probability peak and (ii) the width of the EDF, i.e., its two first moments. The first category includes the hydrides of groups 1, 2, and 13, which feature central atoms with very low electronegativities. They display a single dominant RSRS where element A hosts fewer valence electrons than those corresponding to the neutral atom. When moving away from Li to B, electronegativity increases, and as a direct consequence, additional RSRSs become more relevant, widening the probability distribution. Changing the period from 2 to 4 in the alkaline and alkaline-earth hydrides does not influence the shape of the EDF appreciably. However, doing so in group 13 has interesting consequences. For instance, the peak shifts on going from Al to Ga, which is thus considerably more electronegative than B or Al. The most probable distribution in GaH_3 turns out to be $[Ga^{+}(H_3)^{-}]$.

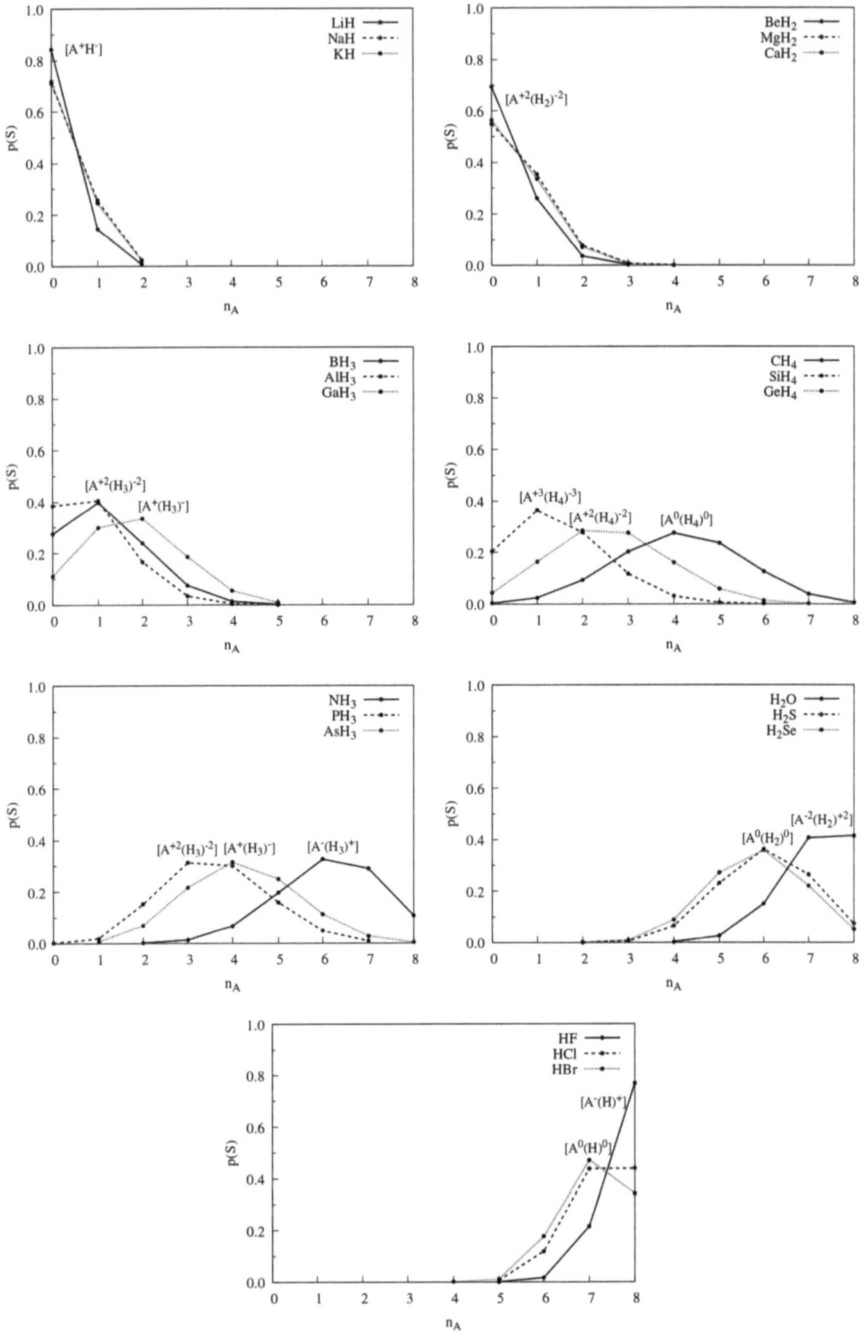

Fig. 6 Two-group QTAIM probabilities wherein one group is composed of the A heteroatom and the other contains the rest of the molecule in the AH_n systems described in the main text. Each plot contains the evolution of the EDF for a given group as the row is changed. The electron count displayed, n_A, is the number of valence electrons in A. Second-, third-, and fourth-row systems are identified by solid, dashed, and dotted lines, respectively. The most probable RSRS for each case is explicitly indicated in square brackets. Results obtained at the DFT B3LYP/def2-TZVP level

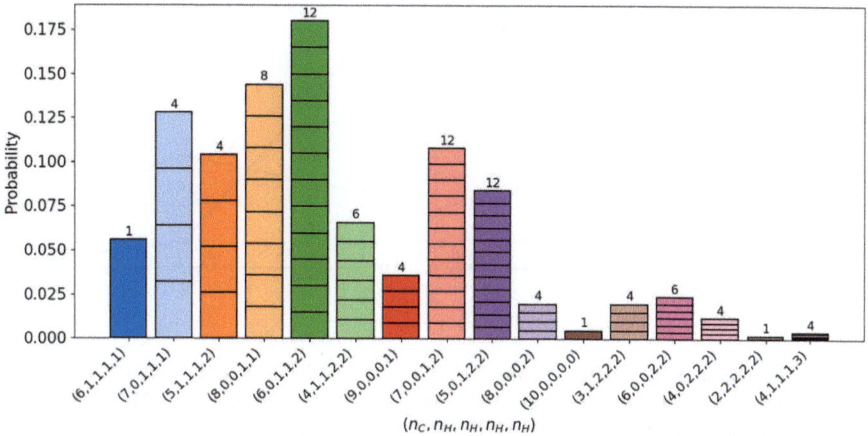

Fig. 7 Irreducible multi-basin EDF analysis in the QTAIM partition for the CH_4 molecule (at the B3LYP/def2-TZVP level). Only the most relevant irreducible categories are shown. Subdivisions display the share of each of the individual distributions to an irreducible category (whose multiplicity is indicated above each bar). The nomenclature of the distributions is $(n_C, n_H, n_H, n_H, n_H)$, where n_C represents the electron population in the atomic basin of C, and n_H the same for the four H atomic basins

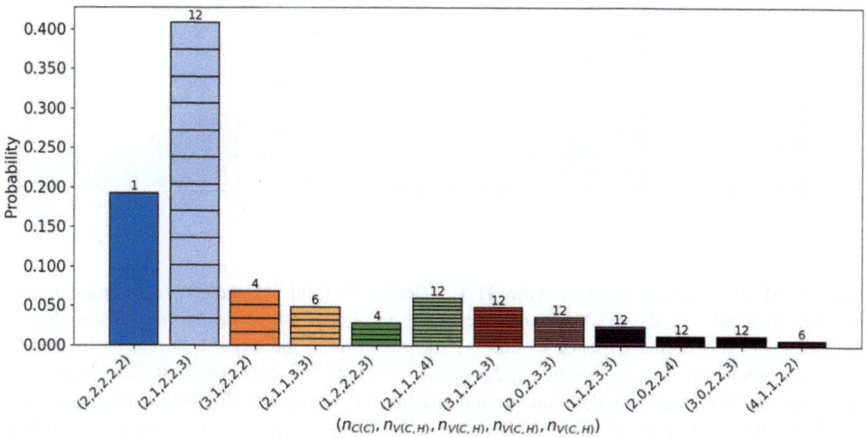

Fig. 8 Irreducible multi-basin EDF analysis in the ELF partition for the CH_4 molecule (at the B3LYP/def2-TZVP level). Only the most relevant irreducible categories are shown. Subdivisions display the share of each of the individual distributions to an irreducible category (whose multiplicity is indicated above each bar). Labels indicate the number of electrons in each basin, $(n_{C(C)}, n_{V(C,H)}, n_{V(C,H)}, n_{V(C,H)}, n_{V(C,H)})$, where C(C) is the core basin of C atom, and V(C,H) the synaptic basins involving the C–H bonds. Results obtained at B3LYP/def2-TZVP level

The second major group, exemplified by methane, shows a more balanced electron distribution where neither the carbon nor hydrogen atomic regions tend to accumulate additional electrons. This scenario illustrates a covalent bond where

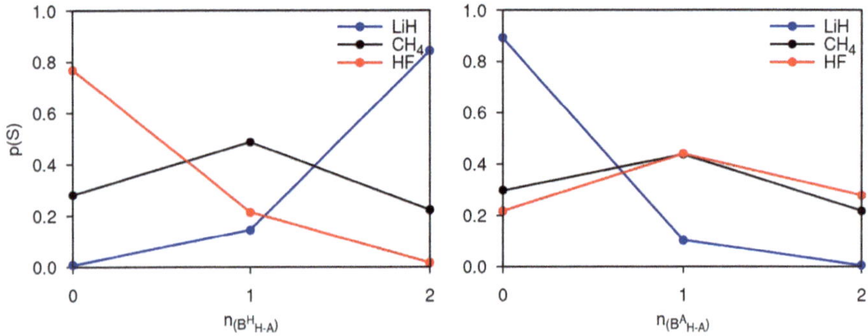

Fig. 9 Two-basin EDF analysis in the QEI partition for the LiH, CH_4 and HF molecules, considering the hydrogen $B^H_{A\text{-}H}$ (left) and the A $B^A_{A\text{-}H}$ (right) subbasins of the V(A,H) bond domain. Results were obtained at the B3LYP/def2-TZVP level of theory

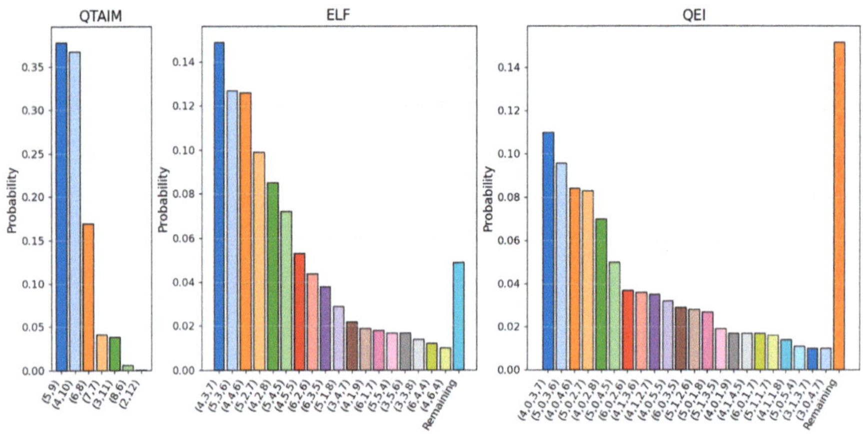

Fig. 10 Multi-basin EDF analysis in the QTAIM (left), ELF (middle), and QEI (right) partitions for the CO molecule. In the QTAIM case, we report $p(n_C, n_O)$, where n_C represents the electron population in the atomic basin of C, and n_O the electrons in the oxygen atomic basin. For the ELF decomposition, $p(n_{C(C)\cup V(C)}, n_{V(C,O)}, n_{C(O)\cup V(O)})$, where C(C) and C(O) are the core basins of C and O atoms, V(C) and V(O) the monosynaptic basins of C and O, and V(C,O) the disynaptic basin involving the C–O bond. Finally, for QEI, we report $p(n_{C(C)\cup V(C)}, n_{B^C_{C\text{-}O}}, n_{B^O_{C\text{-}O}}, n_{C(O)\cup V(O)})$, where $B^C_{C\text{-}O}$ and $B^O_{C\text{-}O}$ are the contribution of the C atom and the O atom, respectively, to the divided disynaptic ELF basin. All data were obtained at the B3LYP/def2-TZVP level

electrons are more evenly shared between the bonded atoms. The resulting symmetrical distribution is centered at the neutral RSRS, represented as $[C^0(H_4)^0]$. A similar behavior is identified in H_2S and H_2Se, although in these instances, the distribution is somewhat more peaked, indicating a slightly more ionic nature compared to CH_4.

Systems in the third category exhibit bonding patterns that contrast with those in the initial group. These create highly polar bonds, and the atom A often gathers

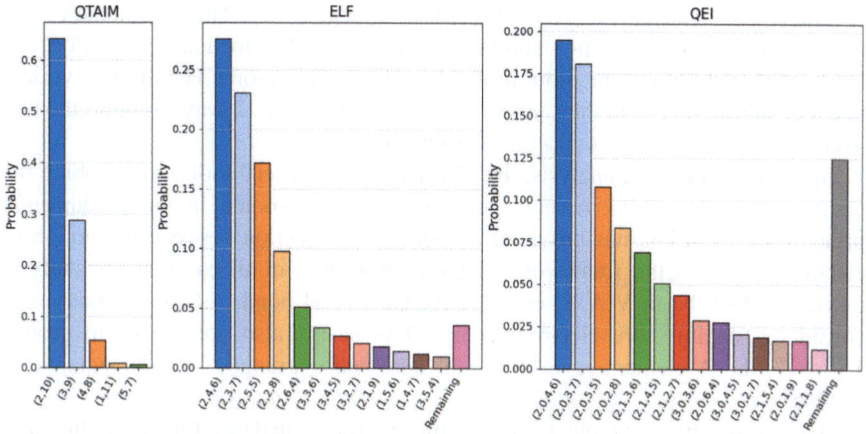

Fig. 11 Multi-basin EDF analysis in the QTAIM (left), ELF (middle), and QEI (right) partitions for the BeO molecule. In the QTAIM case, the straightforward $p(n_{Be}, n_O)$ is given. Three-domain probabilities are provided for the ELF partition, $p(n_{C(Be)}, n_{V(Be,O)}, n_{C(O)\cup V(O)})$, where C(O) and V (O) are the core and monosynaptic basins of O, and V(Be,O) the disynaptic basin involving the Be–O bond. For QEI, four-basin probabilities, $p(n_{C(Be)}, n_{B^{Be}_{Be\text{-}O}}, n_{B^{O}_{Be\text{-}O}}, n_{C(O)\cup V(O)})$, where $B^{Be}_{Be\text{-}O}$ and $B^{O}_{Be\text{-}O}$ refer to the bond subbasins belonging to the contribution of the Be and O atoms, respectively, are provided. Data obtained at the B3LYP/def2-TZVP level

electrons due to its higher electronegativity than hydrogen (H). Notable examples include NH_3, H_2O, and HF. In these compounds, the distributions skew toward higher n_A values. Furthermore, as one moves from nitrogen (N) to fluorine (F), the curves become narrower, suggesting fewer relevant distributions as the electronegativity gap between A and H increases.

The analysis of multi-basin EDFs within the quantum theory of atoms in molecules (QTAIM) framework leads to an extensive discussion. It is highlighted that, rather than concentrating solely on the most likely distribution, attention should also be given to all distributions that contribute to the average electron population in each basin. This discussion is exemplified using methane as a paradigm of covalent hydride (refer to Fig. 7). Here, the analysis of EDFs considers the five atomic basins associated with each atom. Findings reveal that the most probable RSRS corresponds to the neutral distribution, where each atomic domain retains its reference electron counts (with n_A matching Z_A). Notably, since the hydrogen atoms are equivalent, a bunch of distributions shares the same probabilities. For instance, in the second-most probable distribution, the carbon atom gains an electron from any one of the four H regions, and this distribution is four-fold degenerate (due to the equivalence of four H atoms), as shown in light blue in Fig. 7. Consequently, distributions of chemically equivalent domains can be effectively grouped into supra-RSRSs, as illustrated in Fig. 7. We call these irreducible distributions. The combinatorial variety introduces a new perspective where small individual contributions combine to produce high probabilities. In fact, most of the first ten distributions surpass the neutral RSRS in terms of cumulative probabilities. This is similar to several effects in statistical mechanics related to degeneracy: The final statistics will depend on both individual probabilities and multiplicity. In methane, if six electrons are located on C, it will be more probable to find one protonic H and one hydridic one than all the four H in their neutral state since there are 12 ways to build the first distribution and only one to do it in the second.

In chemical bonding analysis, the ELF has been a crucial tool for describing the nature of electron pairs in real space. This picture, rooted in the fundamental principles of VSEPR, has provided an intimate connection between the notion of electron pairs and the framework of QCT. Note that in all cases under study, a disynaptic ELF basin V(A,H) emerges for each A–H bond. As all heteroatoms A considered exhibit core electrons, core C(A) basins also appear. From groups 15 to 17, the A element reveals several lone pairs, arranged in monosynaptic V (A) basins.

As an illustrative example, the CH_4 molecule is revisited (see Fig. 8). In this case, four disynaptic basins representing the C–H bonds are observed, along with the carbon core electron basin, C(C). In agreement with the QTAIM-based EDF analysis, the system shows a significantly high probability for the Lewis distribution—in which each bonding basin accommodates a pair of electrons. The next more probable distribution, approximately six times less likely, indicates the transfer of an electron from one bonding basin, V(C,H), to another analogous one. Despite this probability difference, the cumulative probability becomes twice that of the Lewis structure. Together, they account for more than half of the total probability, indicating a rather homogeneous distribution that translates into a prominent covalent character for the C–H bonds.

How do we reconciliate these two pictures? To do that, as explained, we focus on the finer partition provided by the QEI framework. This methodology enables the decomposition of polysynaptic ELF basins into atomic contributions, which, in the systems considered herein, translates into each V(A,H) ELF basin being split into two subregions: the atomic contribution of A to the V(A,H) bond, $B^{A}_{A\text{-}H}$, and its equivalent for the H atom, $B^{H}_{A\text{-}H}$. Note that the number of potential basins to be considered increases greatly when doing this, especially in cases with many A–H bonds. As a result, not all the basins can always be considered individually. Let us start by splitting just one of the V(A,H) basins, isolating its $B^{A}_{A\text{-}H}$ and $B^{H}_{A\text{-}H}$ components, and performing two-domain analyses for each of them. Figure 9 shows the probability of finding a given number of electrons in each of these subregions. For the sake of clarity, only three systems are depicted: the very polarized LiH and HF molecules, and the covalent CH_4. The $B^{H}_{A\text{-}H}$ basin displays again three different possible behaviors. In methane, it hosts one electron in the most probable distribution, in agreement with the covalent character of the bond. In LiH, the $B^{H}_{A\text{-}H}$ basin has a sharp probability peak when hosting the two electrons corresponding to the bond, compatible with a hydride description. On the contrary, in HF, the $B^{H}_{A\text{-}H}$ basin probability peaks at 0 electrons so that the F atom takes the two bonding electrons, as expected. These results can be complemented by examining the $B^{A}_{A\text{-}H}$ probabilities (Fig. 9). In CH_4, the probability is maximal at one electron, and the distribution is markedly symmetric. In LiH, the $B^{A}_{A\text{-}H}$ basin peaks at zero electrons, as it is also expected. Surprisingly, in HF, the probability peak also takes place when the F basin has one electron. In broad strokes, the reason for such an unexpected result lies in that the $B^{A}_{A\text{-}H}$ basin is small, and the F tends to accumulate electrons in the lone pairs.

Let us look at other simple but somewhat more complex cases. In this context, the CO molecule deserves special attention (see Fig. 10). Looking at the EDFs in the atomic partition, where two regions appear (one for C and another for O atoms), it is proven that the C–O bond has a highly polar nature [81]. Thus, the most likely distribution places an extra electron in the O basin coming from the C atomic region, leading to a $[C]^{+}[O]^{-}$ RSRS. The second most likely RSRS transfers two electrons from C to O, $[C]^{2+}[O]^{2-}$. Only in the third most relevant distribution does the neutral RSRS appear, $[C]^{0}$ $[O]^{0}$, with a probability half that of the other two distributions.

Moving on to ELF-based EDFs, in CO, there are two cores, C(C) and C(O), two monosynaptic V(C) and V(O) basins—corresponding to the lone pairs of C and O atoms, respectively—and one V(C,O) bond basin. Since a full five-basin analysis is difficult to interpret, it is instructive to combine the core and lone pair domains of each atom into a single fragment, leaving the V(C,O) basin alone. This leads to a simpler three-basin EDF picture. Within this framework, the two most likely distributions show a bond basin populated by three electrons, far from the six expected in a triple bond description [82]. Moreover, the bond basin shows considerable variability in its electron number. This suggests that the fluctuation in the population of the monosynaptic regions (lone pairs and core basins) is mediated by the disynaptic ELF basin [83, 84]. The finer QEI partition is able to shed light on these controversial features by further splitting the V(C,O) basin into O and C contributions. The results confirm the strong tendency of O to accumulate electrons at the expense of C atoms, as the $B^{C}_{C\text{-}O}$ basin remains unoccupied until the 7th distribution in probability.

Thus, it can be concluded that the oxygen atom tends to host the electrons involved in the C–O bond and that the overall bonding picture is significantly more complex than is generally reported for this molecule. The fact that the O atom has a large probability of closing its valence shell (thus hosting 10 electrons) reduces its reactivity and might justify why CO typically binds through the C end.

Another case that deserves special attention is that of the diatomic BeO system, whose EDF results are shown in Fig. 11. By looking only at the EDFs in the QTAIM framework, it is easy to conclude that the system is highly ionic. The most likely RSRS—that with 10 electrons in O and 2 in Be, $[Be]^{2+}[O]^{2-}$—is fully ionic. The ELF partition gives rise to two core regions, C(Be) and C(O), the lone pairs of the oxygen atom V(O), and the bond V(Be,O). A three-basin EDF is provided where C (Be) and V(Be,O) are taken individually, while the C(O) and V(O) domains are combined into a single fragment. In contrast to the large charge separation resulting from the atomic partitioning, the ELF bond basin is highly populated in all relevant distributions, with four electrons assigned in the most probable distribution, but varying from two to six in the four subsequent most probable structures. The population fluctuation is completely dominated by the oxygen monosynaptic regions. As in the case of CO, the QEI serves as a suitable tool to rationalize the QTAIM/ELF differences. It is concluded that the electrons involved in the Be–O bond are clearly provided by the $B^{O}_{Be\text{-}O}$.

As shown, electron distribution functions effectively bridge the gap between traditional chemical bonding concepts, such as covalency or ionicity, and the quantum perspective of electrons subject to statistical fluctuations in real space. By introducing the QEI intersection, the ELF and QTAIM become two possible views within a larger picture in which electrons belong to both atoms and bonds. This new strategy is particularly fruitful for understanding the meaning of electronegativity, providing useful insights both in highly ionic situations and in more complex interactions, as observed in the CO and BeO systems.

5.2 A Quantum Chemical Topology Approach to Pyramidanes

Exotic molecules have always been an important playground in which to test bonding models or to propose new concepts in chemical bonding [85–87]. For example, the unusual structural features displayed by some chemical species tend to attract research efforts that sometimes lead to the discovery of novel types of interactions and to the proposal of new concepts that can then be scrutinized by the community [88–90]. In this respect, the appealing geometric structure of the so-called pyramidanes serves as a revealing example.

The family of related chemical structures known as [n]-pyramidanes is characterized by a mostly planar *n*-membered face (the base) attached to an apex atom. When $n=4$, this leads to an unusual pyramidal geometry in which the apex atom lies in a tetracoordinate geometry different from a tetrahedral one, an example of an inverted configuration [86, 91]. Both experimentally and theoretically, several structural designs have been proposed by varying the pyramidal skeleton, ranging from simple C-based homoatomic frameworks (which have not yet been synthesized) to bulkier groups containing boron or silicon derivatives, which have been prepared experimentally [87, 92, 93]. In addition, different *n* values for the base of the pyramid have been considered, from trigonal, tetragonal to hexagonal, with homo- or heteroatomic components and all kinds of functional groups [87, 91]. Here, the structure of the simplest pyramidanes, derived from [4]-pyramidane (not yet synthesized), has been chosen [36]. We also take some raw data from this reference. Its four-membered base is derived from cyclobutadiene, C_4H_4, as shown in Fig. 1 for the case of the Si-[4]-pyramidane molecule. The group placed at the apical vertex is typically a single atom from the tetrel group (C, Si, Ge, Sn, and Pb).

We start by considering an EDF analysis. Even with the simplest QTAIM partition, the number of basins exceeds the maximum which allows any reasonable

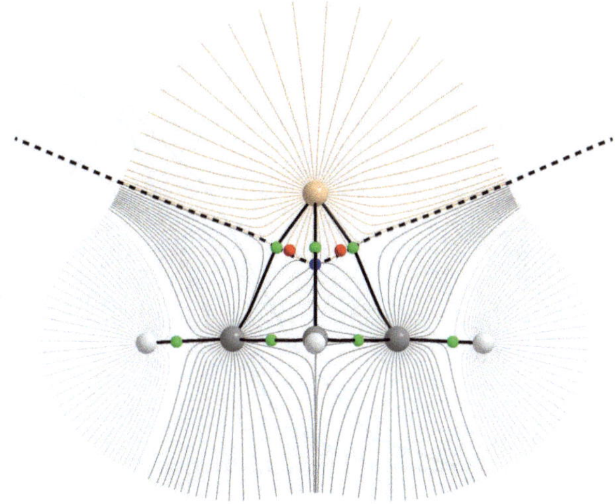

Fig. 12 B3LYP-D3BJ/def2-TZVP gradient lines of ρ defining the atomic partition of the Si-substituted [4]-pyramidane, as seen from one of the diagonal planes. The bond critical points are indicated in green, the ring critical points in red, and the central cage critical point in blue. The electron density threshold is taken as 0.001 au. The figure has been obtained with AIMAll [37]

interpretation of the probability distribution. Gathering basins into appropriate joint groups is thus mandatory. Here we consider the apex atomic region as one domain and group the full cyclobutadienyl $[C_4H_4]$ moiety into a single macro-region, as shown in Fig. 12.

The results are shown in Fig. 13. Note that the pyramidane base, $[C_4H_4]$, is common to all systems. Its electron number can therefore be used as a measure of the electron number fluctuations associated with the bond to the apex atom. When C is the apex element, the probability of finding a given number of electrons in the base peaks at 28 electrons, so we have two neutral fragments, the $E^0[C_4H_4]^0$ RSRS. In addition, the C-substituted pyramidane has a predominantly symmetric probability distribution. This fact illustrates a highly symmetric electron exchange between the two fragments, which is directly related to a highly covalent behavior. In the remaining cases, the maximum probability occurs when the apex atom has donated an electron to the base, i.e., for the $E^+[C_4H_4]^-$ structure (E is the apex element). The Si compound deserves special attention as it has a very asymmetric distribution, skewed toward structures where the base is negatively charged. In this system, the neutral structure is only third in probability, while in the other cases E = Ge, Sn, Pb, it is second. From this analysis, we conclude that the $C[C_4H_4]$ system presents a covalent bonding scenario that shifts toward charge transfer from the apex to the base in the rest of the systems, this being particularly important in the silicon compound.

Statistical analysis via EDFs is compatible with the IQA energy decomposition scheme. The latter allows the classification of each interatomic interaction according to the magnitude of the ionic and covalent contributions. Since interatomic interactions are additive, they can be expressed collectively as fragment interactions. For the sake of consistency, the energetic terms to be discussed correspond to the

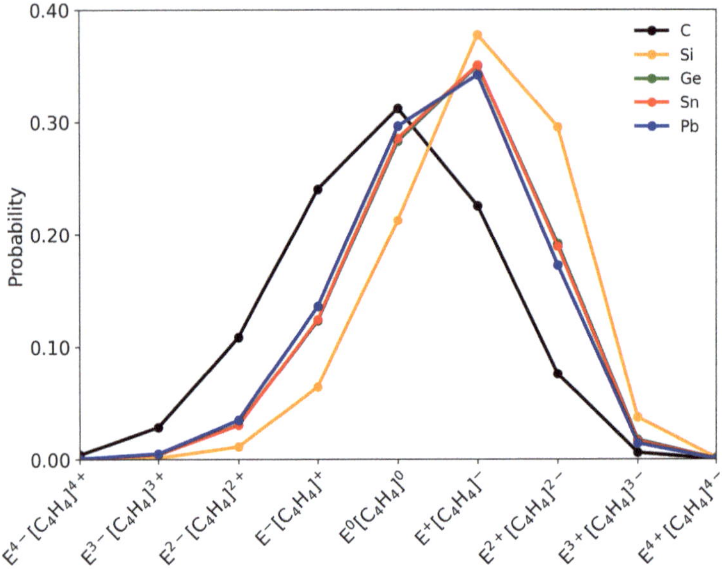

Fig. 13 Two-basin EDFs for the $E[C_4H_4]$ pyramidane systems, being E = C, Si, Ge, Sn, Pb. All systems have been computed at the B3LYP-D3BJ/def2-TZVP level with ZORA relativistic corrections. Note that the data points for E = Ge and Sn are nearly indistinguishable due to significant overlap

interaction between the pyramidane apex element and the pyramidane base, obtained after adding all the pairwise additive interactions between the apex, E, and each atom in the base, as in Eq. 41. This is done separately for the ionic or electrostatic V_{cl}, the exchange-correlation or covalent V_{xc}, and the total interaction E_{total} contributions. The results are shown in Fig. 14, where the apex-base delocalization index is also shown. For the C-[4]-pyramidane, the classical term is destabilizing, while the total interaction energy is favorable due to compensation by the exchange-correlation contribution. Such behavior is characteristic of covalent bonding. Accordingly, the Si-substituted structure exhibits the most attractive V_{cl} contribution, due to the substantial charge transfer from the Si to the base, while the V_{xc} term is the less stabilizing, consistent with its much higher ionic behavior. For E = Ge, Sn, and Pb, an intermediate behavior is shown: The V_{cl} term is stabilizing but lower in absolute value than for E = Si. In line with this, the V_{xc} contribution is larger than for E = Si, but significantly less than for E = C. Overall, these three systems (E = Ge, Sn, Pb) can be considered to establish a mainly ionic apex-base interaction, albeit with a higher degree of covalency than $Si[C_4H_4]$.

As it can be seen, the statistical and the energetic analyses are interrelated and complementary, providing compatible yet independent information about the electronic structure and chemical bonding of a given system.

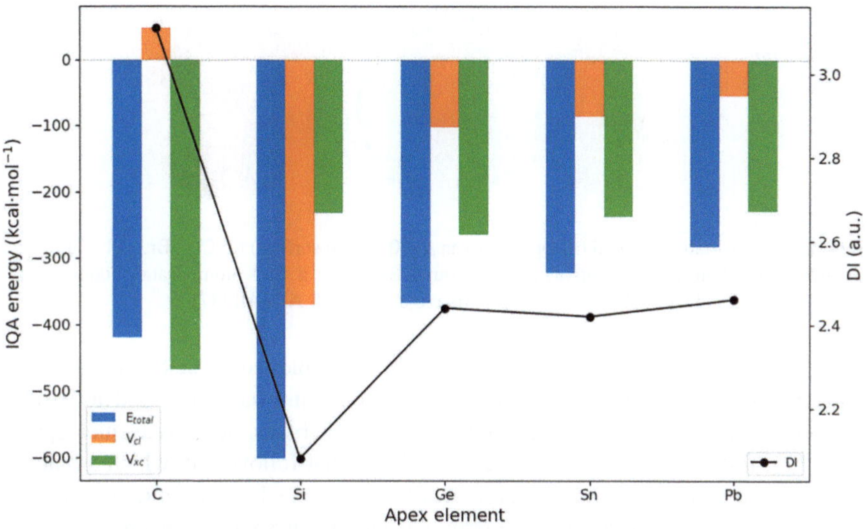

Fig. 14 IQA energetic contributions for the interaction between the apex element and the $[C_4H_4]$ base (blue: total interaction energy, orange: electrostatic contribution, green: exchange-correlation term). The apex-base delocalization index is also shown as a black line. All systems have been computed at the B3LYP-D3BJ/def2-TZVP level with ZORA relativistic corrections

5.3 Scrutinizing the Nature of Halogen Bonds

So far, we have introduced the foundations of a number of real-space-based formalisms that can be used to extract a wealth of information about chemical bonding from state-of-the-art quantum chemical calculations. We have also shown what kind of insights they can provide in strongly bonded systems such as CO and in more exotic cases of intermediate strength such as the pyramidane structures. Now we turn to the main topic of this monograph and show how the same technology can be used to assess non-covalent interactions. Real-space analyses thus become some of the few techniques that can be used across the entire spectrum of interaction strengths.

In the last two decades, researchers have introduced a zoo of X-bonds to extend the hydrogen bond (HB), one of the basic building blocks of supramolecular architecture [94]. Similar to HBs, the stabilization of most of these newly introduced XBs is typically associated with electrostatic interactions, as in the case of the so-called halogen bond. Halogens are highly electronegative elements that tend to attract electron density and become negatively charged, so electrostatic attraction is obviously present in the very-well studied hydrogen-bonded dimer of HF, for example. However, it was soon realized that when a halogen X binds to another electronegative atom, as in homodiatomic X_2 molecules, its lone pairs arrange to form a negatively charged torus or belt perpendicular to the covalent bond, leaving an electron-deficient hole along the axis: the so-called σ-hole (see Fig. 15). Such a (frequently) positively charged region, as revealed by the electrostatic potential

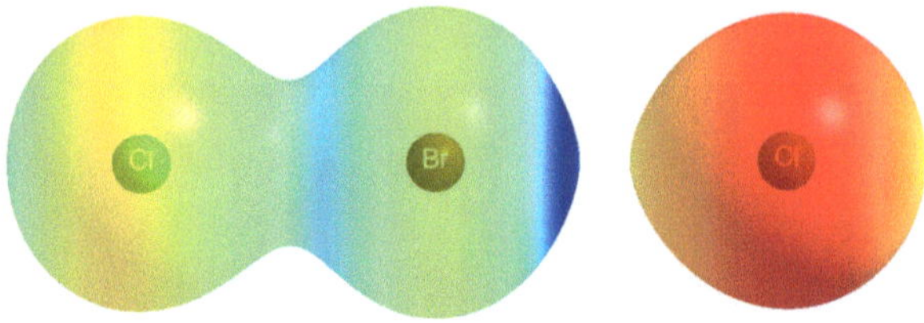

Fig. 15 Electrostatic potential (mapped onto an $\rho = 0.070$ au surface) for $Cl - Br...Cl^-$ showing a σ-hole as the blue (most positive) region around Br facing a Cl^- atom. Data obtained at the M06-2X-D3/x2c-TZVPPall level of theory. Image drawn with AIMAll [37]

(ESP), provides a dipolar structure that can be exploited to stabilize molecular aggregates via secondary electrostatic interactions, and has been used to explain the ability of halogen atoms to act not only as Lewis bases, but also as nucleophilic sites [95, 96]. This is the halogen bond (XB in what follows), in which a halogen atom with the aforementioned charge distribution is attacked by a nucleophile.

Since crystal engineering and supramolecular chemistry have become important areas of research, many works have appeared to shed light on this type of non-covalent interaction, resorting not only to the ESP, but also to a wealth of other techniques, mainly based on an orbital-based perspective [95, 97–103]. Interestingly, many of these studies have revealed that non-electrostatic contributions such as polarization, charge transfer, or dispersion play a crucial role in the stabilization of XBs. Less effort has been paid, however, in assessing such interactions by means of real-space methods [102, 104–112]. Note that the precise meaning of polarization, charge transfer, or dispersion depends crucially on the type of energy decomposition and that all of them are fully incorporated in the IQA exchange-correlation term.

Let us then comment on how XBs are described with our technologies in very simple $X...Y...X^-$ (X, Y = Cl, Br, I) model systems [113]. To simulate the crystalline environment in which XBs are usually found, we include both optimized symmetric $[X_1...Y_2...X_3]^-$ arrangements and linear asymmetric $X_1 - Y_2...X_3^-$ ones with fixed $X_1 - Y_2$ distances (as found in their isolated diatomics), although for the sake of simplicity only the latter will be addressed herein. In addition, we only consider IQA in this section, since in the case of non-covalent interactions the populations of the various molecular fragments are dominated by the parent RSRSs. The σ-hole model, as usually envisaged, naïvely assumes that there is a stabilizing electrostatic interaction between the negatively charged lone pair torus of one fragment and the σ-hole region of another, both located on their respective ESP envelopes, forgetting that electrostatics involves the full 3D electron distribution. Since the latter is correctly taken into account in IQA, it is interesting to check whether the model's insight corresponds to reality or not.

Table 1 gathers the results obtained for the Cl series, taken as a representative example of our model systems. Some raw data have been taken from Ref. [113]. As can be seen, in agreement with other energy decomposition analyses [114], IQA shows that, contrary to the common rationalization of this non-covalent interaction

Table 1 Atomic charges (in au, first column) and IQA energies (kcal mol^{-1}) for the chlorine series. The central panel contains interatomic interactions between atoms Y_2 and X_3^- (labeled as *AB*) and the right panel interfragment interactions, where $\mathcal{G}$ is the $X_1 - Y_2$ group and $\mathcal{H}$ denotes X_3^-. $E_{bind} = E_{def} + E_{inter}^{\mathcal{GH}}$, $E_{def} = E_{def}^{\mathcal{G}} + E_{def}^{\mathcal{H}}$. M06-2X-D3/x2c-TZVPPall data with RESC scalar relativistic corrections for iodine-containing systems

$X_1 - Y_2...X_3^-$ Complex	E_{inter}^{AB}	V_{xc}^{AB}	V_{cl}^{AB}	E_{bind}	E_{def}	$E_{inter}^{\mathcal{GH}}$	$V_{xc}^{\mathcal{GH}}$	$V_{cl}^{\mathcal{GH}}$
$Cl^{-0.30} - Cl^{+0.04}...Cl^{-0.74}$	− 66.78	− 60.61	− 6.17	− 21.61	38.31	− 59.92	− 66.06	6.14
$Cl^{-0.46} - Br^{+0.16}...Cl^{-0.70}$	− 94.12	− 76.97	− 17.15	− 32.64	46.45	− 79.10	− 82.41	3.31
$Cl^{-0.61} - I^{+0.34}...Cl^{-0.73}$	− 115.65	− 75.89	− 39.76	− 35.47	54.86	− 90.33	− 79.89	− 10.44
$Br^{-0.22} - Cl^{-0.10}...Br^{-0.69}$	− 58.30	− 64.18	5.87	− 22.14	36.65	− 58.80	− 70.75	11.95
$I^{-0.06} - Cl^{-0.28}...I^{-0.66}$	− 37.72	− 53.69	15.98	− 16.53	30.46	− 46.99	− 61.05	14.06

in terms of electrostatics, it is the nonclassical part of the interaction energy, i.e., the exchange-correlation contribution that measures covalency, that mainly accounts for the total interaction energy between the atoms involved in the σ-hole interaction. It is interesting to see that as the central atom becomes more polarizable and the attacking halide more polarizing, V_{cl} significantly becomes more and more stabilizing. However, even in highly polarized cases, it is the non-electrostatic contribution V_{xc} that remains dominant. Moreover, when the positions of the more polarizing-more polarizable atoms are interchanged and thus X_3- becomes more polarizable than Y_2, it is found that although the central-terminal atomic interactions are overall favorable, the electrostatic contributions may become destabilizing. The final value of the electrostatic interaction between two moieties is therefore difficult to predict from simple ESP envelope plots.

The polarizable-polarizing character of the interacting $Y_2...X_3$ halogen atoms (and their concomitant electronegativity difference) therefore modulates the strength of their mutual electrostatic interaction, which can even become positive (destabilizing). Interestingly, the signs and magnitude of total electrostatic interactions correlate well with the magnitude and sign of the atomic charges involved (although higher-order multipoles also contribute to the total V_{cl}).

Atomic charges not only provide information about the Coulomb interaction between atoms but can also reveal bonding patterns. By studying their change upon complexation, it is found that a large charge transfer takes place between the edge atoms, with $\Delta Q(X_1) \approx -\Delta Q(X_3)$. They also show that the central atom does not seem to play an important role, as it roughly maintains the charge it carries in the corresponding diatomic. Real-space analyses thus provide clear evidence that the σ-hole model is flawed and in need of revision, while providing new insights into the role of charge transfer. To investigate the latter finding further, Fig. 16 shows how the QTAIM charges evolve as the $Br_2...Cl_3$ distance is changed in $Cl_1 - Br_2...Cl_3^-$. The edge atoms do indeed follow mirror changes in their charge, while the central one shows no significant change in the process. This finding is consistent with the well-known Pimentel-Rundle model [115, 116] for 3c-4e (3 center–4 electrons) bonds, which explains the bonding with purely orbital arguments without invoking electrostatic contributions, thus reinforcing the abovementioned nonclassical character of this type of interactions.

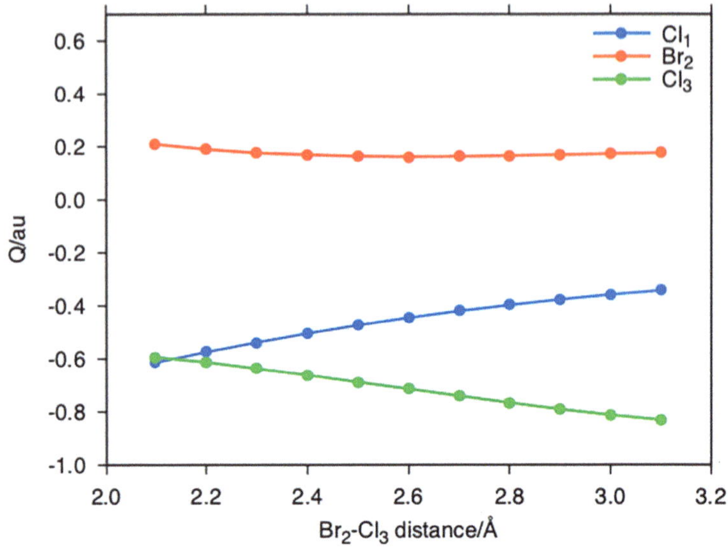

Fig. 16 Evolution of atomic charges as the $Br_2 \ldots Cl_3$ distance is changed in $Cl_1 - Br_2 \ldots Cl_3^-$. Data obtained at the M06-2X-D3/x2c-TZVPPall level of theory

Further insight can be gained if, as in our previous EDF examples, we group several atoms together. To that end, we can consider the molecule as composed of a diatomic featuring a σ-hole and a halide that attacks it (see the right-side block of Table 1). This partitioning is particularly suitable to investigate XB formation and shows how the two interacting moieties deform considerably upon interaction, although the final binding energy becomes negative only after the strongly stabilizing interaction energy is added.

We also note that the determining role of nonclassical contributions was also pointed out by Wolters and Bickelhaupt [117], who also studied these trihalide systems. However, they reported stabilizing electrostatics, which can be attributed to the increased electrostatic stabilization associated with orbital-based energy decomposition analyses that use interpenetrating electron densities. Overall, the real-space analysis of XB in model systems shows that the σ-hole concept and the associated electrostatic description of XB, although useful as a qualitative tool, are far from capturing the essential features of this type of interaction, which are mainly nonclassical in nature.

As shown in the previous showcase analyses, IQA in combination with QTAIM partitioning results in a very well-suited way to elucidate the nature and properties of a non-covalent interaction such as XB on the same basis as for any other type of chemical bond or specific interaction.

Conclusion

This chapter has provided a comprehensive exploration of quantum chemical topology (QCT) and its contemporary applications in analyzing molecular electronic structures. Approached from different perspectives—the electron counting and the energetic analysis—QCT has been an essential instrument to strengthen the knowledge about several challenging chemical interactions.

It has been shown that real-space partitioning techniques provide deep insights into electron distribution and bonding. By leveraging QTAIM atomic basins, we have seen how the partitioning of space based on electron density gradients offers a robust way to define atomic contributions to molecular properties. The ELF framework complements this by identifying chemically meaningful entities such as lone pairs, bonding regions, and multicenter interactions, which are often less apparent in traditional electronic structure calculations. Furthermore, the QTAIM-ELF intersection (QEI) has emerged as an interesting tool for bridging gaps between different bonding models.

Energy decomposition analysis within the IQA framework has further demonstrated the power of real-space methods by depicting a clear distinction between classical electrostatic interactions and quantum mechanical exchange-correlation effects. This approach has been particularly useful in offering a quantitatively rigorous means of assessing the relative contributions of different bonding forces.

Beyond its methodological contributions, we have tried to show in this chapter the broader impact of QCT in advancing our understanding of chemical bonding and molecular behavior. The capacity to analyze electron distributions at different levels of granularity, i.e., from individual atomic basins to more intricate molecular frameworks, illustrates the robustness and adaptability of these methods in capturing both localized and delocalized electronic effects in diverse contexts.

The practical applications explored in this chapter reinforce the significance of QCT as a powerful framework for interpreting molecular electronic structure. These methods allow for an unambiguous description of the wavefunction and reconcile multiple interpretations of bonding through a physically grounded perspective. As computational chemistry continues to evolve, real-space analyses will continue to ensure an exhaustive feedback about the wide range of chemical interactions in the vast array of systems to explore.

Acknowledgements The authors acknowledge the Ministerio de Ciencia e Innovación (Grant No. PID2021-122763NBI00), as well as Fundación Ibercaja and Universidad de Zaragoza (JIUZ2023-CIE-10) for financial support. D.B.-E. acknowledges the Spanish FICyT for a predoctoral grant (PA-23-BP22-168).

References

1. Pauling L (1931) J Am Chem Soc 53(9):3225. https://doi.org/10.1021/ja01360a004
2. Mulliken RS (1928) Phys Rev 32:186. https://doi.org/10.1103/PhysRev.32.186. https://link.aps.org/doi/10.1103/PhysRev.32.186
3. Mulliken RS (1928) Phys Rev 32:761. https://doi.org/10.1103/PhysRev.32.761. https://link.aps.org/doi/10.1103/PhysRev.32.761
4. Lennard-Jones JE (1929) Trans Faraday Soc 25:668. https://doi.org/10.1039/TF9292500668
5. Hoffmann R, Shaik S, Hiberty PC (2003) Account Chem Res 36(10):750. https://doi.org/10.1021/ar030162a
6. Galbraith JM, Shaik S, Danovich D, Braïda B, Wu W, Hiberty P, Cooper DL, Karadakov PB, Dunning THJ (2021) J Chem Educ 98(12):3617. https://doi.org/10.1021/acs.jchemed.1c00919
7. Slater JC (1932) Phys Rev 41:255. https://doi.org/10.1103/PhysRev.41.255.2. https://link.aps.org/doi/10.1103/PhysRev.41.255.2
8. Van Vleck JH, Sherman A (1935) Rev Mod Phys 7:167. https://doi.org/10.1103/RevModPhys.7.167. https://link.aps.org/doi/10.1103/RevModPhys.7.167
9. Hiberty PC, Leforestier C (1978) J Am Chem Soc 100(7):2012. https://doi.org/10.1021/ja00475a007
10. Hiberty PC, Ohanessian G (1985) Int J Quant Chem 27(3):259. https://doi.org/10.1002/qua.560270304. https://onlinelibrary.wiley.com/doi/abs/10.1002/qua.560270304
11. Blanco MA, Martín Pendás Á, Francisco E (2005) J Chem Theory Comput 1(6):1096. https://doi.org/10.1021/ct0501093
12. Popelier PLA (2016) On quantum chemical topology. Springer International Publishing, Cham, pp 23–52. https://doi.org/10.1007/978-3-319-29022-5_2
13. Stone A (2013) The theory of intermolecular forces. Oxford University Press, Oxford. https://doi.org/10.1093/acprof:oso/9780199672394.001.0001
14. Hayes I, Stone A (1984) Mol Phys 53(1):83. https://doi.org/10.1080/00268978400102151
15. Martín Pendás Á, Kohout M, Blanco MA, Francisco E (2012) Beyond standard charge density topological analyses. Springer Netherlands, Dordrecht, pp 303–358. https://doi.org/10.1007/978-90-481-3836-4_9
16. Becke AD (1988) J Chem Phys 88(4):2547. https://doi.org/10.1063/1.454033
17. Salvador P, Ramos-Cordoba E (2013) J Chem Phys 139(7):071103. https://doi.org/10.1063/1.4818751
18. Hirshfeld FL (1977) Theor Chem Accounts 44:129. https://doi.org/10.1007/BF00549096
19. Slater JC (1964) J Chem Phys 41(10):3199. https://doi.org/10.1063/1.1725697
20. Francisco E, Martín Pendás Á, Blanco MA (2006) J Chem Theory Comput 2(1):90. https://doi.org/10.1021/ct0502209. PMID: 26626383
21. Silvi B, Savin A (1994) Nature 371:683. https://doi.org/10.1038/371683a0
22. Dillen J (2015) J Comput Chem 36(12):883. https://doi.org/10.1002/jcc.23869. https://onlinelibrary.wiley.com/doi/abs/10.1002/jcc.23869
23. Malcolm NOJ, Popelier PLA (2003) Faraday Discuss 124:353. https://doi.org/10.1039/B211650M
24. Born M, Oppenheimer R (1927) Annalen der Physik 389(20):457. https://doi.org/10.1002/andp.19273892002. https://onlinelibrary.wiley.com/doi/abs/10.1002/andp.19273892002
25. Bader R (1990) Atoms in molecules: A quantum theory. International series of monographs on chemistry. Clarendon Press, London
26. Kato T (1957) Commun Pure Appl Math 10(2):151. https://doi.org/10.1002/cpa.3160100201. https://onlinelibrary.wiley.com/doi/abs/10.1002/cpa.3160100201
27. Becke AD, Edgecombe KE (1990) J Chem Phys 92(9):5397. https://doi.org/10.1063/1.458517
28. Polo V, Andres J, Berski S, Domingo LR, Silvi B (2008) J Phys Chem A 112(31):7128. https://doi.org/10.1021/jp801429m. PMID: 18557601

29. Andrés J, Berski S, Silvi B (2016) Chem Commun 52:8183. https://doi.org/10.1039/C5CC09816E
30. Kohout M (2004) Int J f Quant Chem 97(1):651. https://doi.org/10.1002/qua.10768. https://onlinelibrary.wiley.com/doi/abs/10.1002/qua.10768
31. Schmider H, Becke A (2000) J Mol Struct THEOCHEM 527:51. https://doi.org/10.1016/S0166-1280(00)00477-2
32. Martín Pendás Á, Blanco MA, Costales A, Sánchez PM, Luaña V (1999) Phys Rev Lett 83:1930. https://doi.org/10.1103/PhysRevLett.83.1930. https://link.aps.org/doi/10.1103/PhysRevLett.83.1930
33. Dye JL (2009) Account Chem Res 42(10):1564. https://doi.org/10.1021/ar9000857
34. Saha R, Das P, Chattaraj PK (2022) ChemPhysChem 23(23):e202200329. https://doi.org/10.1002/cphc.202200329. https://chemistry-europe.onlinelibrary.wiley.com/doi/abs/10.1002/cphc.202200329
35. Imagawa T, Giarrana L, Andrada DM, Morgenstern B, Nakamoto M, Scheschkewitz D (2023) J Am Chem Soc 145(8):4757. https://doi.org/10.1021/jacs.2c13530
36. Vidal L, Barrena-Espés D, Echeverría J, Munárriz J, Martín Pendás Á (2024) ChemPhysChem 25(22):e202400329. https://doi.org/10.1002/cphc.202400329. https://chemistry-europe.onlinelibrary.wiley.com/doi/abs/10.1002/cphc.202400329
37. Keith TA (2019) TK Gristmill Software. Overland Park KS, USA. aim.tkgristmill.com
38. García-Revilla M, Popelier PLA, Francisco E, Martín Pendás Á (2011) J Chem Theory Comput 7(6):1704. https://doi.org/10.1021/ct2001842. PMID: 26596434
39. Outeiral C, Vincent MA, Martín Pendás Á, Popelier PLA (2018) Chem Sci 9:5517. https://doi.org/10.1039/C8SC01338A
40. Mayer I (1983) Chem Phys Lett 97(3):270. https://doi.org/10.1016/0009-2614(83)80005-0. https://www.sciencedirect.com/science/article/pii/0009261483800050
41. Julg A, Julg P (1978) Int J Quant Chem 13(4):483
42. Ángyán JG, Rosta E, Surján PR (1999) Chem Phys Lett 299(1):1. https://doi.org/10.1016/S0009-2614(98)01255-X. https://www.sciencedirect.com/science/article/pii/S000926149801255X
43. Hohenberg P, Kohn W (1964) Phys Rev 136:B864. https://doi.org/10.1103/PhysRev.136.B864. https://link.aps.org/doi/10.1103/PhysRev.136.B864
44. Landeros-Rivera B, Gallegos M, Munárriz J, Laplaza R, Contreras-García J (2022) Phys Chem Chem Phys 24:21538. https://doi.org/10.1039/D2CP01517J
45. Dobson JF (1991) J Chem Phys 94(6):4328. https://doi.org/10.1063/1.460619
46. Pettersen EF, Goddard TD, Huang CC, Couch GS, Greenblatt DM, Meng EC, Ferrin TE (2004) J Comput Chem 25(13):1605. https://doi.org/10.1002/jcc.20084. https://onlinelibrary.wiley.com/doi/abs/10.1002/jcc.20084
47. Raub S, Jansen G (2001) Theor Chem Accounts 106:223. https://doi.org/10.1007/s002140100268
48. Barrena-Espés D, Munárriz J, Martín Pendás Á (2024) J Chem Phys 160(14):144106. https://doi.org/10.1063/5.0199318
49. Klessinger M (1970) Angewandte Chemie. International Edition in English 9(7):500. https://doi.org/10.1002/anie.197005001. https://onlinelibrary.wiley.com/doi/abs/10.1002/anie.197005001
50. Mullay J (1987) Propell Explos Pyrotech 12(4):121. https://doi.org/10.1002/prep.19870120403. https://onlinelibrary.wiley.com/doi/abs/10.1002/prep.19870120403
51. Ferro-Costas D, Pérez-Juste I, Mosquera RA (2014) J Comput Chem 35(13):978. https://doi.org/10.1002/jcc.23574. https://onlinelibrary.wiley.com/doi/abs/10.1002/jcc.23574
52. Ferro-Costas D, Mosquera RA (2013) J Chem Theory Comput 9(11):4816. https://doi.org/10.1021/ct400630f
53. Francisco E, Martín Pendás Á, Blanco MA (2007) J Chem Phys 126(9):094102. https://doi.org/10.1063/1.2709883

54. Martín Pendás Á, Francisco E, Blanco MA (2007) J Phys Chem A 111(6):1084. https://doi.org/10.1021/jp064600h
55. Martín Pendás Á, Francisco E, Blanco MA (2007) Phys Chem Chem Phys 9:1087. https://doi.org/10.1039/B616310F
56. Martín Pendás Á, Francisco E, Blanco MA (2007) J Chem Phys 127(14):144103. https://doi.org/10.1063/1.2784392
57. Francisco E, Martín Pendás Á, Blanco MA (2008) Comput Phys Commun 178(8):621. https://doi.org/10.1016/j.cpc.2007.11.015. https://www.sciencedirect.com/science/article/pii/S0010465507004882
58. Francisco E, Martín Pendás Á (2014) Comput Phys Commun 185(10):2663. https://doi.org/10.1016/j.cpc.2014.05.009. https://www.sciencedirect.com/science/article/pii/S0010465514001635
59. Martín Pendás Á, Francisco E (2019) ChemPhysChem 20(21):2722. https://doi.org/10.1002/cphc.201900641. https://chemistry-europe.onlinelibrary.wiley.com/doi/abs/10.1002/cphc.201900641
60. Cancès E, Keriven R, Lodier F, Savin A (2004) Theor Chem Accounts 111:373. https://doi.org/10.1007/s00214-003-0509-4
61. Francisco E, Martín Pendás Á, Blanco MA (2011) Theor Chem Accounts 128:433. https://api.semanticscholar.org/CorpusID:97555235
62. Kurth S, Perdew JP (2000) Int J Quant Chem 77(5):814. https://doi.org/10.1002/(SICI)1097-461X(2000)77:5%3C814::AID-QUA3%3E3.0.CO;2-F
63. Francisco E, Martín Pendás Á, García-Revilla M, Álvarez Boto R (2013) Comput Theor Chem 1003:71. https://doi.org/10.1016/j.comptc.2012.09.009. https://www.sciencedirect.com/science/article/pii/S2210271X12004756. Reduced density matrices: A simpler approach to many-electron problems?
64. Jeziorski B, Moszynski R, Szalewicz K (1994) Chem Rev 94(7):1887. https://doi.org/10.1021/cr00031a008
65. Kitaura K, Morokuma K (1976) Int J Quant Chem 10(2):325. https://doi.org/10.1002/qua.560100211. https://onlinelibrary.wiley.com/doi/abs/10.1002/qua.560100211
66. Ziegler T, Rauk A (1977) Theor Chimica Acta 46:1. https://doi.org/10.1007/BF02401406
67. Bickelhaupt FM, Baerends EJ (2000) Kohn-Sham density functional theory: Predicting and understanding chemistry. Wiley, New York, pp 1–86. https://doi.org/10.1002/9780470125922.ch1. https://onlinelibrary.wiley.com/doi/abs/10.1002/9780470125922.ch1
68. Zhao L, von Hopffgarten M, Andrada DM, Frenking G (2018) WIREs Comput Mol Sci 8(3):e1345. https://doi.org/10.1002/wcms.1345. https://wires.onlinelibrary.wiley.com/doi/abs/10.1002/wcms.1345
69. Levine DS, Head-Gordon M (2017) Proc Natl Acad Sci 114(48):12649. https://doi.org/10.1073/pnas.1715763114. https://www.pnas.org/doi/abs/10.1073/pnas.1715763114
70. Martín Pendás Á, Blanco MA, Francisco E (2006) J Chem Phys 125(18):184112. https://doi.org/10.1063/1.2378807
71. Martín Pendás Á, Blanco MA, Francisco E (2007) J Comput Chem 28(1):161. https://doi.org/10.1002/jcc.20469. https://onlinelibrary.wiley.com/doi/abs/10.1002/jcc.20469
72. Symons BCB, Williamson DJ, Brooks CM, Wilson AL, Popelier PLA (2019) ChemistryOpen 8(5):553. https://doi.org/10.1002/open.201900096. https://chemistry-europe.onlinelibrary.wiley.com/doi/abs/10.1002/open.201900096
73. Gallegos M, Costales A, Martín Pendás Á (2021) ChemPhysChem 22(8):775. https://doi.org/10.1002/cphc.202000975. https://chemistry-europe.onlinelibrary.wiley.com/doi/abs/10.1002/cphc.202000975
74. Martín Pendás Á, Casals-Sainz JL, Francisco E (2019) Chem A Eur J 25(1):309. https://doi.org/10.1002/chem.201804160. https://chemistry-europe.onlinelibrary.wiley.com/doi/abs/10.1002/chem.201804160

75. Martín Pendás Á, Francisco E, Suárez D, Costales A, Díaz N, Munárriz J, Rocha-Rinza T, Guevara-Vela JM (2023) Phys Chem Chem Phys 25:10231. https://doi.org/10.1039/D2CP05540F
76. Frisch MJ, Trucks GW, Schlegel HB, Scuseria GE, Robb MA, Cheeseman JR, Scalmani G, Barone V, Petersson GA, Nakatsuji H, Li X, Caricato M, Marenich AV, Bloino J, Janesko BG, Gomperts R, Mennucci B, Hratchian HP, Ortiz JV, Izmaylov AF, Sonnenberg JL, Williams-Young D, Ding F, Lipparini F, Egidi F, Goings J, Peng B, Petrone A, Henderson T, Ranasinghe D, Zakrzewski VG, Gao J, Rega N, Zheng G, Liang W, Hada M, Ehara M, Toyota K, Fukuda R, Hasegawa J, Ishida M, Nakajima T, Honda Y, Kitao O, Nakai H, Vreven T, Throssell K, Montgomery JA Jr, Peralta JE, Ogliaro F, Bearpark MJ, Heyd JJ, Brothers EN, Kudin KN, Staroverov VN, Keith TA, Kobayashi R, Normand J, Raghavachari K, Rendell AP, Burant JC, Iyengar SS, Tomasi J, Cossi M, Millam JM, Klene M, Adamo C, Cammi R, Ochterski JW, Martin RL, Morokuma K, Farkas O, Foresman JB, Fox DJ (2016) Gaussiañ16 Revision C.01. Gaussian Inc. Wallingford CT
77. Neese F (2022) WIREs Comput Mol Sci 12(5). https://doi.org/10.1002/wcms.1606
78. Martín Pendás A, Francisco E (2025) Promolden. A qtaim/iqa code. Available from the authors upon request
79. Francisco E, Martín Pendás A, Blanco M (2008) Comput Phys Commun 178(8):621
80. Noury FFS, Krokidis X, Silvi B (1997) The topmod package. https://www.lct.jussieu.fr/pagesperso/silvi/topmod.html
81. Bader RFW, Bandrauk AD (1968) J Chem Phys 49(4):1653. https://doi.org/10.1063/1.1670292
82. Silvi B (2004) Phys Chem Chem Phys 6:256. https://doi.org/10.1039/B311272A
83. Shaik S, Rzepa HS, Hoffmann R (2013) Angewandte Chemie. International Edition 52(10): 3020. https://doi.org/10.1002/anie.201208206. https://onlinelibrary.wiley.com/doi/abs/10.1002/anie.201208206
84. Frenking G, Hermann M (2013) Angewandte Chemie. International Edition 52(23):5922. https://doi.org/10.1002/anie.201301485. https://onlinelibrary.wiley.com/doi/abs/10.1002/anie.201301485
85. Kroto HW, Heath JR, O'Brien SC, Curl RF, Smalley RE (1985) Nature 318:162. https://doi.org/10.1038/318162a0
86. Sun Q, Mück-Lichtenfeld C, Kehr G, Erker G (2023) Nature Rev Chem 7:732. https://doi.org/10.1038/s41570-023-00525-7
87. Luder DJ, Terefenko N, Sun Q, Eckert H, Mück-Lichtenfeld C, Kehr G, Erker G, Wiegand T (2024) Chem A Eur J 30(11):e202303701. https://doi.org/10.1002/chem.202303701. https://chemistry-europe.onlinelibrary.wiley.com/doi/abs/10.1002/chem.202303701
88. Minyaev RM, Minkin VI (2008) Russ J Gener Chem 78:732. https://doi.org/10.1134/S1070363208040348
89. Casals-Sainz JL, Jara-Cortés J, Hernández-Trujillo J, Guevara-Vela JM, Francisco E, Martín Pendás Á (2019) Chem A Eur J 25(52):12169. https://doi.org/10.1002/chem.201902369. https://chemistry-europe.onlinelibrary.wiley.com/doi/abs/10.1002/chem.201902369
90. Sowlati-Hashjin S, Šadek V, Sadjadi S, Karttunen M, Martín Pendás Á, Foroutan-Nejad C (2022) Nature Commun 13:2069. https://doi.org/10.1038/s41467-022-29504-0
91. Lee VY, Gapurenko OA, Ito Y, Meguro T, Sugasawa H, Sekiguchi A, Minyaev RM, Minkin VI, Herber RH, Gornitzka H (2016) Organometallics 35(3):346. https://doi.org/10.1021/acs.organomet.5b00924
92. Minyaev RM, Gribanova TN (2000) Russ Chem Bull 49:783. https://doi.org/10.1007/BF02494697
93. Torstensen K, Ghosh A (2024) ACS Organic Inorganic Au 4(1):102. https://doi.org/10.1021/acsorginorgau.3c00041
94. Hobza P, Řezáč J (2016) Chem Rev 116(9):4911. https://doi.org/10.1021/acs.chemrev.6b00247

95. Cavallo G, Metrangolo P, Milani R, Pilati T, Priimagi A, Resnati G, Terraneo G (2016) Chem Rev 116(4):2478. https://doi.org/10.1021/acs.chemrev.5b00484
96. Politzer P, Murray JS, Clark T (2013) Phys Chem Chem Phys 15:11178. https://doi.org/10.1039/C3CP00054K
97. Wolters LP, Schyman P, Pavan MJ, Jorgensen WL, Bickelhaupt FM, Kozuch S (2014) WIREs Comput Mol Sci 4(6):523. https://doi.org/10.1002/wcms.1189
98. Kolář MH, Hobza P (2016) Chem Rev 116(9):5155. https://doi.org/10.1021/acs.chemrev.5b00560
99. Costa PJ (2017) Phys Sci Rev 2(11):20170136. https://doi.org/10.1515/psr-2017-0136
100. Wolters LP, Smits NWG, Guerra CF (2015) Phys Chem Chem Phys 17(3):1585. https://doi.org/10.1039/c4cp03740e
101. Novák M, Foroutan-Nejad C, Marek R (2015) Phys Chem Chem Phys 17(9):6440. https://doi.org/10.1039/c4cp05532b
102. Bora PL, Novák M, Novotný J, Foroutan-Nejad C, Marek R (2017) Chem Eur J 23(30):7315. https://doi.org/10.1002/chem.201700179
103. Thirman J, Engelage E, Huber SM, Head-Gordon M (2018) Phys Chem Chem Phys 20(2):905. https://doi.org/10.1039/c7cp06959f
104. Syzgantseva OA, Tognetti V, Joubert L (2013) J Phys Chem A 117(36):8969. https://doi.org/10.1021/jp4059774
105. Yahia-Ouahmed M, Tognetti V, Joubert L (2015) Comput Theor Chem 1053:254. https://doi.org/10.1016/j.comptc.2014.09.004
106. Yahia-Ouahmed M, Tognetti V, Joubert L (2016) Theor Chem Acc 135(2). https://doi.org/10.1007/s00214-015-1796-2
107. Eskandari K, Lesani M (2015) Chem Eur J 21(12):4739. https://doi.org/10.1002/chem.201405054
108. Jahromi HJ, Eskandari K, Alizadeh A (2015) J Mol Model 21(5). https://doi.org/10.1007/s00894-015-2660-y
109. Buralli G, Petelski A, Peruchena N, Sosa G, Duarte D (2017) Molecules 22(11):2034. https://doi.org/10.3390/molecules22112034
110. Guevara-Vela JM, Ochoa-Resendiz D, Costales A, Hernández-Lamoneda R, Martín PendásÁ (2018) ChemPhysChem 19(19):2512. https://doi.org/10.1002/cphc.201800474
111. Orangi N, Eskandari K, Thacker JCR, Popelier PLA (2019) ChemPhysChem 20(15):1922. https://doi.org/10.1002/cphc.201900250
112. Alkorta I, Silva AF, Popelier PLA (2020) Molecules 25(11):2674. https://doi.org/10.3390/molecules25112674
113. Jiménez-Grávalos F, Gallegos M, Martín Pendás Á, Novikov AS (2021) J Comput Chem 42(10):676. https://doi.org/10.1002/jcc.26488. https://onlinelibrary.wiley.com/doi/abs/10.1002/jcc.26488
114. Martín Pendás A, Casals Sainz JL, Francisco E (2017) Quantitative determination of the nature of intermolecular bonds by EDA analysis. The Royal Society of Chemistry, pp 178–221. https://doi.org/10.1039/BK9781782621737-00178
115. Pimentel GC (1951) J Chem Phys 19(4):446. https://doi.org/10.1063/1.1748245
116. Rundle RE (1957) J Phys Chem 61(1):45. https://doi.org/10.1021/j150547a010
117. Wolters LP, Bickelhaupt FM (2012) ChemistryOpen 1(2):96. https://doi.org/10.1002/open.201100015

Struct Bond (2026) 190: 263–296
https://doi.org/10.1007/430_2025_102

Published online: 13 August 2025

The Interpenetration Index and Its Applications in Chemistry

Jorge Echeverría and Santiago Alvarez

Contents

Abstract We introduce the *interpenetration index* (p_{AB}) as a unified metric for quantifying and comparing interatomic interactions across an extensive range of chemical systems. Building on the idea of representing each atom by three concentric spheres (core, valence, and van der Waals), we define the interpenetration index that measures the overlap of the outermost (van der Waals) regions of two nearby atoms. This approach yields a size-corrected parameter that places all interactions – from van der Waals contacts in noble gas dimers to multiple bonds in main-group,

J. Echeverría (✉)
Departamento de Química Inorgánica, Instituto de Síntesis Química y Catálisis Homogénea (ISQCH), CSIC-Universidad de Zaragoza, Zaragoza, Spain
e-mail: jorge.echeverria@unizar.es

S. Alvarez
Department de Química Inorgànica i Orgànica, Secció de Química Inorgànica, and Institut de Química Teòrica i Computacional, Universitat de Barcelona, Barcelona, Spain

transition-metal, and actinide complexes – on a single, continuous scale. Through computational and experimental case studies, we show that p_{AB} can successfully distinguish weak dispersion-driven associations, moderate hydrogen and σ-hole bonds, and strong single or multiple covalent bonds. Specifically, van der Waals systems map onto low or near-zero penetrations, while covalent and ionic bonds exhibit penetrations of 100% or higher. The method also accounts for more complex scenarios, such as proton transfer in hydrogen bonds and high-order metal–metal or actinyl bonds with unusually large overlaps. This framework offers a simple yet robust way to interpret and predict structure–property relationships in diverse chemical contexts, emphasizing that bonding should be viewed as a continuum rather than discrete categories.

Keywords Chemical bonding · Interpenetration index · Noncovalent interactions · Penetration index · van der Waals crust

1 Introduction

Chemists commonly depict atoms as hard spheres that either can exist independently or are linked together by "sticks" representing chemical bonds. This simple picture, while an approximation, has proven remarkably effective for describing and predicting the structures, stabilities, and reactivities of a vast array of chemical entities. The size of the atomic spheres (i.e. atomic radii) and the length of the sticks (bond lengths) together constitute the bedrock of structural chemistry. They provide parameters that let us rationalize why certain molecules adopt particular geometries, how they pack in crystals, or how they form supramolecular assemblies.

Despite its enduring usefulness, the simplistic "ball-and-stick" model glosses over the reality that the electron density around an atom is continuous and decays exponentially with the distance to the nucleus. There is no abrupt boundary telling us precisely where an atom "ends". Instead, chemists rely on conventions – for example, specifying the distance that encloses 99% of the electron density or the isosurface at which the electron density takes on a defined value (e.g. 0.001 e/bohr^3) [1]. These conventions allow us to carve out an "effective radius" for each atom.

1.1 Concentric Spheres: Core, Valence, and van der Waals Regions

In this chapter, we focus on a conceptual model in which each atom is represented by three concentric spheres of diminishing electron density [2] (Fig. 1a, b):

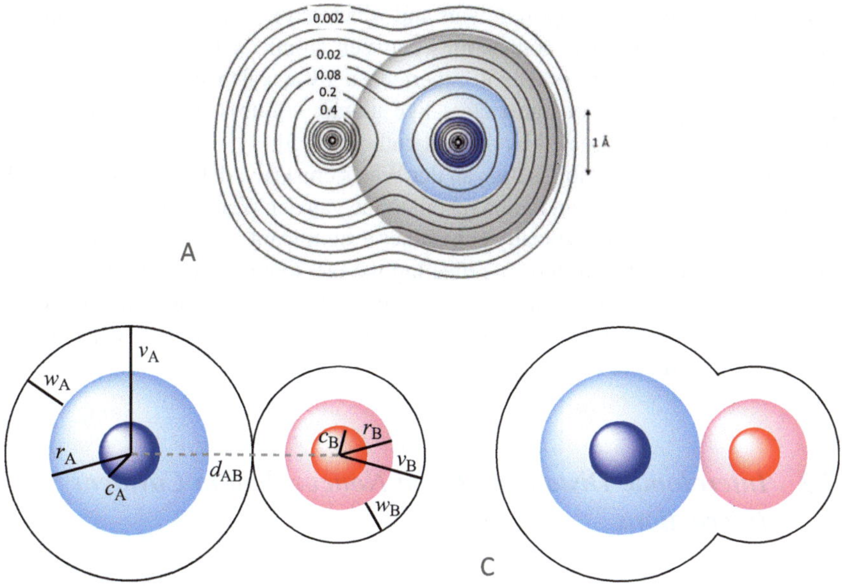

Fig. 1 (**a**) Calculated electron density map for the Cl_2 molecule superimposed on the idealized core (dark blue), valence (light blue), and van der Waals (grey) spheres of one of the Cl atoms defined by the corresponding radii. Isodensity lines shown in electrons/bohr^3. (**b**) Definition of the core (dark colour), valence (light colour), and van der Waals (transparent) spheres with radii *c*, *r*, and *v*, and of the van der Waals crust of width *w* for atoms A and B, and visual description of the case of two contacting van der Waals spheres with no interpenetration. (**c**) Schematic description of the case in which two atoms are at a distance equal to the sum of the covalent radii ($d_{AB} = r_A + r_B$). (Adapted from Ref. [2] with permission from the RSC)

1. A core sphere of radius *c*, primarily harbouring the deeply bound core electrons
2. A valence sphere of radius *r*, which includes the electrons in the atom's highest occupied shell (typically the one engaged in chemical bonding)
3. A van der Waals (vdW) sphere of radius *v*, accounting for the diffusely distributed electron density at larger distances from the nucleus

From a quantum chemical standpoint, these three regions loosely map onto electron shells characterized by principal quantum numbers 1 to $n - 1$ (core), n (valence), and $n + i$ (extended region, with $i \geq 1$) – though the details can be more nuanced depending on the specific atom and its orbitals. Experimentally, the boundaries r and v may be determined from empirical data such as X-ray crystallography (covalent radii) and gas-phase or molecular beam scattering (van der Waals radii). Interestingly, comparison of available core radii with covalent radii shows a good linear correlation according to $c = 1.09r - 0.65$ ($r^2 = 0.944$); that is, we can roughly estimate the core radius of an atom as its covalent radius minus 0.65 Å.

In the following, we will use the empirical sets of atomic covalent [3] and van der Waals [4] radii, derived from crystallographic and spectroscopic gas-phase structural data, that were recently proposed by us and which have been broadly adopted by the chemical community. The connections between atomic radii and other properties, such as the electrostatic potential experienced by a valence electron, a constant electron density value, or parameters like atomic softness or polarizability, have been thoroughly investigated by several authors [1, 5].

Central to our discussion is the van der Waals crust (w), defined as the radial shell extending from the covalent radius (r) out to the van der Waals radius (v), with width $w = v - r$ [6]. This crust typically constitutes a significant fraction of the total atomic volume (often 60–90% or more), reflecting the fact that electron density, while low in this outer region, still occupies an extensive radial domain. Note that noble gases and late transition metals often exhibit comparatively wider crusts, whereas alkali metals show narrower ones.

1.2 Interatomic Distances: From Noncovalent to Covalent Interactions

When two atoms approach one another, the extent of overlap of their outer electron-density regions determines whether the interaction is weak (e.g. van der Waals or London dispersion) or strong (covalent or ionic bonding). Two conventional metrics help demarcate these extremes:

1. The van der Waals distance, set by the sum of van der Waals radii ($v_A + v_B$). If two atoms just "touch" at this distance, they are generally considered as noncovalently interacting (Fig. 1b).
2. The covalent distance, generally approximated by the sum of covalent radii ($r_A + r_B$). When the valence spheres of two atoms are "in contact" at this distance, we typically speak of a covalent bond (Fig. 1c).

Of course, real chemical interactions exist on a continuum between these two boundaries, and even the same pair of elements can exhibit shorter or longer distances depending on their formal oxidation states, molecular environment, or electronic structure. Simply comparing the interatomic distance d_{AB} with ($v_A + v_B$) or ($r_A + r_B$) can be insufficient if we want a universal parameter that discriminates among a wide variety of interaction strengths.

In our previous work [2], we showed that the interpenetration index works nicely for ionic solids, despite the use of covalent radii in Eq. (1). In an Appendix A to this chapter, we further show that the differences between ionic and covalent atomic radii are strongly reduced when radii sums are considered, as in the denominator of Eq. (1), and that in many cases the sums of both types of radii are nearly identical or differ by a small constant amount (typically 0.1–0.3 Å) for a given family of atom pairs.

1.3 The Interpenetration Index

To address this challenge, we adopt the interpenetration index (p_{AB}), originally formulated by us [2], which quantifies how far two atoms intrude into each other's van der Waals crusts. Mathematically:

$$p_{AB} = 100 \cdot (v_A + v_B - d_{AB})/(v_A + v_B - r_A - r_B) \qquad (1)$$

As depicted in Fig. 1b, c, two reference scenarios can be easily established:

- $p_{AB} = 0\%$ if $\Delta_{AB} = v_A + v_B$, meaning the atoms merely contact at their van der Waals surfaces
- $p_{AB} = 100\%$ if $d_{AB} = r_A + r_B$, corresponding to a classical single covalent bond (valence spheres touching)

Thus, the interpenetration index p_{AB} effectively normalizes the interatomic distance for the sizes of the interacting atoms, enabling meaningful comparisons across the periodic table and for bonds of varying strengths. When p_{AB} is negative (i.e. $\Delta_{AB} > v_A + v_B$), it implies minimal or no meaningful overlap – often no strong interaction is present, although weak van der Waals attractions may still be operating at small negative penetrations. Conversely, $p_{AB} > 100\%$ suggests a substantial penetration into the valence spheres, often indicative of stronger or multiple bonds.

The overarching goal of this chapter is to illustrate how p_{AB} can be used to classify and compare the whole spectrum of chemical interactions, including:

- Weak, noncovalent contacts typical of supramolecular chemistry or crystal packing
- Moderate overlaps found in hydrogen bonds or other directional, partially covalent interactions, such as σ-hole interactions
- Strong covalent or ionic bonds, where valence shells significantly overlap or fully merge
- Multiple bonds, in which the overlap extends beyond single-bond limits and can produce higher p_{AB} values
- Positronic bonding, whereupon two atoms or anions are held at a close distance via interaction with a bridging positron

By unifying these diverse cases within the same quantitative scale, we hope to provide a broad and intuitive framework that clarifies the continuum from purely nonbonding interactions to highly covalent ones. This insight can be particularly valuable in areas such as reaction mechanism analysis, where the subtle formation or breaking of contacts needs to be viewed as a dynamic process rather than a simple transition from "nonbonded" to "bonded".

Following a comment by Grochala [7], who warned about the problems that might arise when using such an interpenetration index for pairs of atoms with dissimilar sizes and different van der Waals crust widths, we analysed this case [8] and concluded that the index p_{AB} carries information on both the left-right (p_A) and right-left (p_B) penetrations and that those differences depend only on the widths of the van der Waals crusts of A and B, not on their atomic radii. It was concluded that for most applications, it is preferable to use the interpenetration index p_{AB}, which is conceptually clearer, implies no loss of information on the individual penetration indices p_A and p_B, and is simpler to use. Therefore, we will use only *interpenetration indices* p_{AB} throughout this work, not *penetration indices* p_A and p_B, even if for simplicity we may occasionally use the word "penetration" to refer to p_{AB}.

In the sections that follow, we will detail the derivation and computational strategies behind the interpenetration index, compare its predictions to established structural data, and propose a road map for interpreting a variety of chemical phenomena – ranging from fundamental bond formation to interatomic contacts in complex systems. Recent examples of application of the interpenetration indices to analyse noncovalent bonding interactions by other authors will also be highlighted, demonstrating the positive reception this parameter has had in the chemical community during the last 2 years.

2 Noble Gas Dimers

We begin by examining the prototypical van der Waals interactions manifested in weakly bound noble gas diatoms (Ng_2). To that end, we conducted DFT calculations over a range of Ng···Ng distances (where "Ng" designates any noble gas) and performed energy decomposition analyses (further details on the ALMO-EDA scheme used can be found in Ref. [2]). Rather than focusing on interatomic distances directly, we recast the results in terms of interpenetration indices by using a modified Lennard-Jones potential in which the distance variable is replaced by $100 - p_{AB}$ (Eq. 2). In this formulation, ε and σ represent the depth of the potential well at the minimum and the value of $100 - p$ at which the energy crosses the zero line, respectively. The exponent n is a fitting parameter that equals 6 in the standard Lennard-Jones form. Setting to zero the derivative of the energy with respect to p yields Eq. (3), which defines the interpenetration index at the energy minimum (p_{min}). By setting the energy in Eq. (2) to zero, the penetration at the crossing point, p_0, is given by Eq. (4):

$$E = 4\varepsilon\left[\left(\frac{\sigma}{100-p}\right)^{2n} - \left(\frac{\sigma}{100-p}\right)^{n}\right] \tag{2}$$

$$p_{min} = 100 - \sigma * 2^{\frac{1}{n}} \tag{3}$$

$$p_0 = 100 - \sigma \tag{4}$$

The resulting energy curves are shown in Fig. 2. While the high-energy region beyond p_0 is not perfectly captured by the fits, this discrepancy does not affect our

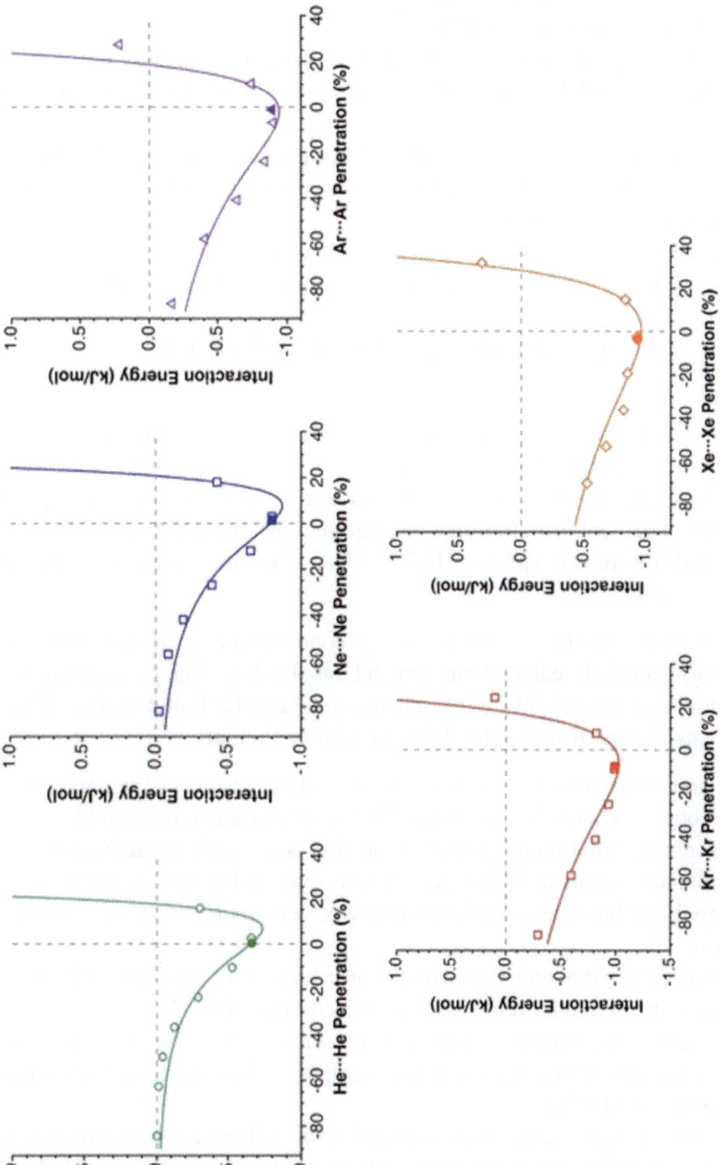

Fig. 2 The interaction energy of Ng_2 adducts represented as a function of the interpenetration index for He to Xe. The solid symbols correspond to the optimized Ng–Ng distance. (Reproduced from Ref. [2] with permission from the RSC)

discussion since the minimum-energy and pre-dissociation zones are well reproduced. Accordingly, the following observations can be made:

1. Interaction energies across the noble gas dimers are between −0.7 and −1.1 kJ/mol, averaging −0.91(±0.09) kJ/mol.
2. In all cases, the penetration at the energy minimum, p_{min}, is close to 0% (ranging from −6.2% to +7.5%), consistent with typical van der Waals interactions (Fig. 1b).
3. The width of the potential well broadens down the group, as shown by decreasing n values. This trend reflects changes in the vibrational force constant and the corresponding Raman frequencies.
4. The optimized dimers' interpenetration indices correlate with those extracted from the Lennard-Jones fit through a linear relationship given in Eq. (5):

$$p_{opt} = 1.0264 * p_{min} - 5.6574 \quad \left(r^2 = 0.97\right) \tag{5}$$

According to high-level calculations, the He_2 dimer has a minimal energy at 2.96 Å [9], corresponding to an interpenetration index of −4.3%. Given the small dissociation energies and shallow wells, a few-percent variation is unsurprising. Other computations [10] confirm minimal interatomic distances for He, Ne, and Ar dimers – and for mixed species (HeNe, HeAr, NeAr) – yielding penetrations between −7% and +5%.

Figure 3 illustrates how various energy contributions extracted from energy decomposition analysis calculations depend on the Ng···Ng penetration in Ng_2 (further details on the ALMO-EDA scheme used can be found in Ref. [2]). The five noble gas dimers display remarkably similar behaviour:

1. Orbital and dispersion attractions dominate at long distances (penetrations of −30% or lower), where electrostatic and Pauli terms remain negligible.
2. Near the energy minimum, London dispersion is the main attractive term.
3. Pauli repulsion surges at higher penetrations, preventing deeper overlaps.
4. Pauli repulsion becomes significant at lower penetrations than the electrostatic attraction.
5. Both Pauli repulsion and electrostatic attraction vary exponentially with the penetration index and correlate well with each other (Fig. 4a, b).
6. Consequently, electrostatics alone cannot stabilize Ng_2 dimers comprised of neutral noble gas atoms, because it is accompanied by a stronger Pauli repulsion at any penetration value.
7. The combined Pauli + electrostatics contribution follows an exponential penetration dependence that is largely independent of the noble gas involved (Fig. 4c). This factor limits achievable penetrations to only a few percentage units.
8. Finally, the above point implies that electronegativity and atom size have a negligible role in determining how far two Ng atoms penetrate each other's van der Waals crusts.

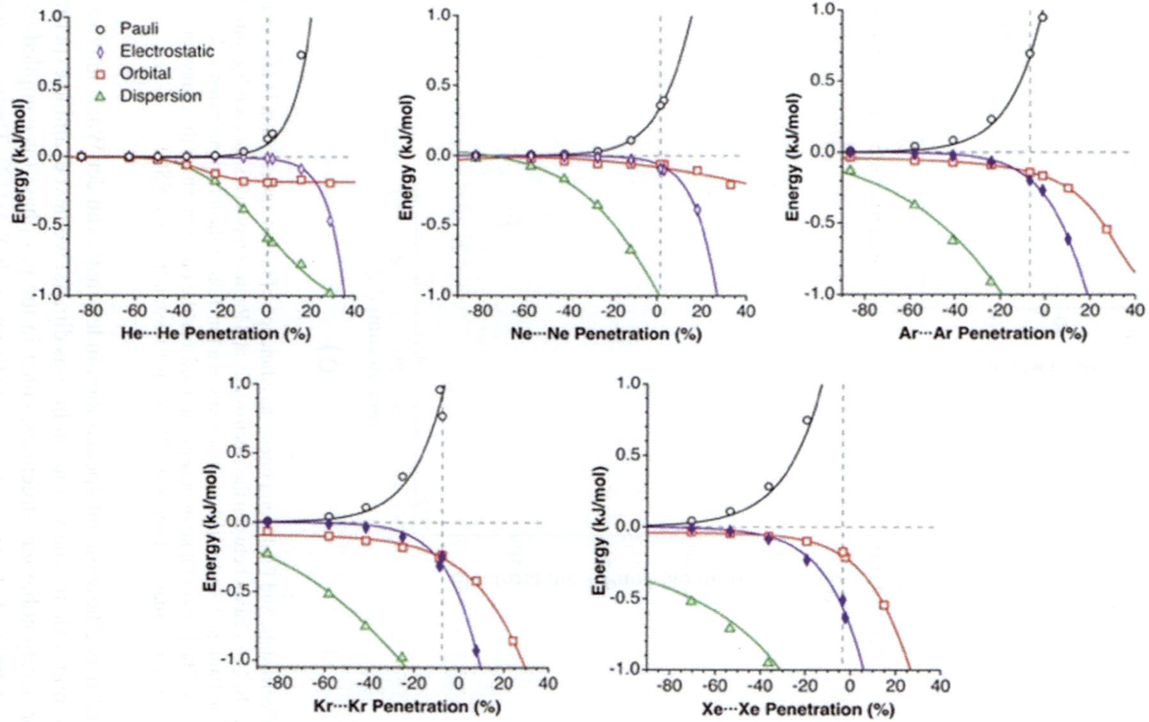

Fig. 3 Evolution of the components of the interaction energy as a function of the interpenetration index for noble gas dimers. The vertical dashed lines indicate the position of the energy minima. (Reproduced from Ref. [2] with permission from the RSC)

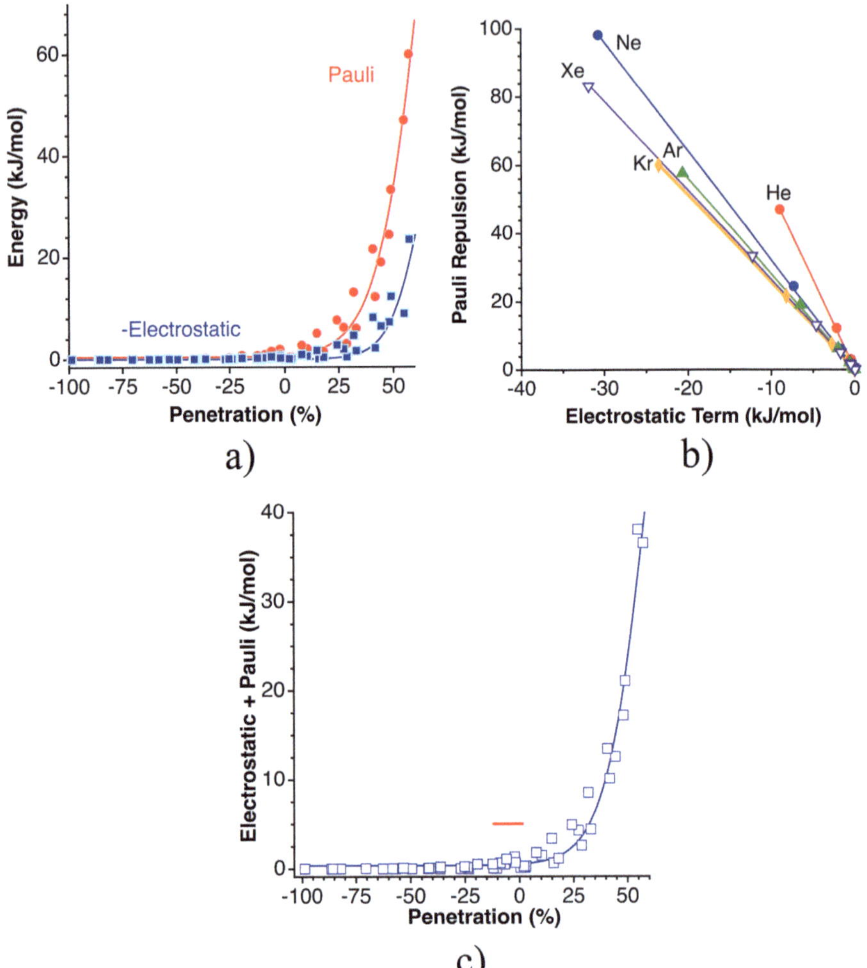

Fig. 4 (**a**) Electrostatic and Pauli components as a function of the interpenetration index for all the Ng_2 adducts. (**b**) Correlation between the electrostatic and Pauli components for each Ng_2 dimer. (**c**) The sum of the Pauli and electrostatic components as a function of the interpenetration index for all the Ng_2 dimers. The horizontal straight segment indicates the range in which the dimers are stable towards dissociation. (Adapted from Ref. [2] with permission from the RSC)

Having established how the interpenetration index maps van der Waals interactions onto a common scale for the various noble gas dimers, we next examine experimental data for purely noble gas structures, either in the gas phase (as compiled by the Mogadoc database) [11] or in the solid state (from the ICSD) [12]. These penetration indices are summarized in Fig. 5. While no Mogadoc entries for He_2 or Xe_2 appear, certain literature sources do report their distances [13]. Meanwhile, the short-lived Rn_2 and Og_2 dimers have been theoretically studied at ~4.42 Å and 4.28 Å,

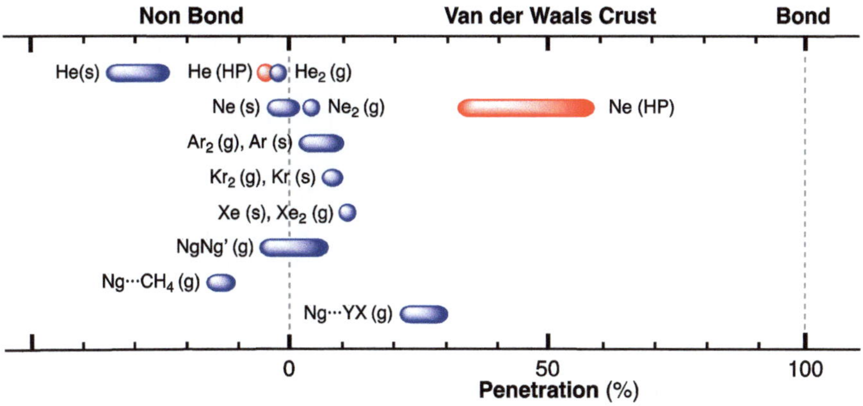

Fig. 5 Ranges of interpenetration indices found for pairs of noble gas atoms in the gas and solid states. Blue bars correspond to structures at atmospheric pressure, and red bars to structures under pressure (HP). Ng stands for noble gases in general, and X and Y for halogens. (Reproduced from Ref. [2] with permission from the RSC)

respectively. Because no reliable van der Waals radii exist for these heaviest noble gases [4], each dimer's interpenetration index is arbitrarily taken as 0% by assigning half the optimized distance (2.21 Å for Rn, 2.14 Å for Og) as its radius.

Experimentally observed Ng···Ng penetrations thus cluster around 0%. Since the van der Waals radii used to compute penetration indices were principally derived from Ng···O contacts (rather than Ng···Ng), it follows that the measured Ng···Ng penetrations show slight deviations from exact zero, though still consistent with predominantly van der Waals interactions. Intriguingly, mixed noble gas pairs (Ng···Ng′) also group near zero, generally falling within small positive or negative values. As a point of contrast, Fig. 5 also shows that noble gas–methane penetrations are even smaller, whereas noble gas–halogen or noble gas–interhalogen interactions surpass 20%, clearly indicating that polarization supplements the dispersion in those cases. Lastly, it is noteworthy that solid-state Ng···Ng penetrations often register lower values than their gas-phase counterparts, likely reflecting the higher coordination numbers in crystal structures, which prevent deeper pairwise approaches.

3 Hydrogen Bonds

The many types of hydrogen bonds found across chemistry and biology are commonly classified into three general categories [14]. In this scheme, hydrogen bonds whose energies lie between about 4 and 15 kcal/mol – comparable to those found between water molecules – are considered moderate or "classical". Those exceeding this range are regarded as strong, and those falling below it are termed weak [15].

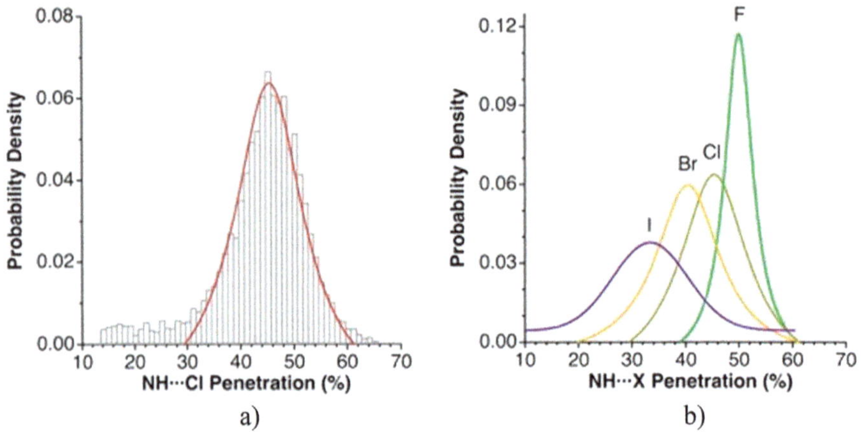

Fig. 6 (**a**) Distribution of the N–H···Cl interpenetration index for hydrogen bonds with alkylammonium cations, showing the fitting to a Lorentzian function. (**b**) Distribution of the N–H···X interpenetrations in hydrogen bonds between alkylammonium cations and halogen atoms. (Reproduced from Ref. [2] with permission from the RSC)

To initiate our exploration of how a hydrogen atom and its acceptor interpenetrate in a D–H···A hydrogen bond, we examined [2] the penetration probability density for a variety of donor (D = O, N) and acceptor (A = N, O, O=C, S=C, Cl, Br) combinations. Within the set of so-called classical hydrogen bonds, let us focus on N–H···Cl contacts. Their penetration distribution (Fig. 6a) peaks sharply at 42%, with a lower tail at smaller values corresponding to weaker interactions. Restricting the dataset to chloride anions paired with alkylammonium cations ($R_3NH\cdots Cl$, 6,224 occurrences) yields a Lorentzian fit (neglecting $p < 30\%$) with a maximum at 45.2% (Fig. 6a). Analogous data for the other halides show maxima at 49.9%, 45.2%, 39.9%, and 33.4% for F, Cl, Br, and I, respectively (Fig. 6b). These values diminish linearly down the group ($p_{HX} = 61.3 - 5.48n_X$; $r^2 = 0.99$), where n_X is the period of X.

Grabowski and Sokalski computed dissociation energies for a broad selection of hydrogen-bonded molecular pairs encompassing X···H contacts (X = F, O, N, C, H) [16]. Their results closely correlate with the penetration indices (Fig. 7), even though these hydrogen bonds involve a variety of atoms.

The large macrocycle 18-crown-6 and related derivatives (e.g. sila-crown species) [17] are well known for binding cations. In systems containing the hydronium cation (H_3O^+), each structure provides three crystallographically distinct H···O contacts (in 11 data sets from six compounds) with interpenetration indices spanning the range 39–60%. In every adduct, the shortest contact penetrates above 48%, placing it among the highest values recorded for O···H–O hydrogen bonds. This likely reflects the ionic character of the donor. By contrast, in potassium clathrates of the same crown ether, the K···O interpenetration often exceeds 80%, with a maximal probability at ~90% – a consequence of combining donor-acceptor and ion-dipole forces. Similarly, the protonated forms of trimethyl-triazacyclononane give

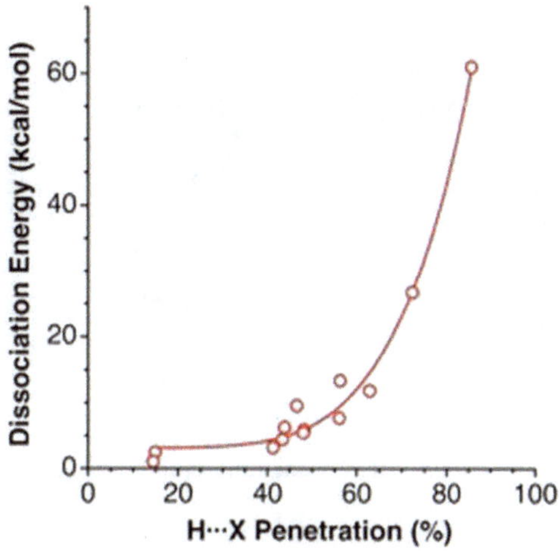

Fig. 7 Dissociation energies of hydrogen-bonded complexes comprising H···O, H···N, H···F, H···C, and H···H contacts reported by Grabowski and Sokalski, represented as a function of the corresponding penetration indices. (Reproduced from Ref. [2] with permission from the RSC)

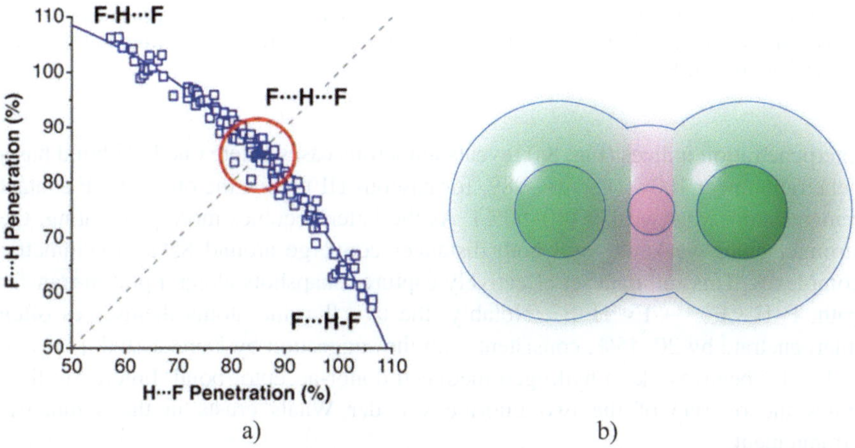

Fig. 8 (**a**) Correlation between the penetrations of the two F–H contacts in the F–H–F⁻ anion. In this figure and in Fig. 9a, the dashed lines correspond to symmetric structures with two identical X–H penetrations, where X = F and O, respectively. (**b**) Representation of the valence and van der Waals spheres in the F–H–F⁻ anion. (Reproduced from Ref. [2] with permission from the RSC)

intramolecular N–H$^+$···N hydrogen bonds with 33–42% penetration (compared to ~105% for the N–H$^+$ bond), values akin to those for non-protonated N–H···N hydrogen bonds.

One of the strongest hydrogen bonds is the F–H···F interaction [14]. Disregarding cases where one or both fluorines are covalently bound to other elements, the FHF$^-$ anion frequently adopts a symmetric structure with F–H penetrations near 80 ± 1% in the gas phase [18], in its potassium salt [19], in cocrystals with IF_5 [20], and in solution [21]. It can also appear in organic crystal structures as an asymmetric arrangement with one short and one long F–H distance. A scatterplot of the two

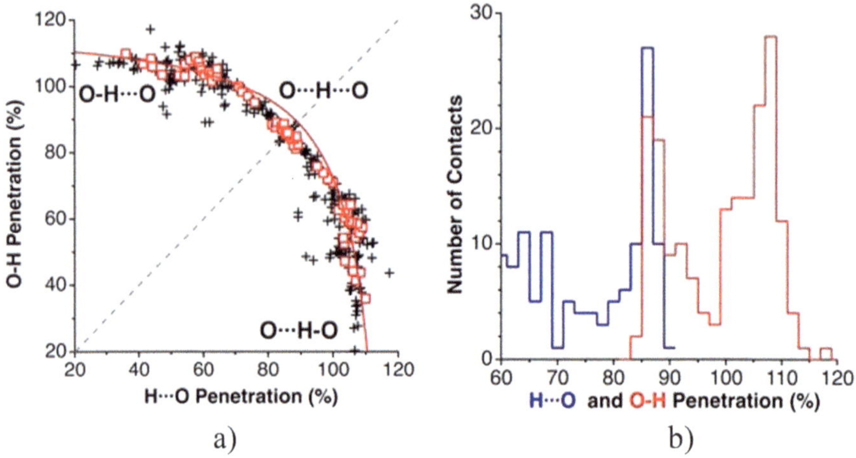

Fig. 9 (**a**) Correlation between the large and small H···O penetrations in $H_2O \cdots H^+ \cdots OH_2$ groups (squares correspond to crystal structures with R ≤ 2.5%). (**b**) Distribution of the penetration indices in the long and short O–H contacts in $H_2O \cdots H^+ \cdots OH_2$ groups. (Reproduced from Ref. [2] with permission from the RSC)

interpenetration indices (Fig. 8a) reveals numerous cases where one F–H bond has a penetration near 100% (versus ~98% for gaseous HF) while the other H···F contact remains somewhat weaker (40–60%). As the latter becomes more penetrating, the stronger bond weakens, until both distances converge around 85% in symmetric complexes. This full data set effectively captures snapshots along a proton-transfer path, $F\text{–}H \cdots F^- \rightarrow F \cdots H\text{–}F^-$. Notably, the two fluorine atoms themselves often interpenetrate by 20–45%, consistent with the suggestion by Dereka et al. [21] that $F\text{–}H \cdots F^-$ behaves like a hydrogen-mediated donor-acceptor bond. Figure 8b illustrates the overlap of the two fluorine van der Waals crusts in the symmetric arrangement.

The $H_2O^+\text{–}H \cdots OH_2$ cation forms hydrogen bonds that rival the $F\text{–}H \cdots F^-$ system in strength, along with F–H···O [14]. It occurs in crystalline compounds containing chloride [22], bromide [23], or nitrate [24] and with organic anions. As in the $F \cdots H \cdots F^-$ case, the two O–H penetrations display a wide range and appear correlated: when one water approaches the hydronium cation more closely, the corresponding O–H bond elongates (Fig. 9a) until both distances become identical. The variability seen in Fig. 9a likely arises from limited precision in locating hydrogen atoms by X-ray diffraction. Selecting only structures with R ≤ 2.5% shows that the O–H bond remains essentially unchanged for small H···O penetrations (<60%), but then starts to lengthen, mirroring the $F\text{–}H \cdots F^-$ system (Fig. 8a). If both O–H distances are plotted jointly for these cations, a linear fit ($r^2 = 0.95$) is achieved: $p_{H \cdots X} = 149.0\text{–}0.735\ p_{X\text{–}H}$ (X = F or O). Interestingly, the dual-peaked distribution in Fig. 9b – centred near 110% and 88% – implies that there may be

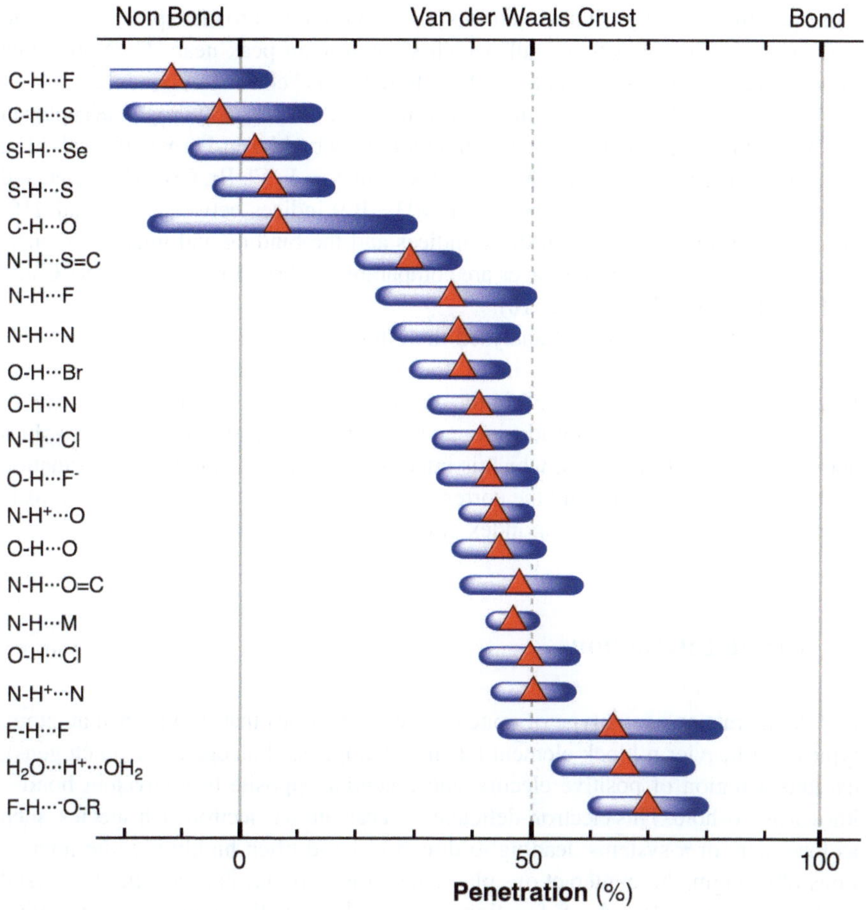

Fig. 10 Ranges of interpenetration indices presented by several types of hydrogen bonds (blue bars) represented by the width of the Lorentzian curve obtained by least squares fitting of the probability density distribution, and position of the maximum of the probability density (red triangles). (Reproduced from Ref. [2] with permission from the RSC)

three energy minima in the proton-transfer process $H_2O–H^+ + OH_2 \rightarrow H_2O + H^+OH_2$, one of which is a symmetric intermediate.

A broad overview of a host of hydrogen bonds is given in Fig. 10. The C–H···F, C–H···O, and C–H···S contacts, often classed as weak hydrogen bonds, indeed show the lowest penetration indices. The numerous C–H···S hydrogen bonds have recently been examined in a detailed structural study, concluding that these interactions favour angles larger than 155° and distances between 2.83 and 3.13 Å [25]. From the viewpoint of van der Waals crust interpenetration, we find a distribution peaking near −7% and spanning up to 25% – consistent with the −2% to +15% range reported by Fargher et al. with a narrower angular criterion [25]. Likewise, Keil et al. [26] identified intramolecular Si–H···SeR_2 contacts at 2.71–2.91 Å, corresponding to 6–19% interpenetrations well within the broader spread (−20% to +30%) of H···Se contacts overall, which show a main peak near 2%. Notice that these interactions could alternatively be referred to as "chalcogen bonds".

A new type of hydrogen bonding involving a Be atom as a hydrogen acceptor in $(OC)_3$Be**:** has been recently studied computationally by Martín-Fernández et al. [27], using as hydrogen donors the HX molecules, with X = F, Cl, Br, CN, NC, CCH, and OH. These results of these authors give *p*(H···Be) indices between 11% and 41% and good correlations between those indices and the binding and interaction energies. Those interpenetration indices are comparable to those for the N–H···S=C and N–H···F hydrogen bonds (Fig. 10).

Datta and co-workers [28] analysed the effect of pressure on the N···H–O–CH (=O) (formic acid) penetration index for pyridine, pyridazine, and pyrazine and found a dramatic increase of 20% in the penetration index at a critical pressure (3, 5, and 15 GPa for pyridine, pyridazine, and pyrazine, respectively) reaching values above 90% in the three cases, which is interpreted as an indication of the formation of the covalent N–H bond and the corresponding ion pair, accompanied by a drastic decrease in the H-O penetration index at the critical pressure.

4 σ-Hole Interactions

σ-Hole interactions are a type of noncovalent interactions that occur when an atom, typically a heavier p-block element (such as halogens, chalcogens, or pnictogens), exhibits a region of positive electrostatic potential opposite to a covalent bond – known as a σ-hole. This electron-deficient area can attract electron-rich species, such as lone pairs or π-systems, leading to directional and often highly specific interactions [29]. Sigma-hole interactions play a key role in molecular recognition, crystal engineering, and design of functional materials, usually complementing other noncovalent forces like hydrogen bonding.

Bickelhaupt and co-workers [30] have argued that σ-hole intermolecular interactions should not be viewed merely as electrostatic attractions between two isolated molecular charge distributions, but rather as phenomena involving a substantial orbital overlap. Our present use of penetration indices aligns with their conclusion that HOMO-LUMO interactions – i.e. interpenetration of valence spheres – play a major stabilizing role in σ-hole contacts. To illustrate this, we have applied penetration indices to three families of computationally studied adducts, comparing their binding energies and other parameters such as the reported energy decomposition analysis (EDA) terms and the electron density at the bond critical point (BCP) determined via the atoms in molecules (AIM) method.

These three families are summarized in Fig. 11. The first (**1a**) involves a four-coordinated group 14 atom (E) binding a Lewis base (either halide or a group 16 compound), forming either anionic or neutral adducts – so-called tetrel bonds – in line with Frontera and co-workers' studies [31]. The second family (**1b**), examined by Shaik and co-workers, features a halogen or interhalogen species (Y–X)

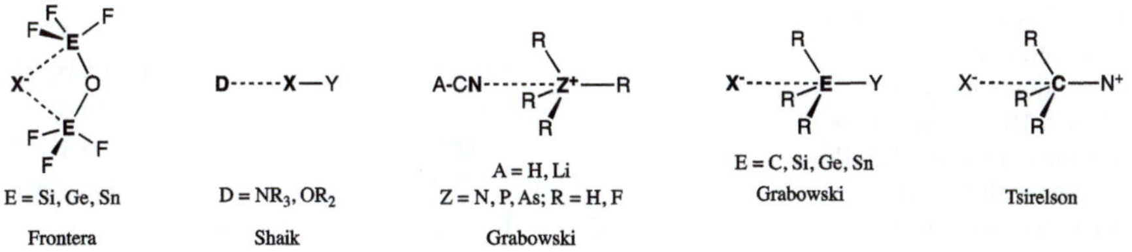

1a **1b** **1c** **1d** **1e**

Fig. 11 σ-Hole interactions studied by different groups and discussed in this section in terms of their penetration indices. The σ-hole-bearing atom and the donor atom are in boldface, and X and Y represent halogens in all cases. (Reproduced from Ref. [2] with permission from the RSC)

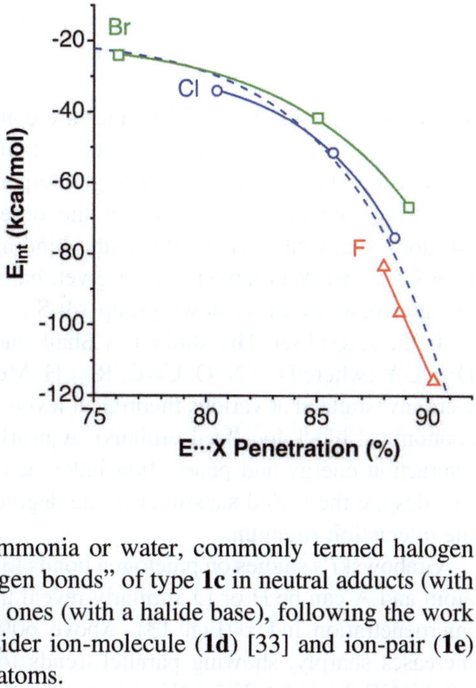

Fig. 12 Dependence on the penetration indices of the $X^{-}\cdots E$ binding energy ($X = F$, Cl, Br; $E = Si$, Ge, Sn) for σ-hole bonds in adducts **1a** studied by Frontera and co-workers. The dashed line corresponds to the least squares fitting for all the compounds. (Reproduced from Ref. [2] with permission from the RSC)

interacting with a Lewis base like ammonia or water, commonly termed halogen bonds [32]. We also focus on "pnictogen bonds" of type **1c** in neutral adducts (with NCH or NCLi as the base) or anionic ones (with a halide base), following the work by Grabowski [33]. Finally, we consider ion-molecule (**1d**) [33] and ion-pair (**1e**) [34] interactions involving group 14 atoms.

Frontera and co-workers' nine $X^{-}\cdots E$ adducts ($X = F$, Cl, Br; $E = Si$, Ge, Sn) show a strong dependence of interaction energy on the penetration index, particularly when comparing systems involving the same halide (Fig. 12). This trend can be fitted to a power law (Eq. 6):

$$E_{\text{int}} = E_0 - a \cdot p^n \tag{6}$$

where E_0, a, and n are fitting parameters. While E_0 in principle represents the energy at very low penetration (i.e. at large negative p), its numerical value has been obtained from a limited data set and is highly sensitive to the interaction energies

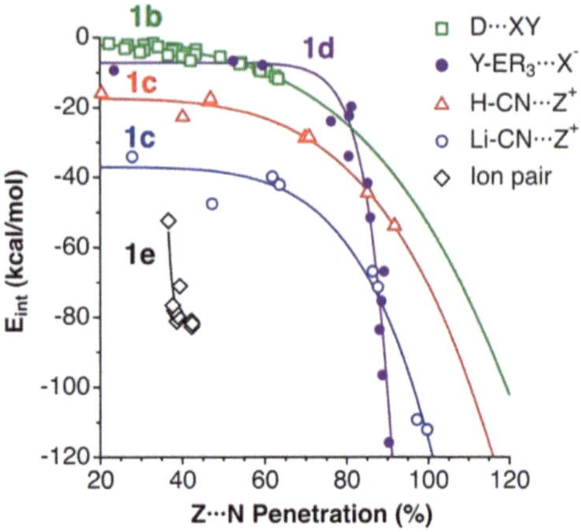

Fig. 13 Dependence on the penetration index of the interaction energy of adducts **1b** at their energy minima (squares), of R_3Pn-X^+ ions with NC–H (triangles) and NC–Li (empty circles; X, R = H or F; Z = N, P, As) **1c**, R-$EY_3 \cdots X^-$ adducts **1d** (solid circles), and ion pairs **1e** (filled squares). (Adapted from Ref. [2] with permission from the RSC)

at low penetrations. When all halides are combined, the same functional form still gives a reasonable description of the binding energies for the nine E–X pairs, showcasing the advantage of using interpenetration indices. Within this series, increasing penetration proceeds in the order Br $\leq$ Cl $<$ F for a given group 14 atom, while the magnitude of the binding energy decreases down the group: F $\gg$ Cl $>$ Br. Alternatively, for a given halogen, both penetration and interaction energy rise as we move down group 14: Si $<$ Ge $<$ Sn.

In the second set (**1b**), studied by Shaik and co-workers [32], adducts of the type D···X–Y (where D = N, O, C=C; R = H, Me; X = halogen; Y = halogen or NO_2) were investigated at various theoretical levels (full valence bond theory or the more economical block-localized orbitals). A nearly linear correlation emerges between interaction energy and penetration index across diverse pairs (Fig. 13), indicating that despite the varied substituents, the degree of overlap has a strong influence on the interaction strength.

Grabowski's studies on pnictogen bonds [33] (Z···NC–A) where Z is a pnictogen atom and A can be H or Li similarly reveal that binding energies increase with the interpenetration index (Fig. 13). Above 60% penetration, the interaction energy increases sharply, showing parallel trends for A = Li or A = H. Two systems, $PF_4^+ \cdots$NC–Li and $AsF_4^+ \cdots$NC–Li, even reach penetrations close to 100%, adopting near-trigonal bipyramidal geometries of the ZF_4N group. Indeed, the penetration index correlates with the N–Z–R angle (**1c**), tracing an SN^2-type pathway. Relative to set **1b**, these pnictogen-containing adducts cover a comparable penetration range but with larger binding energies due to their ionic nature. Conversely, compared to Frontera's halide systems (**1a**) [31], the same range of energies spans a broader range of penetration – again reflecting the possibility of trigonal bipyramidal structures, clearly exemplified by species **1d** (Fig. 13). Despite these differences, all three groups of adducts follow a similar penetration–energy relationship akin to Eq. (6). Meanwhile, family **1e**, where oppositely charged ions interact, exhibits stronger binding energies at modest penetration indices.

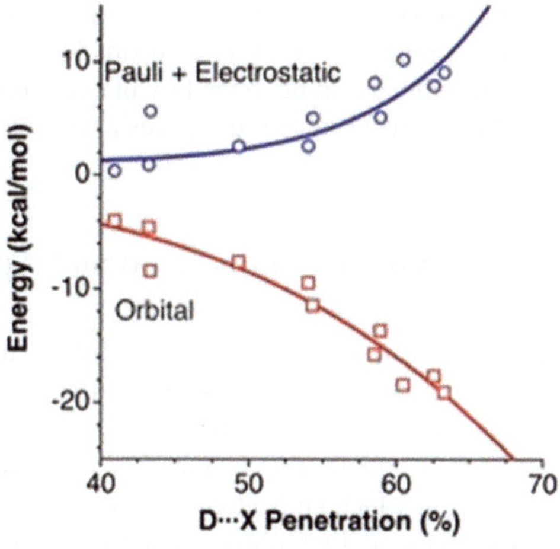

Fig. 14 The frozen and orbital interaction EDA components calculated by Wang et al. [35] for adducts **1b**, represented as a function of the D···X penetration. (Reproduced from Ref. [2] with permission from the RSC)

Further insights arise from EDA data, as illustrated in Fig. 14: the repulsive (or slightly attractive) frozen component (electrostatic + Pauli) and the attractive orbital term both track with the penetration index. Three outliers – interactions between a fluorine molecule and $Me_nH_{3-n}N$ – likely deviate due to additional N–CH_3···F contacts at shorter distances, which may display C···F or C···H interactions or both. Indeed, in the gas-phase structure of $Me_3N \cdots F_2$, C···F separations correspond to a 23% penetration index, and these distances grow shorter as the number of methyl substituents increases, implying multiple partial contacts.

Summarizing, the range of halogen bond penetration indices studied by several authors falls mostly between 11% and 65%, suggesting that these values typify halogen bonds. In contrast, tetrel and pnictogen bonds often display higher penetrations, reflecting ion-dipole components and formation of hypercoordinated structures. Notably, the 26 distinct σ- and π-interactions analysed by Murray and Politzer [36], which include halogen, chalcogen, pnictogen, and hydrogen bonds, also adhere to an energy–penetration trend comparable to that in Fig. 12.

A particular case studied by Velasquez et al. [37] considers the weak interactions between acetone or chloride and haloalkenes, in a variety of topologies that include C–X···O halogen bonds, but also C–H···O hydrogen bonds and πC···O interactions. All these interactions are stabilizing, with interaction energies of −2 to −5 kcal/mol. While monohalogenated ethene molecules prefer to establish hydrogen bonds with Lewis bases, di- and tetrahalogenated ethenes can form lone pair–π interactions via their C=C frameworks, with the π* orbital of the alkene acting as an electron acceptor. For these interactions, an O···C=C attack angle close to 107° seems to maximize the interaction strength, which confers these lone pair–π contacts some degree of directionality. The O···C=C penetrations were found to be within the range of 7–23%, and similar values (18–27%) were obtained for the Cl···C interactions.

Shen et al. [38] have studied the electrophilicity and nucleophilicity of dithiocarbamate square-planar complexes [M(dtc)$_2$], where M is Ni, Pd, or Pt, for the S···X interactions (X = HCl, ICl). The interpenetration indices, p(M/S···X), were found to be in the range of 44–78%, comparable to the penetrations characteristic of strong hydrogen bonds, some secondary interactions, and **2c1e** bonds (see Figs. 10 and 21, and Sect. 7 for data and references to one-electron bonds).

5 Secondary Donor-Metal and Agostic Interactions

We had earlier shown [2] that secondary metal-donor interactions of coordinated ligands present M···C interpenetration indices in the range of 0–60% for σ-coordinated arenes to Rh, Ir, Ni, Pd, Pt, Ag, or Au and M···O in the range of 37–75% for nitrato ligands. Later on, Nebra and co-workers [39] reported an Ag(III) complex in which an N atom of a bipyridine ligand occupies a basal position of the square-pyramidal coordination sphere, while its second N atom establishes a secondary interaction at a penetration of 85%. A CSD search for square pyramidal d^8 complexes with a bipyridine or phenanthroline ligand occupying a basal (regular M–N bond) and an axial (secondary interaction) coordination position will allow us to put that new datum in context. The lengths of all the M–N basal bonds (M = Co, Ni, Cu, Pd, Ag, Ir, Pt, Au) correspond to penetrations of between 91% and 100%, characteristic of two-electron covalent bonds. In contrast, the secondary axial M···N interactions present penetration indices covering a much wider range, 58–98%, and thus such secondary interactions cover a similar range of penetrations to the 2c1e bonds and regular 2c2e covalent bonds combined. The wide range of penetrations found is the result of the geometrical constraints imposed by the bidentate nature of the ligand (**2**, Scheme 1): The large metal atoms favour a significant decrease of the N···M–N angle (α) and a poorer N···M penetration. In fact, this is a general trend shown by the structural data, but with a significant dispersion around the least squares line. A comparison of the trends for the atomic covalent radius and the

Scheme 1 Chemical structures of **2a** and **2b**

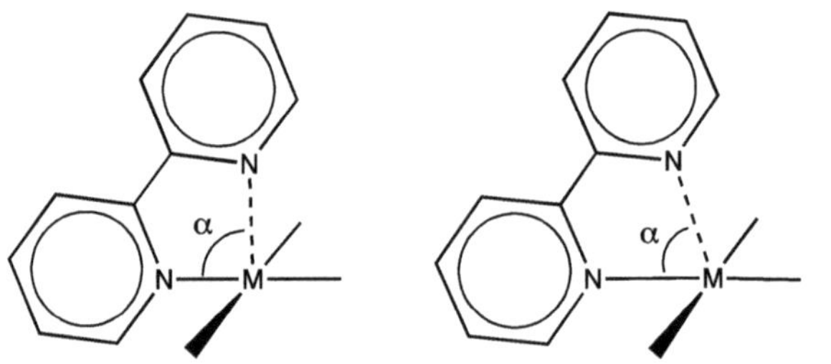

Small M
α ≥ 76°
Co, Ni, Cu

2a

Large M
α ≤ 76°
Au, Pt, Pd, Ir, Ag

2b

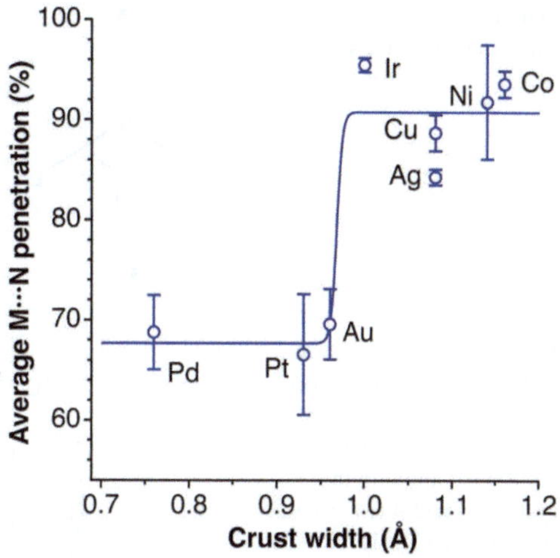

Fig. 15 Effect of the width of the van der Waals crust of the metal atom on the penetration indices for the secondary N···M bonds in square pyramidal complexes with bipyridine or phenanthroline ligands. The error bars indicate ±1 esd

secondary penetrations (Eqs. 7 and 8) shows that the latter decrease as the atomic size increases (from Co to Pt), but for a further increase of the atomic radius, the penetration increases again. The reason for such non-linear behaviour is that the increase of the metal size per se implies a higher penetration (Eq. 1) and thus counters the effect of the decrease of the angle α.

$$\text{Atomic radius}: \quad \text{Co} < \text{Ni} < \text{Cu} < \text{Pt} \approx \text{Au} < \text{Pd} < \text{Ir} < \text{Ag} \tag{7}$$

$$\text{M}\cdots\text{N penetration}: \quad \text{Co} > \text{Ni} > \text{Cu} > \text{Pt} < \text{Pd} < \text{Au} < \text{Ag} < \text{Ir} \tag{8}$$

A surprising result is that the range of penetration values can be classified in two sets depending on the width of the van der Waals crust of the metal atom, rather than on its size (Fig. 15).

Velasquez et al. [40] studied intermolecular interactions involving Hg-coordinated azides and found a good correlation between the N···N or the N···Hg penetration indices and the interaction energies, depending on which is the dominant interaction. The secondary N···Hg interactions were shown to predominate in those interaction topologies allowing for short N···Hg contacts. Also, a good correlation of the electron density at the N···N bond critical point and the interpenetration index, and a still better correlation with the Laplacian of that electron density, was found. In QTAIM, typically we cannot compare the BCP properties between different atom pairs (N–N vs. N–Hg, for example). However, one of the strengths of the interpenetration index is that it allows a direct comparison of different atom pairs.

Stoccoro et al. [41] analysed agostic Pt···H–C interactions by means of the penetration indices observed for the axial Pt···H–C interaction in the complex [MePtCl(dmso–κ*S*)(1-^{i}Pr-py)] (50–55% for the Pt···H *ipso* contact calculated in the gas phase and in solution, as well as X-ray diffraction results, and 41% for the

Scheme 2 Chemical structures of **3** and **4**

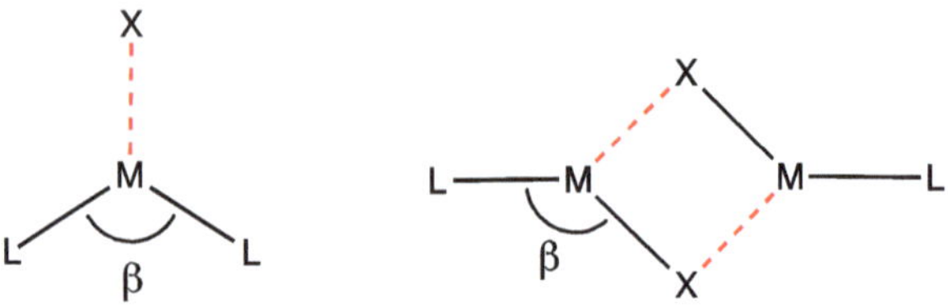

3 **4**

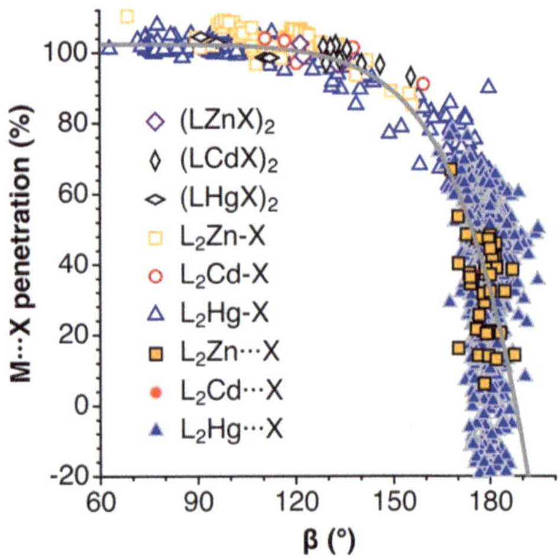

Fig. 16 Plot of the M–X and M···X interpenetration indices between L–M–X monomers as a function of the β bond angle in the systems **3** and **4** (M = Cu, Ag, Au, Zn, Cd, Hg; X = Cl, Br, I; L is a non-halogen two-electron donor ligand). The solid line represents a least squares fitting of the full data set to the expression $p_{MX} = p_0 - A \cdot 10^{-19} \cdot \beta^9$. (Reprinted with permission from Ref. [42])

Pt···C ipso interaction). Such a close interaction produces a local potential energy minimum, which contributes to the stabilization of the chiral conformation.

We have also explored [42] the secondary interactions of a halide (Cl^-, Br^-, and I^-) with d^{10} metal ions (Cu^+, Ag^+ and Au^+, Zn^{2+}, Cd^{2+} and Hg^{2+}) in the association reactions $[ML_2] + X^- \longrightarrow [MXL_2]$ (**3**), where L is any monodentate ligand, and in the dimerization reactions 2 [XML] $\longrightarrow$ $[(\mu\text{-}X)_2M_2L_2]$ (**4**) (Scheme 2). An exciting finding is that the analysis of all CSD crystal structures shows the existence of a single reaction path that correlates the L–M–L or X–M–L bond angle and the M···X interpenetration index, all the way from very weak contacts (penetrations as low as −20%) to fully formed single bonds (penetrations of around 100%), comprising well over 4,300 crystal structures (Fig. 16).

6 Covalent Bonds: Single and Multiple

6.1 Main-Group Element-Element Bonds

In this section, we aim to show how interpenetration indices can be extended to describe covalent bonds spanning single to triple bond orders between main-group elements. To illustrate this, we investigated the penetration indices for some of the

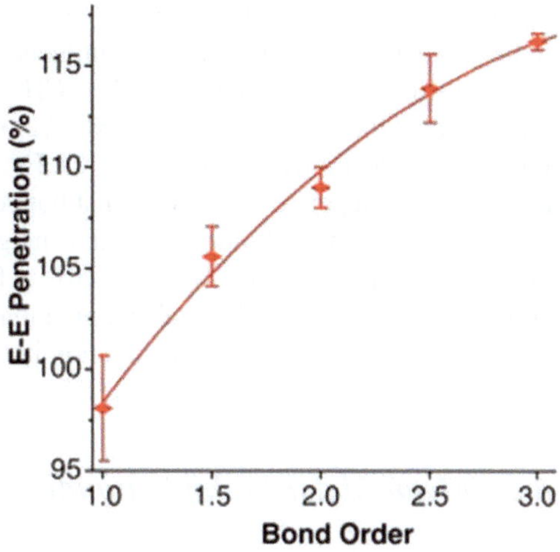

Fig. 17 Average penetration indices and esd found for element-element bonds with different bond orders in some prototypical compounds tabulated in Ref. 2 in the gas phase. (Reproduced from Ref. [2] with permission from the RSC)

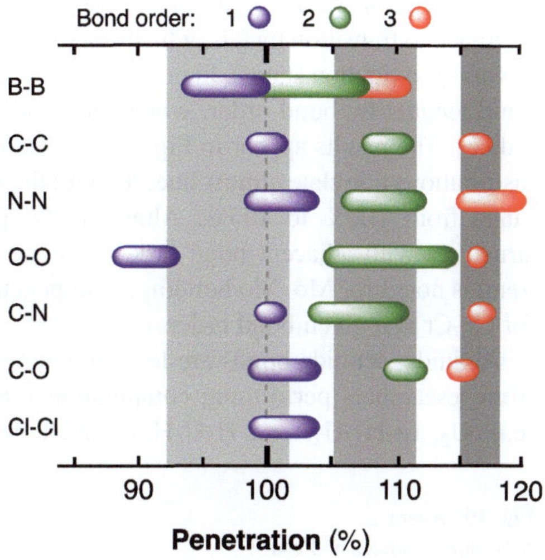

Fig. 18 Progression of the interpenetration indices (average ± two standard deviations) of several single, double, and triple element-element bonds, as deduced from the analysis of crystal structural data from the CSD. The shaded areas correspond to the ranges of values found for the gas-phase bonds (Ref. 2). (Reproduced from Ref. [2] with permission from the RSC)

most common bonds involving second-row elements. We first examined a set of simple gas-phase molecules, which display a clear correlation between bond order and penetration index (Fig. 17).

While the strong bond order–interpenetration index correlation for a handful of prototypical molecules is encouraging, it may not reflect the full range of possibilities. Consequently, we carried out a similar analysis for the same element-element bonds retrieved from the CSD crystal structure database. As shown in Fig. 18, and except for the relatively low penetration in peroxides, all single bonds lie near 100% penetration, double bonds cluster around 108%, and triple bonds form around 117%. These findings align with the ranges deduced for our limited set of gas-phase species (Fig. 17).

In a recent study, Bickelhaupt and co-workers [43] emphasized how the atomic radii of interacting atoms strongly influence bond dissociation energies across a series of small R–R′ molecules (R, R′ = CH_3, NH_2, OH, F, SiH_3, PH_2, SH, Cl). They reported both the energy gap and the SOMO overlap between the constituent radicals for three Me–R′ molecules (R′ = Me, Cl, F) at their energy minima, and our present analysis of the reported data shows that both correlate well with the interpenetration index: the penetration diminishes as the energy gap widens, and the overlap rises in tandem with penetration. Though only a few data points were available, this approach appears promising for computational and theoretical studies of chemical bonding, warranting further investigation in the future.

6.2 *Metal-Metal Multiple Bonding*

Having demonstrated that penetration indices enable meaningful comparisons of bonds between diverse main-group elements, a natural step is to extend the same approach to transition metals, which can exhibit bond orders up to five (e.g. Cr or Mo in various oxidation states). In prior work [44], we analysed the distribution of Cr–Cr bond lengths by bond order, which can readily be converted into interpenetration indices. The results appear in Fig. 19: the midpoints of the Cr–Cr penetration index distributions correlate almost linearly with the bond order ($r^2 = 0.989$), covering the range from 100% to 145%. Although the penetration distributions overlap for structures with adjacent bond orders, a clear progression still emerges. A similar trend is noted for Mo–Mo bonding, with penetration values higher for Mo–Mo than for Cr–Cr at a given bond order.

Actinide–actinide bonds are less common, but U–U bonding has drawn interest from researchers performing computational studies on relatively small molecules (e.g. U_2, (μ-H)$_2U_2$, and $H_2U_2H_2$), which have been observed via gas-phase or

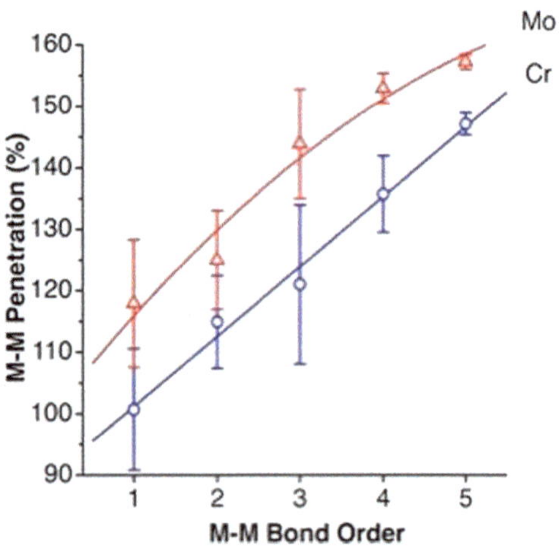

Fig. 19 Average penetration indices and esd for M–M bonds of a given bond order (M = Cr, Mo). (Reproduced from Ref. [2] with permission from the RSC)

matrix-isolation techniques. Although six bonding electron pairs could theoretically form in U_2, high-level calculations assign it a quintuple bond [45]. Despite this assignment, the penetration index at the optimized U–U distance reaches a remarkably high 199%. Earlier DFT studies [46] of $(\mu\text{-H})_2U_2$ and $H_2U_2H_2$ yield similarly elevated U–U penetrations (216% and 210%, respectively), even though the latter case suggests a maximum bond order of four. By contrast, crystallographically characterized U–U bonds exhibit penetrations around 125–146%, paralleling Cr–Cr and Mo–Mo double or triple bonds (Fig. 19). Meanwhile, Th–Th bonds in 75 structural data entries from 17 compounds in the CSD show penetrations of 105–129%, aligning with thorium's reduced capacity for multiple bonding given its fewer valence electrons. Overall, the absence of bridging ligands in U_2 and its unusually large penetration index point to intriguing avenues for further exploration. In particular, we cannot at present calibrate the size of the U cores, but certainly significant core-valence region overlap is expected at such large penetrations and maybe even a small core-core interpenetration.

6.3 Multiple Bonding in Actinyl Groups

One of the most characteristic features in actinide chemistry is the actinyl group, formed by two bonds between an actinide centre and terminal oxygen atoms in a trans arrangement – that is, a nearly linear O≡An≡O unit, typically treated as having triple-bond character [47]. Analogous structures can occur with transition metals, for instance in molybdenyl or osmyl species, which can be compared to CO_2. By analysing the structural data for various actinyl complexes, we aim at determining whether the penetration index offers a universal bonding metric for all actinides while also comparing An–O bonds to similarly short M=O bonds across the periodic table.

Figure 20 summarizes the results of this analysis. Each of the four types of oxyl bonds considered occupies a relatively narrow penetration index range (under 25%), with the actinyl groups displaying clearly higher penetrations than the other M=O bonds, suggesting triple-bond character. Another point of interest is that the standard deviation for the peak distribution of CO_2 – observed across different crystallographic environments – exceeds that of the other M=O families, even though those families encompass a wide variety of metals/actinides or substituents (as in ketones). This finding highlights the utility of penetration indices as a concise structural descriptor for analogous bonds involving atoms of different sizes. Despite some experimental uncertainty in determining CO_2 penetrations, their values still lie near the high end of the C=O distribution for acyclic ketones.

Next, we focus more narrowly on the actinyl region and sort the data by individual elements, fitting each set of histogram data to Lorentzian functions for clarity (Fig. 20b). The interpenetration indices for each actinide cluster in distinct ranges – and even show clear separations by formal oxidation state (Table 1) – underscoring the sensitivity of penetration indices to the electronic and structural variation within actinyl bonds.

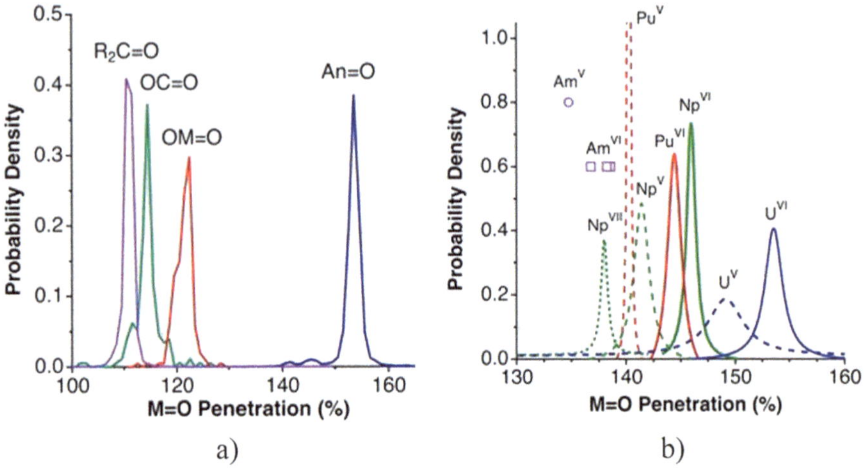

Fig. 20 (**a**) Distribution of interpenetration indices in ethers, CO_2, transition-metal *trans*-MO_2, and actinyl *trans*-AnO_2 groups. (**b**) Breakdown of the actinyl peak by element and oxidation state. (Reproduced from Ref. [2] with permission from the RSC)

Table 1 Ranges of M–O bond distances and penetration indices of actinyl groups, average penetration indices (p_{av}), and penetration indices for the maximum probability density (p_{max}), obtained from a fitting of the probability density histograms to a Lorentzian function. *N* is the number of crystallographically independent data (reproduced from Ref. [2] with permission from the RSC)

	N	M–O (Å)	*p* (%)	p_{av} (%)	p_{max} (%)
All U	6,212	1.52–2.47	109–169	153.5 (10)	153.6
U^{VI}	2,078	1.52–2.47	109–169	153.5 (10)	153.5
U^{V}	60	1.76–1.90	145–154	149.4 (19)	149.1
Np^{VII}	4	1.89–2.06	128–138	135.2 (35)	138.0
Np^{VI}	113	1.68–1.78	144–150	146.0 (6)	146.0
Np^{V}	96	1.73–1.90	137–147	141.4 (8)	141.4
Pu^{VI}	74	1.71–1.78	142–146	144.4 (5)	144.4
Pu^{V}	6	1.81–1.82	140–141	140.2 (2)	140.3
Am^{VI}	3	1.74–1.77	136–139	137.9 (9)	–
Am^{V}	1	1.81	135	–	–

7 Positronic Bonding

In recent years, there has been an increasing interest in the possibility of forming molecules by joining two anions via a bridging positron, such as p + [$2H^-$], usually referred to as *positron bonding* [48]. In this case, the energy minimum is found for the two hydrides at 3.260 Å, implying a negligible interpenetration of their van der

Waals crusts (interpenetration index of −48%). The positron probability density is located in between the two hydrides, indicating that the otherwise repulsive anions are kept at proximity due to an essentially Coulombic interaction with the bridging positron. This type of arrangement is often named a two-centre one-positron *bond* (**2c1p**), which suggests comparison with the related **2c1e** bond in H_2^+ at 1.060 Å, or a penetration index of 75%, well within the range found for other electron-deficient bonds [2].

The same dihalide systems with two positrons attached (**2c2p**) [49] show an unsurprising behaviour: (a) the positronic density is located mostly between the two halides and (b) the X···Y penetration becomes now slightly positive in most cases (Table 2) due to the stronger Coulombic attraction between the two positrons and each of the halide anions.

In the **3c2p** triangular $2e^+[3H^-]$ system [50], the positronic density sitting at the centre of the triangle keeps the hydride ions at a distance similar to that in $e^+[2H^-]$, since the stronger attraction provided by the extra positron is mostly compensated by the higher repulsion between three hydrides.

All the analogous systems in which positrons are replaced by electrons, used by many authors for comparison, behave as expected from our earlier analysis of electron-deficient and electron-precise chemical bonds [2]: **2c1e** bonds for $A\text{–}B^+$ diatomic ions (A, B = H, Li, Na, K, Rb) have penetration indices in the range of 59–82%; **3c2e** systems (H_3^+ and Li_3^+) have penetration indices of 86% and 75%, respectively; **3c4e** bonds, as in the asymmetric FHF^- anion, show penetration indices (65 structural data from the CSD) of 35 (2)% for the H-bridged F···F pair, 76 (7)% for the long F···H bond, and 93 (6)% for the short F–H bond. Finally, the covalent **2c2e** A–B bonds of the AB-neutral molecules have penetrations between 102% and 113%. A last example worth mentioning is that of the $p^+[LiH]$ molecule [53] that could be described as a **2c2e1p** system and whose rather high penetration index (p_{LiH} = 86.1%) is intermediate between those of the **2c2e** LiH molecule (p_{LiH} = 99.7%) and the **2c1e** cation LiH^+ (p_{LiH} = 65.1%).

Table 2 Penetration indices for two atoms bridged by one or two positrons, calculated from the interatomic distances given in the literature

		d (Å)[a]	p_{XY}	*d* (Å)	p_{XY}	*d* (Å)	p_{XY}
X	Y	**2c1p**: $e^+[X^-Y^-]$ [51, 52]		**2c2p**: $2e^+[X^-Y^-]$ [49]		**3c2p**: $2e^+[3X^-]$ [50]	
F	F	3.088	−9.4	2.66	14.6		
F	Cl	3.545	−15.7	3.04	14.2		
F	Br	3.709	−25.1	3.31	0.6		
Cl	Cl	3.869	−14.3	3.37	16.9		
Cl	Br	4.019	−23.2	3.54	9.6		
Br	Br	4.149	−32.5	3.88	−12.1		
H	H	3.260	−48.3			3.175	−43.5
NC	CN	3.9	−55.1				
OCN	NCO	3.5	−9.5				

[a] Very similar values have been reported by Archila-Peña et al. [49] for the halides

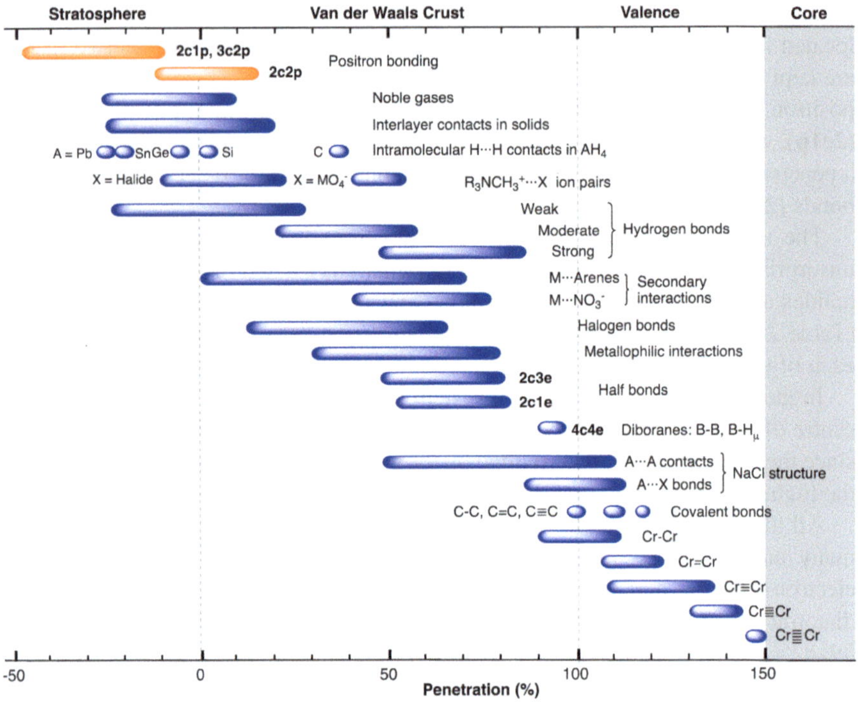

Fig. 21 Ranges of interpenetration indices presented by different bond types, showing a continuous distribution from the strongly covalent metal–metal quintuple bonds (lowest rightmost bar) to the very low penetration of pure van der Waals interactions and the still lower penetration between positron-bridged atoms (uppermost two bars). (Adapted from Ref. [2] with permission from the RSC)

The incorporation of the ranges of interpenetration indices for positron-bridged system to the bond-type chart presented in Ref. [2] (Fig. 21) extends the continuum of interatomic interactions towards the lowest atomic overlaps, with values similar or even smaller than those of the noble gas diatomic adducts held together by van der Waals interactions.

8 Conclusions

In this work, we have introduced and demonstrated the utility of the interpenetration index (p_{AB}) as a unifying metric for describing a wide spectrum of interatomic interactions – ranging from noncovalent, van der Waals contacts in noble gas dimers to the strongest multiple bonds in both main-group and transition-metal systems. By defining atomic radii in terms of three concentric spheres (core, valence, and van der Waals), and then quantifying the overlap of the outer, van der Waals "crusts", we have shown that even highly disparate interactions can be placed on a single, continuous scale.

The validity of this approach has been illustrated through several case studies. Noble gas dimers, for instance, largely occupy low or near-zero penetration regimes, as expected for weakly bound species dominated by dispersion. Hydrogen bonds, which can vary considerably in strength, fall within intermediate penetration ranges but also reveal a nuanced correlation between penetration and bond strength. σ-Hole interactions, including halogen, pnictogen, and tetrel bonds, display broader coverage of the penetration scale – reflecting significant valence-shell overlap in these partially covalent, directionally stabilized contacts. At the high end of the scale, single and multiple covalent bonds in both main-group and transition-metal compounds emerge with penetrations of 100% or beyond, in line with deep valence-shell overlap. Notably, the method also accommodates unusual systems such as U–U or actinyl bonds, revealing exceptionally high penetrations that underscore their complex bonding character. Recent works that employ interpenetration indices include the finding of a moderate correlation of the electron density at the bond critical points with the penetration index in the C–H···π contacts found in the crystal packing of a chiral compound [54], irrespective of whether the carbon atoms are sp- or sp^2. In a different direction, Kounalis et al. [55] discussed a hemilabile ligand and, correcting the Ti–P, Ti–N, and Ti–Cl bond lengths for their van der Waals radii employing the penetration index method (85%), further confirmed the remarkably long nature of the Ti–P bonds.

Overall, the interpenetration index provides a consistent, size-corrected measure of interatomic distances that is readily comparable across the entire periodic table. This framework can clarify where a given interaction lies on the continuum from purely nonbonded to strongly bonded, with potential applications in crystal engineering, reaction mechanism analysis, and computational screening of novel species. Future work may further refine this metric by integrating more detailed electronic descriptors, but as presented, the penetration index already offers a powerful tool for chemists seeking a unified perspective on molecular structure and bonding.

Acknowledgements This work has been supported by the Spanish Ministry of Science, Innovation and Universities (grants PGC2018-093863-B-C21, PID2021-122464NB-I00, PID2022-140244NB-I00 and María de Maeztu grant CEX2021-001202-M). The authors also acknowledge the Generalitat de Catalunya-AGAUR (grant 2017-SGR-1289) and the allocation of computer time at CSUC.

Appendix A: Analogy of Ionic and Covalent Atomic Radii

In this and precedent work, we have used the set of covalent and van der Waals radii published by ourselves because (a) they are obtained from experimental structural data, (b) they have the widest coverage of the natural elements of the periodic table, and (c) both sets have consistent periodic behaviours. Similar results for penetration indices should be obtained using alternative sets of atomic radii, but this possibility has not been studied in depth so far. A particular approach would replace the covalent radii (r_A and r_B in Eq. 1) by their ionic counterparts, since solid-state chemists are more familiar with this version of atomic size in ionic solids. Some conceptual aspects of the analogies and differences between ionic and covalent radii have been discussed previously in our seminal publication (Chem Sci, 2023, 14, 11647), providing some examples of atom pairs for which the sums of the ionic radii are nearly identical to the covalent radii sum.

It must be stressed that the interpenetration index depends on the sum of the atomic radii, not on the individual sizes of the atoms (Eq. 1 in the main text), and for that reason, we wish to provide in this section further evidence of the analogies between ionic and covalent radii sums. First, we look at the atom pairs in which both the electropositive and the electronegative atoms belong to the *s* and *p* blocks, considering the ions with the typical charges that obey the octet rule, as specified in Table A.1. The 92 atom pairs considered give a nice correlation (Fig. A.1a), very close to the ideal behaviour represented by $\Sigma r_{ionic} = \Sigma r_{cov}$. We have chosen to represent the correlation by Eq. (A.1), where the parameter *a* measures an average deviation from the identity between the ionic and covalent radii sums, scarcely 0.10 Å in the main group atom pairs.

Table A.1 Fitting parameter of Eq. (A.1) for different sets of atom pairs (Figs. A.1 and A.2)

Set	Cation	Anion[a]	*a*	*N*[b]
Main group				
MG	Groups 1–3, Si–Pb, As–Bi	**X**	0.10	92
Lanthanides				
Ln	Ln^{3+}	Cl^-, O^{2-}	−0.26	30
Actinides				
An	Fr^+, Ra^{2+}, Pu^{3+}, Am^{3+}, Cm^{3+}, Cf^{3+}	Cl^-, O^{2-}	−0.04	12
An′	Ac^{3+}, Th^{4+}, Pa^{3+}, U^{4+}, Np^{4+}	Cl^-, O^{2-}	−0.28	10
3d transition metals				
A	Sc^{3+}, Ti^{4+}, V^{5+}	**X**	−0.20	27
B	Cr^{3+}, Mn^{2+}, Fe^{2+}, Co^{3+}	**X**	0.05	42
	Sc^{3+}, Ti^{4+}, V^{5+}	**Y**		
C	Ni^{2+}, Cu^+, Zn^{2+}	**X**	0.27	35
	Cr^{3+}, Mn^{2+}, Fe^{2+}, Co^{3+}	**Y**		
D	Ni^{2+}, Cu^+, Zn^{2+}	**Y**	0.53	6
4d transition metals				
E	Ru^{3+}, Rh^{3+}	**Z**	−0.21	8
F	Y, Zr, Nb, Mo, Tc	**Z**	−0.01	20
G	Pd^{2+}, Cd^{2+}	**Z**	0.25	8
H	Ag^+	**Z**	0.46	4

[a] **X** = F^-, Cl^-, Br^-, I^-, O^{2-}, S^{2-}, Se^{2-}, Te^{2-}, N^{3-}; **Y** = P^{3-}, As^{3-}; **Z** = F^-, Cl^-, Br^-, O^{2-}

[b] Number of atom pairs in the set

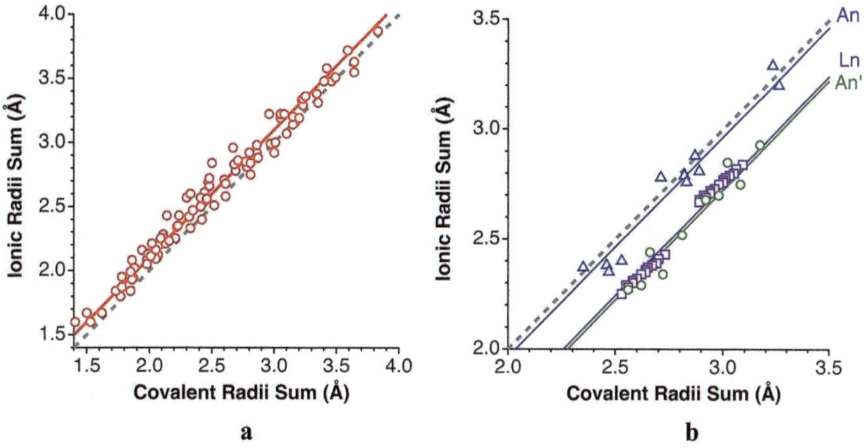

Fig. A.1 (**a**) Correlation between the sums of ionic radii and of covalent radii for main-group element pairs (see Table A.1). (**b**) Analogous correlation for the lanthanide atom pairs (**Ln**) with a Cl^- or an O^{2-} anion, and similarly for two sets of actinides, **An** and **An′** (Table A.1). The continuous lines correspond to least squares fittings to Eq. (A.1) and the dashed lines to the ideal behaviour $\Sigma r_{\text{ionic}} = \Sigma r_{\text{cov}}$

$$\Sigma r_{\text{ionic}} = a + \Sigma r_{\text{cov}} \tag{A.1}$$

The pairs formed by a lanthanide trivalent cation with a chloride or an oxo ion also present an excellent correlation between the two radii sums (set **Ln** in Fig. A.1b), although in this case, the ionic sum for the trivalent cations is systematically 0.26 Å shorter than the covalent radii sum (Table A.1). Some of the actinides, Ac–Np, present a similar behaviour (**An′** set, Table A.1 and Fig. A.1b), but the **An** set, including Fr, Ra, Pu, Am, Cm, and Cf, has practically coincident values of their ionic and covalent radii sums (Fig. A.1b and Table A.1).

The 3*d* transition metals can be grouped in four sets (sets **A**–**D** in Table A.1 and Fig. A.2a) depending on the differences between the two radii sums. Interestingly, in the most populated group, set **B**, the two radii sums differ by only about 0.05 Å. The absolute values of those differences for sets **A** and **C** atom pairs are about 0.20–0.27 (Table A.1). Finally, there is only a small number of atom pairs (Ni^{2+}, Cu^{+}, Zn^{2+} with phosphide and arsenide, set **D**) for which the sum of ionic radii can be higher than the covalent sum by up to 0.59 Å. Among the 4*d* transition metal series (Fig. A.2b), we find also a nearly perfect agreement of the ionic and covalent radii sums for the set **F** of atom pairs, a somewhat higher deviation (±0.2 Å) for sets **E** and **G**, and a large deviation for the smallest set **H**. No correlation whatsoever can be found for the 5*d* transition metal-Z anion pairs.

In summary, for the 294 atom pairs collected in Table A.1, the average absolute value of the deviations of the ionic radius sum from the covalent radius is 0.17 (10) Å, with a maximum value of 0.59 Å. Disregarding sets **D** and **H**, we are left with 274 atom pairs for which the average deviation is 0.15(9) Å and the maximum value

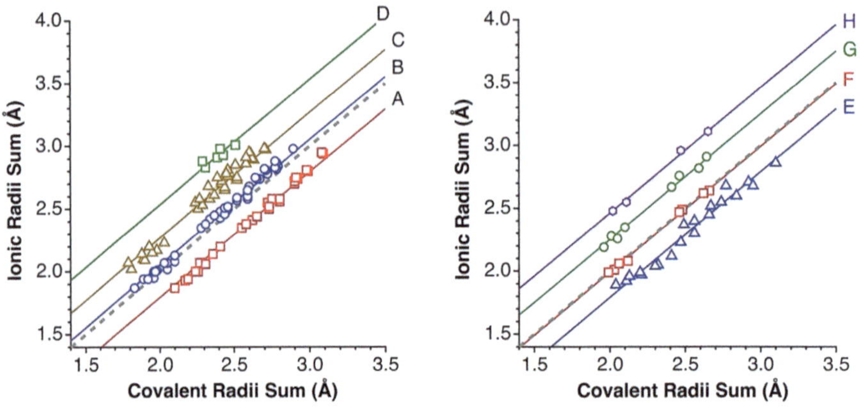

Fig. A.2 (**a**) Correlation between the sums of ionic radii and of covalent radii for 3*d* transition metal-electronegative atom pairs. For the definition of the sets of atom pairs, see Table A.1. (**b**) Analogous correlations for the 4*d* transition metals. The continuous lines correspond to least squares fittings with a fixed slope of 1.00 and the dashed lines to the ideal behaviour $\Sigma r_{ionic} = \Sigma r_{cov}$

is 0.36 Å. We can therefore conclude that in most cases, the sums of the covalent and ionic radii for an element pair differ by an amount which is within the uncertainty of the atomic radii. Besides, for several families of atom pairs, correction of the ionic radii sum with the parameter *a* (Table A.1) would yield practically the same results than the covalent radii sum, as far as the interpenetration indices are concerned.

References

1. Rahm M, Hoffmann R, Ashcroft NW (2016) Atomic and ionic radii of elements 1-96. Chem Eur J 22:14625–14632
2. Echeverría J, Alvarez S (2023) The borderless world of chemical bonding across the Van der Waals crust. Chem Sci 14:11647–11688
3. Cordero B et al (2008) Covalent radii revisited. Dalton Trans 21:2832–2838
4. Alvarez S (2013) A cartography of the Van der Waals territory. Dalton Trans 42:8617–8636
5. Szarek P, Grochala W (2014) Most probable distance between the nucleus and HOMO electron: the latent meaning of atomic radius from the product of chemical hardness and polarizability. J Phys Chem A 118(44):10281–10287
6. Gil DM, Echeverría J, Alvarez S (2022) Tetramethylammonium cation: directionality and covalency in its interactions with halide ions. Inorg Chem 61(24):9082–9095
7. Grochala W (2023) A focus on penetration index – a new descriptor of chemical bonding. Chem Sci 14:11597–11600
8. Echeverría J, Alvarez S (2024) A further focus on penetration indices of misfit van der Waals crusts. Chem Sci 15(31):12166–12168
9. van Mourik T, van Lenthe JH (1995) Benchmark full configuration interaction calculations on the helium dimer. J Chem Phys 102(19):7479–7483
10. Zhou N et al (2014) Improved parameterization of interatomic potentials for rare gas dimers with density-based energy decomposition analysis. J Chem Phys 140(21):214117

11. Vogt J et al (2015) New features in the MOGADOC database. Struct Chem 26(5):1725–1727
12. Zagorac D et al (2019) Recent developments in the inorganic crystal structure database: theoretical crystal structure data and related features. J Appl Crystallogr 52(5):918–925
13. Feltgen R et al (1982) Unique determination of the He2 ground state potential from experiment by use of a reliable potential model. J Chem Phys 76(5):2360–2378
14. Steiner T (2002) The hydrogen bond in the solid state. Angew Chem Int Ed 41(1):48–76
15. Desiraju GR (1999) The weak hydrogen bond. Oxford University Press, New York
16. Grabowski SJ, Sokalski WA (2005) Different types of hydrogen bonds: correlation analysis of interaction energy components. J Phys Org Chem 18:779–784
17. Dankert F et al (2017) A structural study of alkaline earth metal complexes with hybrid disila-crown ethers. Dalton Trans 46(27):8727–8735
18. Kawaguchi K, Hirota E (1987) Diode laser spectroscopy of the $\nu 3$ and $\nu 2$ bands of FHF− in 1300 cm−1 region. J Chem Phys 87(12):6838–6841
19. Ibers JA (1964) Refinement of Peterson and Levy's neutron diffraction data on KHF_2. J Chem Phys 40(2):402–404
20. Mahjoub AR, Leopold D, Seppelt K (1992) Iodine pentafluoride – hydrogen fluoride co-crystallisates, directed weak interactions. Eur J Solid State Inorg Chem 29:635–647
21. Dereka B et al (2021) Crossover from hydrogen to chemical bonding. Science 371(6525): 160–164
22. Lundgren JO, Olovsson I, Studies HB, XV. (1967) The crystal structure of hydrogen chloride dihydrate. Acta Cryst 23:966–971
23. Attig R, Williams JM (1976) The $H_5O_2^+$ ion in hydrogen bromide dihydrate. Angew Chem Int Ed 15:491–492
24. Lebrun N et al (2001) A new crystalline phase of nitric acid dihydrate. Acta Cryst C Cryst Struct Commun 57:1129–1131
25. Fargher HA et al (2022) C–H···S hydrogen bonding interactions. Chem Soc Rev 51(4): 1454–1469
26. Keil H et al (2022) Si–H···Se chalcogen–hydride bond quantified by diffraction and topological analyses. Inorg Chem 61:6319–6325
27. Martín-Fernández C, Elguero J, Alkorta I (2024) Beryllium as a base: complexes of $Be(CO)_3$ with HX (X = F, Cl, Br, CN, NC, CCH, OH). ChemPhysChem 25(21):e202400608
28. Das S, Pratik SM, Datta A (2024) Crystals of organic acid-base complexes defy the ΔpK_a rule under compression. J Phys Chem C 128(40):17175–17183
29. Politzer P et al (2017) The σ-hole revisited. Phys Chem Chem Phys 19(48):32166–32178
30. Hansen T et al (2022) C-X bond activation by palladium: steric shielding versus steric attraction. Chem Eur J 28:3202201093
31. Bauzà A, Mooibroek TJ, Frontera A (2013) Tetrel-bonding interaction: rediscovered supramolecular force? Angew Chem Int Ed 52:12317–12321
32. Mo Y, Danovich D, Shaik S (2022) The roles of charge transfer and polarization in non-covalent interactions: a perspective from ab initio valence bond methods. J Mol Model 28:274
33. Grabowski SJ (2019) Pnicogen and tetrel bonds – tetrahedral Lewis acid centres. Struct Chem 30:1141–1152
34. Bartashevich E, Matveychuk Y, Tsirelson V (2019) Identification of the tetrel bonds between halide anions and carbon atom of methyl groups using electronic criterion. Molecules 24:1083, 12 p
35. Wang C et al (2014) On the nature of the halogen bond. J Chem Theor Comput 10:3726–3737
36. Murray JS, Politzer P (2020) Interaction and polarization energy relationships in σ-hole and π-hole bonding. Crystals 10:76, 16 p
37. Velasquez JD et al (2024) Unveiling the potential of haloalkenes as electron density acceptors. Cryst Growth Des 24(13):5775–5780

38. Shen Z et al (2024) Influence of noncovalent interaction on the nucleophilicity and electrophilicity of metal centers in $M^{II}(S_2{>}CNEt_2)_2$ (M = Ni, Pd, Pt). J Phys Chem A 128(50): 10796–10807
39. Demonti L et al (2021) Cross-coupling through Ag(I)/Ag(III) redox manifold. Chem Eur J 27(62):15396–15405
40. Velasquez JD et al (2024) Azido-mediated intermolecular interactions of transition metal complexes. Phys Chem Chem Phys 26(8):6683–6695
41. Stoccoro S et al (2024) Chirality generated by hindered rotations in platinum(II)-pyridine complexes. Eur J Inorg Chem 27(3):e202300649
42. Echeverria J, Alvarez S (2024) Widening the scope of structural correlations by means of the van der Waals crust penetration indices: the dimerization of groups 11 and 12 L-M-X halo complexes. Cryst Growth Des 24(11):4743–4747
43. Blokker E et al (2021) The chemical bond: when atom size instead of electronegativity difference determines trend in bond strength. Chem Eur J 27:15616–15622
44. Falceto A, Alvarez S (2016) Comparison of the Cr-Cr quadruple and quintuple bonding mechanisms. Struct Bond 170:149–164
45. Gagliardi L, Roos BO (2005) Quantum chemical calculations show that the uranium molecule U_2 has a quintuple bond. Nature 433:848–851
46. Souter PF et al (1997) Experimental and theoretical evidence for the formation of several uranium hydride molecules. J Am Chem Soc 119:1682
47. Autillo M et al (2022) Periodic trends within actinyl(VI) nitrates and their structures, vibrational spectra, and electronic properties. Inorg Chem 61:15607–15618
48. Charry J, Varella MTN, Reyes A (2018) Binding matter with antimatter: the covalent positron bond. Angew Chem Int Ed 57:8859–8864
49. Archila-Peña D et al (2024) Two-positron-bonded dihalides: Ps_2XY (X, Y = F, Cl, Br). Chem Eur J 30:e202402618
50. Charry J et al (2022) The three-center two-positron bond. Chem Sci 13:13795–13802
51. Moncada F et al (2020) Covalent bonds in positron dihalides. Chem Sci 11:44–52
52. Cassidy JP et al (2024) Many-body theory calculations of positronic-bonded molecular dianions. J Chem Phys 160:084304
53. Mella M et al (2000) Positron and positronium chemistry by quantum Monte Carlo. V. The ground state potential energy curve of e^+LiH. J Chem Phys 113:6154–6159
54. Nakamura M, Kashiwagi Y, Morisue M (2024) Comprehensive analysis of C-H···π(alkyne) interactions in the crystal packing of diastereomers of 1,2-di(7′-methoxynaphth-1′-yl)-3,6-di (4″-n-propylphenylethynyl)benzene. CrystEngComm 26(29):3964–3972
55. Kounalis E et al (2024) Pendulum-like hemilability in a Ti-based frustrated Lewis Trio. Chem Sci 15(15):5555–5563

Struct Bond (2026) 190: 297–314
https://doi.org/10.1007/430_2025_95

Published online: 24 April 2025

Graphically Revealing Weak Interactions in Dynamic Environments Using amIGM Method

Tian Lu

Contents

Abstract Visualization of weak interactions is now a very popular kind of analysis method, which allows chemists to recognize the existence, strength, and type of interactions in different regions of a chemical system easily and intuitively. Among these analysis methods, the independent gradient model (IGM) and its variant with better graphical effect, namely IGM based on Hirshfeld partition of the molecular electronic density (IGMH), are particularly useful. They can clearly reveal various interactions between specific fragments and have become widely popular in recent years. This chapter first briefly reviews the definition and characteristics of the two methods and then introduces a new variant of IGM, called modified IGM (mIGM), which, like IGM, only requires atomic coordinate information to perform the analysis and thus the computational cost is fairly low, while its graphical quality in exhibiting weak interactions is almost as good as the markedly more expensive

T. Lu
Beijing Kein Research Center for Natural Sciences, Beijing, People's Republic of China
e-mail: http://www.keinsci.com

IGMH. This chapter then introduces the averaged mIGM (amIGM) method, which extends mIGM to visual analysis of the weak interactions involved in molecular dynamics trajectories. A series of examples fully demonstrate that amIGM can very effectively display all kinds of weak interactions in dynamic environments, and its image quality and flexibility are all superior to the previously proposed averaged noncovalent interaction (aNCI) method with similar uses.

Keywords Independent gradient model · Molecular dynamics · Multiwfn · Noncovalent interaction · Weak interaction

1 Introduction

Interactions in chemical systems can be mainly divided into covalent interactions and noncovalent interactions. Noncovalent interactions can be further divided into ionic bonds and weak interactions; the interaction strength of the latter is usually in the range of several to tens of kJ/mol. Especially since the twenty-first century, intermolecular and intramolecular weak interactions have received more and more attention from chemists, which have a crucial impact on, for example, the conformational distribution of flexible molecules [1], preferential configurations of molecular complexes [2], formation of supramolecules [3, 4], protein-ligand recognition [5], molecular adsorption [6], structure and physical properties of molecular crystals and liquids [7, 8], and selectivity of many chemical reactions [9]. In 2010, Weitao Yang and coworkers proposed the NCI method to visually display noncovalent interactions in chemical systems; this method utilizes colored isosurfaces of reduced density gradient (RDG) to reveal the regions involving notable weak interactions and meantime distinguish strengths and types of interactions [10]. More or less inspired by this work, a series of methods aiming at graphically exhibiting interactions were born since then, including averaged NCI (aNCI) [11], interaction region indicator (IRI) [12], independent gradient model (IGM) [13], IGM based on Hirshfeld partition of the molecular electronic density (IGMH) [14, 15], and so on, some of which have been widely used in recent computational chemistry literature [6, 9, 16–20]. The ideas, definitions, implementations, and practical applications of these methods have been comprehensively reviewed by us recently [21], and interested readers are recommended to read it to fully understand these methods.

The unique advantage of the IGM analysis method is that it allows chemists to flexibly define fragments, permitting the positions and types of weak interactions between the fragments and within every fragment to be conveniently and graphically revealed. In addition, IGM analysis only relies on atomic coordinates; therefore, its computational cost is fairly low. One of the main purposes of this chapter is to extend the popular and useful IGM analysis to the study of the average weak interactions in dynamic environments simulated by molecular dynamics (MD) techniques. We name the new method as averaged IGM (aIGM). Unfortunately, the IGM map often has a relatively poor graphical effect, and as a result, the aIGM method is not fully ideal. Although the IGMH analysis proposed by us earlier has a prominently better graphical effect than IGM, it is too expensive to be adopted in the analysis of a large amount of frames in MD trajectories, because it was defined based on the molecular electron density generated by quantum chemistry calculations. To circumvent this difficulty, we also propose a new variant of IGM named modified IGM (mIGM), which is as cheap as IGM because only atomic information is involved, while its graphical effect is very close to IGMH. Further extending the mIGM to the analysis of dynamic environments results in another new method, amIGM.

In the rest of this chapter, Sect. 2 will present a brief recap of IGM and IGMH to provide readers with an adequate background knowledge to understand how mIGM works. In Sect. 3, the definition and some illustrative examples of mIGM will be given. Section 4 will introduce the idea of aIGM and amIGM. Section 5 will present a series of practical applications of amIGM to demonstrate its importance and great value. Section 6 will make a summary.

2 Brief Recap of IGM and IGMH Methods

The IGM method defined a three-dimensional real space function $\delta g(\mathbf{r}) = g^{\mathrm{IGM}}(\mathbf{r}) - g(\mathbf{r})$ [13, 15, 21], where

$$
\begin{aligned}
g(\mathbf{r}) &= \left| \sum_i \nabla \rho_i^{\mathrm{free}}(\mathbf{r}) \right| \\
g^{\mathrm{IGM}}(\mathbf{r}) &= \sum_i \left| \nabla \rho_i^{\mathrm{free}}(\mathbf{r}) \right|
\end{aligned}
\tag{1}
$$

in which ρ_i^{free} stands for spherically averaged density of atom i in its free (isolated) state, and i loops over all atoms in the system. ρ^{free} data for all elements were built into the IGM analysis code Multiwfn [22, 23], so δg can be rapidly evaluated if a molecular structure file is provided. The isosurfaces of δg are capable of revealing the regions where interatomic interactions in the system are mainly involved. Interatomic regions with a relatively strong interaction tend to have a large δg value; they are only visible in the δg isosurface map with a relatively high isovalue. Regions with a relatively weak interaction usually have a small δg value, and hence the interactions can only be exhibited in the IGM map when the isovalue is set to be low enough. IGM also defined $\delta g^{\mathrm{inter}}$ and $\delta g^{\mathrm{intra}}$ functions to reveal interfragment and intrafragment interactions separately [13, 21],

$$
\begin{aligned}
\delta g^{\mathrm{inter}}(\mathbf{r}) &= g^{\mathrm{IGM,inter}}(\mathbf{r}) - g^{\mathrm{inter}}(\mathbf{r}) \\
\delta g^{\mathrm{intra}}(\mathbf{r}) &= \delta g(\mathbf{r}) - \delta g^{\mathrm{inter}}(\mathbf{r})
\end{aligned}
\tag{2}
$$

where

$$g^{\text{IGM,inter}}(\mathbf{r}) = \sum_A \left| \sum_{i \in A} \nabla \rho_i^{\text{free}}(\mathbf{r}) \right|$$
$$g^{\text{inter}}(\mathbf{r}) = \left| \sum_A \sum_{i \in A} \nabla \rho_i^{\text{free}}(\mathbf{r}) \right| \quad (3)$$

with A and i being the index of fragments and atoms, respectively. Two or more fragments can be arbitrarily defined according to the character of the actual system and research purpose. Inspired by the NCI method, the IGM method maps the sign $(\lambda_2)\rho$ function with different colors on the isosurfaces of δg or g^{inter} or δg^{intra} to distinguish the characteristics of the revealed interactions. In the sign$(\lambda_2)\rho$, $\rho(\mathbf{r}) = \sum_i \rho_i^{\text{free}}(\mathbf{r})$ corresponds to promolecular density, and sign(λ_2) denotes the sign of the second largest eigenvalue of the Hessian matrix of ρ. See the aforementioned review for the underlying idea of this function [21].

As an example, the IGM map of the trimer of capped phenylalanine is shown in Fig. 1, and the geometry was taken from the L7 noncovalent test set [24], which was optimized at TPSS-D/TZVP level. The meaning of the mapped colors of sign$(\lambda_2)\rho$ is indicated on the color bar. The three monomers are, respectively, defined as the three fragments in the IGM analysis. The raw data were calculated by Multiwfn 3.8(dev) [22, 23], the map was rendered by VMD 1.9.3 software [25], and grid spacing of 0.15 Bohr was used, similarly hereinafter. As can be seen, the colored isosurfaces of the g^{inter} defined in IGM successfully revealed the hydrogen bond (H bond) and van der Waals (vdW) interaction regions, which are colored by blue and green, respectively. However, all isosurfaces look quite bulgy, and the bulgy isosurfaces tend to extend into the areas very close to the nuclei, which are obviously not the characteristic regions suitable for exhibiting weak interactions, thus resulting in unreasonable colors on the isosurfaces. As highlighted by the purple arrows, blue and red colors are clearly visible in some vdW interaction regions, which may cause severe misunderstanding about the nature of interactions.

Figure 1 also shows IGMH map. It can be seen that the isosurfaces of the g^{inter} defined by IGMH are much thinner than those of IGM, and the coloring effect is quite reasonable, completely avoiding the aforementioned problems in the IGM map. In IGMH, the ρ_i^{free} involved in IGM is replaced with the atomic density obtained by partitioning molecular density via the Hirshfeld weight function of atom i [15], that is, $\rho_i^{\text{IGMH}}(\mathbf{r}) = \rho(\mathbf{r}) w_i(\mathbf{r})$, where ρ is the molecular density derived from the molecular wavefunction generated by a quantum chemistry calculation, and $w_i(\mathbf{r}) = \rho_i^{\text{free}}/\rho^{\text{pro}}$, in which ρ^{pro} is the aforementioned promolecular density. It is important to note that the Cartesian components of the $\nabla \rho_i^{\text{IGMH}}$ term involved in IGMH are not evaluated in the mathematically correct way as

$$\frac{\partial \rho_i^{\text{IGMH}}}{\partial \mu} = \frac{\partial (\rho w_i)}{\partial \mu} = w_i \frac{\partial \rho}{\partial \mu} + \rho \frac{\partial w_i}{\partial \mu} \quad (\mu = x, y, z) \quad (4)$$

but evaluated in the following specific way [14]:

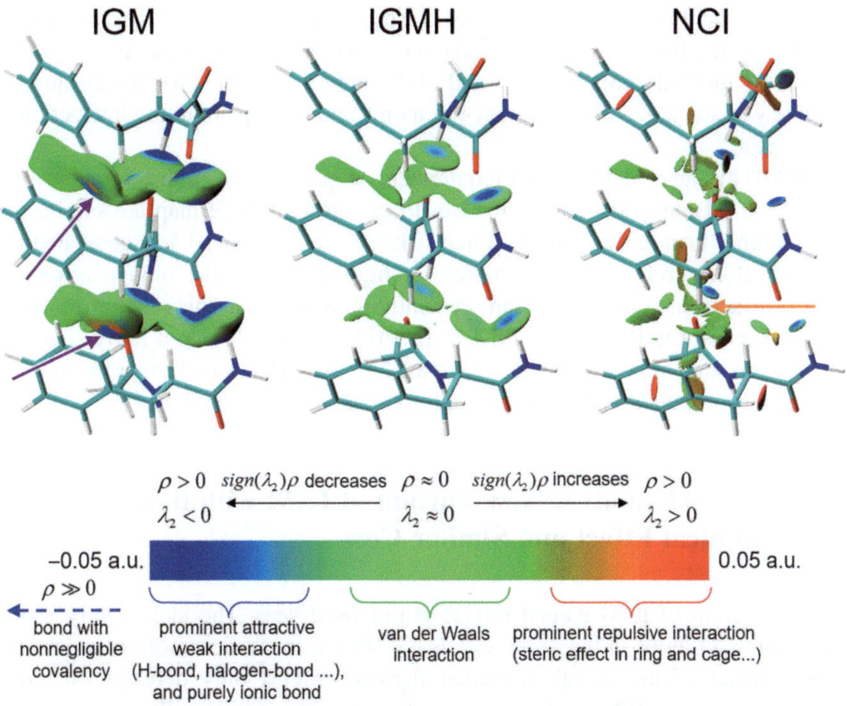

Fig. 1 Weak interactions in the trimer of capped phenylalanine revealed by IGM, IGMH, and NCI methods. The isosurfaces in the IGM and IGMH maps correspond to δg^{inter} of 0.0035 a.u., and the isosurfaces in the NCI map correspond to RDG of 0.5. The meaning of the color corresponding to the mapped function sign(λ_2)ρ is indicated on the color bar. Only for the NCI map are the lower and upper color limits set to −0.04 and 0.02 a.u., respectively, which is the commonly used color scale range in NCI analysis. Grid spacing of 0.15 Bohr is employed for all analyses

$$\frac{\partial \rho_i^{\text{Hirsh}}}{\partial \mu} = w_i \frac{\partial \rho}{\partial \mu} - \rho \frac{\partial w_i}{\partial \mu} \quad (\mu = x, y, z) \tag{5}$$

Also note that the density used in the evaluation of sign(λ_2)ρ in the IGMH analysis comes from a quantum chemistry calculation without using the promolecular approximation.

The NCI method mentioned above is also very popular in revealing weak interactions, while IGMH has many important advantages compared to it:

1. IGMH is able to distinguish intrafragment and interfragment interactions by δg^{intra} and δg^{inter}. In addition, by properly setting the isovalue, one can choose to only visualize relatively strong interactions, or all interactions, in the map. In contrast, the NCI method always exhibits all interactions in a single map, which often contains a large number of isosurfaces, making it difficult to visually study the interactions of actual interest. Indeed, as can be seen from the NCI map in

Fig. 1, there are many very scattered isosurfaces; obviously, this map is not as clear as the IGMH map in revealing the intermolecular interactions.

2. IGMH method defines δG^{atom} and δG^{pair} indices [15, 21], which can quantitatively characterize the importance of atoms and atom pairs in the interactions of interest.
3. IGMH has an evidently lower requirement on grid quality. At the commonly used 0.15–0.2 Bohr grid spacing, the isosurfaces in the IGMH map are significantly smoother than those in the NCI map. The NCI map in Fig. 1 was generated with 0.15 Bohr grid spacing; however, as highlighted by the orange arrow, the isosurfaces in some regions are evidently uneven, while this phenomenon does not exist in the IGM and IGMH maps. Although employing a fairly small grid spacing such as 0.08 Bohr can make the NCI isosurfaces look smooth enough, it can be computationally demanding for a relatively large system.

3 mIGM Method: A New Variant of IGM with Better Graphical Effect and Similar Cost

A new variant of IGM named mIGM is proposed here. The idea is very simple: calculating all terms in the same way as IGMH, except that promolecular density is used instead of the actual molecular density. mIGM only depends on atomic coordinates like IGM, so their computational costs are basically the same. The mIGM map showing the intermolecular interactions in the trimer of capped phenylalanine is given in Fig. 2. It can be seen that the overall graphical characteristics of mIGM with an isovalue of 0.006 a.u. are basically the same as the IGMH map with

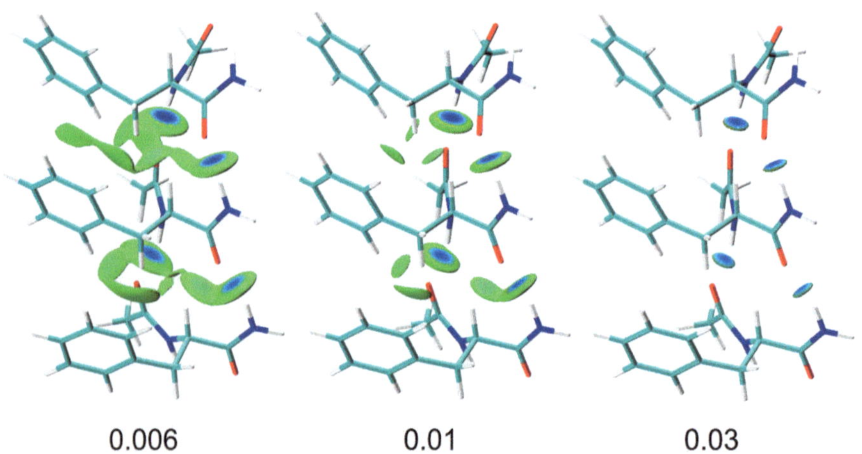

Fig. 2 Weak interactions in the trimer of capped phenylalanine revealed by mIGM method. Isosurfaces of δg^{inter} corresponding to different isovalues are respectively shown. The color bar is the same as in Fig. 1

an isovalue of 0.0035 a.u. in Fig. 1, so mIGM is ideal for use when IGMH analysis is too expensive. Furthermore, as the isovalue of δg^{inter} of mIGM changes, the variation of isosurfaces is completely in line with expectations. At the isovalue of 0.03 a.u., only the significant H bonds are visible. When the isovalue reduces to 0.01 a.u., the prominent vdW interactions stemming from close contacts between atoms are also shown. In the case of isovalue of 0.006 a.u., weaker vdW interactions distributed over a larger area can also be seen. It is noteworthy that, as IGMH is defined based on actual molecular density and thus the change in electron density distribution in the process of formation of weak interactions can be better captured, the more rigorous IGMH is recommended over mIGM whenever IGMH analysis is inexpensive for the system under study.

The key working equation of the mIGM method is explicitly given here. After derivation, the Cartesian components of $\nabla\rho_i^{\text{IGMH}}$ in the IGMH method are calculated as follows:

$$\frac{\partial\rho_i^{\text{IGMH}}}{\partial\mu} = -\frac{\rho_i^{\text{free}}}{\rho^{\text{pro}}}\frac{\partial\rho}{\partial\mu} - \frac{\rho}{\rho^{\text{pro}}}\frac{\partial\rho_i^{\text{free}}}{\partial\mu} + \frac{\rho\rho_i^{\text{free}}}{(\rho^{\text{pro}})^2}\sum_j\frac{\partial\rho_j^{\text{free}}}{\partial\mu} \quad (\mu = x, y, z) \tag{6}$$

In the mIGM method, the ρ in the above equation is replaced with ρ^{pro}, so the Cartesian derivatives of the atomic density involved in mIGM are evaluated as

$$\begin{aligned}\frac{\partial\rho_i^{\text{mIGM}}}{\partial\mu} &= \frac{\rho_i^{\text{free}}}{\rho^{\text{pro}}}\frac{\partial}{\partial\mu}\left(\sum_j\partial\rho_j^{\text{free}}\right) - \frac{\partial\rho_i^{\text{free}}}{\partial\mu} + \frac{\rho_i^{\text{free}}}{\rho^{\text{pro}}}\sum_j\frac{\partial\rho_j^{\text{free}}}{\partial\mu} \quad (\mu = x, y, z) \\ &= -\frac{\partial\rho_i^{\text{free}}}{\partial\mu} + \frac{2\rho_i^{\text{free}}}{\rho^{\text{pro}}}\sum_j\frac{\partial\rho_j^{\text{free}}}{\partial\mu}\end{aligned} \tag{7}$$

Note that $\nabla\rho_i^{\text{IGMH}}$ and $\nabla\rho_i^{\text{IGM}}$ are viewed as zero in the regions very far from any atom ($\rho^{\text{pro}} = 0$).

To better illustrate the reasonableness and usefulness of mIGM, more examples of mIGM are given in Fig. 3. Figure 3a shows the host-guest complex formed by a C_{60} fullerene and a fullerene catcher; the geometry was taken from Ref. [26]. It can be seen that the two broad isosurfaces very vividly revealed the two evident π–π stacking regions. Crystalline tellurium consists of parallel helical chains of Te atoms [27]. In order to reveal the interactions between the chains, mIGM analysis is performed based on 3 × 3 × 4 supercell structure; see Fig. 3b; the atoms in the central chain and the atoms in other chains are, respectively, defined as a fragment. It can be seen that this map successfully exhibits the interchain interactions; the marginally blue color over the isosurfaces implies that the interaction may be somewhat stronger than the common vdW interaction. Indeed, the Mayer bond order [28] between the corresponding closely contacted atoms calculated by Multiwfn based on the wavefunction produced at the PBE/DZVP-MOLOPT-SR-GTH level by CP2K [29] is 0.137, which suggests that the interchain Te-Te

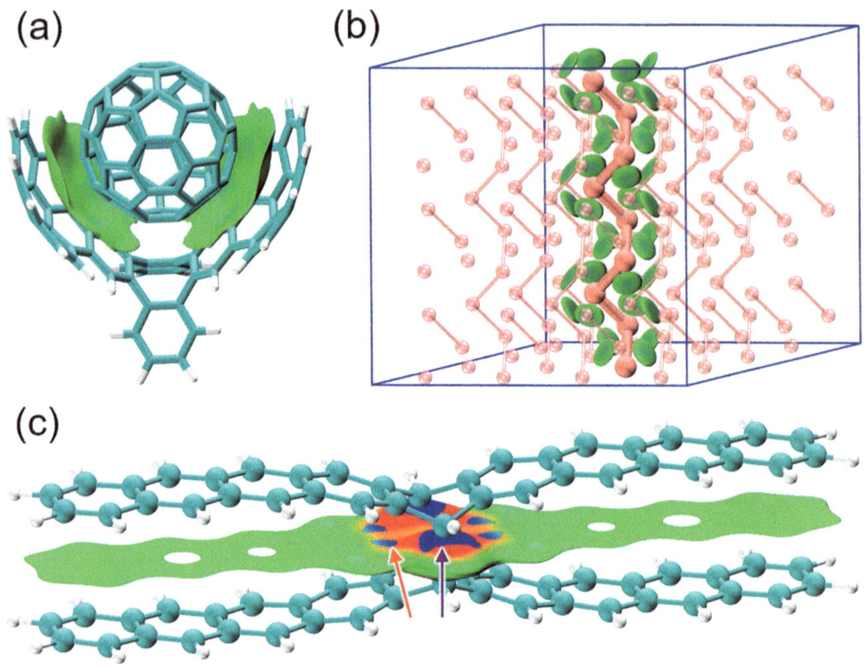

Fig. 3 Weak interactions revealed by the mIGM method. The color bar is the same as in Fig. 1. (**a**) C_{60} and fullerene catcher, with δg^{inter} isovalue of 0.0035 a.u. (**b**) Crystalline tellurium supercell, with δg^{inter} isovalue of 0.01 a.u. The first fragment is the central chain; all surrounding chains correspond to the second fragment. (**c**) Nonacene covalent dimer, with δg^{inter} isovalue of 0.0035 a.u.

interaction shows slight covalency. Figure 3c shows the covalent dimer of nonacene; the two nonacenes are defined as the two fragments in the mIGM analysis. This map shows three pieces of information at the same time: (1) There is a significant π–π interaction on both sides of the dimer (green regions). (2) There is obvious covalent bonding in the middle of the dimer (dark blue regions indicated by the purple arrow). (3) Severe steric hindrance exists around the covalent bonds (red region). Note that the small blue areas indicated by the orange arrow are an artificial phenomenon of mIGM; in fact, there is no prominent C–C attractive interaction, and the blue color cannot be notably detected in the more rigorous IGMH map. This issue comes from the fact that mIGM is not defined based on the actual electron density like IGMH, and the electron density changes considerably around covalent bond regions during formation of the molecule from fragments. However, this shortcoming of mIGM is generally negligible in the analysis of weak interactions.

mIGM has been implemented in the very popular wavefunction analysis code Multiwfn [22, 23] since the version got updated on January 8, 2025. Multiwfn can be freely downloaded at http://sobereva.com/multiwfn. Performing mIGM analysis in Multiwfn is very convenient; the user just needs to load a file containing geometry information (formats such as xyz, pdb, mol2, fch, and mwfn [30]) into Multiwfn and then input commands like the following ones in turn:

```
20 //Enter main function 20 (visual study of weak interactions)
-10 //mIGM analysis
2 //Number of fragments
1-13 //Atomic indices in fragment 1
14-26 //Atomic indices in fragment 2
4 //Set grid spacing
0.15 //0.15 Bohr
```

Then Multiwfn starts to calculate grid data of δg, g^{inter}, δg^{intra}, and sign(λ_2)ρ. Once the calculation is finished, one can choose to export them as .cub files. After that, one can use the VMD plotting scripts provided in the Multiwfn package to directly visualize mIGM map in VMD. Readers are advised to search "mIGM" in the Multiwfn manual for detailed information and examples.

4 aIGM and amIGM Methods: Extending IGM and mIGM to Analysis of Dynamic Environments

In reality, molecules are not fixed but are constantly moving, so weak interactions are also constantly changing. The movement of atoms and molecules can be simulated by the very mature MD technique. The aforementioned IGM, IGMH, and mIGM can only be used to visually analyze the interactions in a specific structure. In order to allow IGM and mIGM to study the average weak interactions in a dynamic environment, inspired by the idea of aNCI [11, 21], the aIGM and amIGM methods are proposed here. In these methods, detectable interfragment interactions are revealed by isosurfaces of averaged δg^{inter} ($\delta\overline{g}^{\text{inter}}$), which is simply the time average of the δg^{inter} calculated for every frame in an MD trajectory. For aIGM and amIGM, the δg^{inter} is calculated in the form of IGM and mIGM, respectively. The averaged sign(λ_2)ρ is used to color the isosurfaces of $\overline{g}^{\text{inter}}$, which is evaluated with the same way as aNCI [11, 21], namely ρ corresponds to time-averaged promolecular density, and λ_2 corresponds to the second largest eigenvalue of time-averaged Hessian of promolecular density.

The computational cost of aIGM and amIGM analyses is linearly proportional to the number of considered frames in the MD trajectory. In practical applications, the greater the change in position and characteristic of the interested weak interactions during the MD process, the more frames are required to be considered; otherwise, the average weak interactions cannot be fully revealed due to insufficient sampling. Considering the computational cost and accuracy, hundreds to thousands of frames are generally considered.

In principle, IGMH can also be analogously extended to studying average weak interactions and can be named as averaged IGMH (aIGMH). However, performing ab initio molecular dynamics (AIMD) is often quite time consuming even if the system only contains dozens of atoms, and calculating grid data of IGMH for several hundreds of MD frames is also expensive, which severely limits the practical use of aIGMH. So, we do not explore aIGMH in this work.

aIGM and amIGM analyses have been available in Multiwfn since the version got updated on January 30, 2021, and January 8, 2025, respectively. The user needs to load an MD trajectory in .xyz format into Multiwfn and then input commands like the following ones in turn:

```
20 //Enter main function 20 (visual study of weak interaction)
-12 //amIGM analysis
2 //Number of fragments
1-13 //Atomic indices in fragment 1
14-26 //Atomic indices in fragment 2
1,1000 //Range of considered frames is 1 to 1000
4 //Set grid spacing
0.15 //0.15 Bohr
```

Then Multiwfn starts to calculate grid data of δg, g^{inter}, δg^{intra}, RDG, and sign $(\lambda_2)\rho$ for every frame and then take the time average to yield the data used for plotting amIGM and aNCI maps. Once the calculation is finished, one can choose to export them as .cub files; after that, a VMD plotting script provided in the Multiwfn package can be used to directly visualize amIGM map in VMD. Readers are advised to search "amIGM" in the Multiwfn manual for detailed information and full examples.

It is noteworthy that the $\delta\overline{g}^{\text{inter}}$ in aIGM and amIGM may also be defined in the following way:

$$\delta\overline{g}^{\text{inter}}(\mathbf{r}) = \overline{g}^{\text{IGM,inter}}(\mathbf{r}) - \overline{g}^{\text{inter}}(\mathbf{r}) \tag{8}$$

where $\overline{g}^{\text{IGM,inter}}$ and $\overline{g}^{\text{inter}}$ are calculated as follows:

$$\begin{aligned} \overline{g}^{\text{IGM,inter}}(\mathbf{r}) &= \sum_A \left| \left\langle \sum_{i \in A} \nabla \rho_i(\mathbf{r}) \right\rangle \right| \\ \overline{g}^{\text{inter}}(\mathbf{r}) &= \left| \left\langle \sum_A \sum_{i \in A} \nabla \rho_i(\mathbf{r}) \right\rangle \right| \end{aligned} \tag{9}$$

in which the $\langle\ \rangle$ symbol stands for the time average of considered trajectory frames. Our current tests showed that this definition of $\delta\overline{g}^{\text{inter}}$ is not as ideal as the aforementioned definition, namely the corresponding isosurfaces are unsmooth and the shapes are not very reasonable, so this definition was not finally adopted.

5 Illustrative Examples of amIGM Method

In the next subsections, we will present five representative examples of amIGM to exhibit its characters and practical value in unveiling weak interactions in realistic dynamic environment simulated by MD technique. amIGM will also be compared with aIGM and aNCI to show its unique advantages.

5.1 Average Interaction Between Water Molecules in Liquid Phase

We carried out MD simulation for a box containing 511 water molecules with the GROMACS 2018.8 program [31] using the SPC/E water model [32] at room temperature with experimental density (NVT ensemble). A water molecule approximately at the box center in the initial structure was fixed during the MD process. The time length of the simulation was 1 ns, and we evenly extracted 1,000 frames for amIGM and aIGM analyses; the resulting maps are given in Fig. 4. It can be seen that the isosurfaces along with the mapped colors in the amIGM map perfectly reveal the regions that can form H bonds; it shows that O–H can act as a H-bond donor and the lone pair region of O can act as a H-bond acceptor. Moreover, as expected, the distribution of the amIGM isosurfaces is completely in line with molecular symmetry. The aIGM analysis in this instance conveys basically the same information as the amIGM analysis; however, like IGM, the isosurfaces in the aIGM map look too bulgy and somewhat inelegant. As a comparison, we arbitrarily extracted a frame from the trajectory to perform mIGM analysis; the resulting map is also shown in

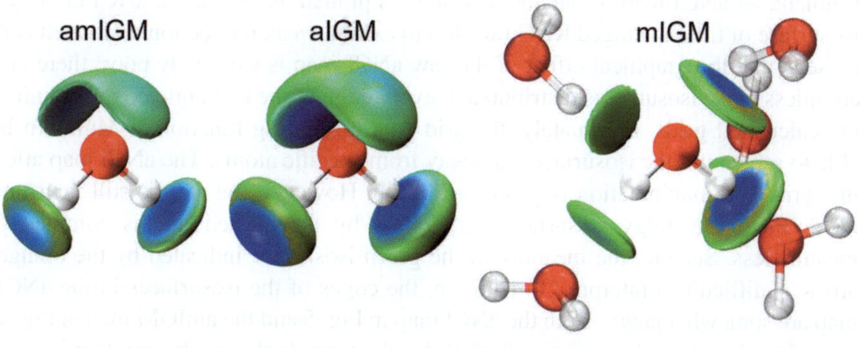

Fig. 4 For a pure water system at room temperature, the average weak interactions between a selected water and surrounding waters revealed by amIGM and aIGM methods, as well as the weak interactions of a frame revealed by mIGM method. The color bar is the same as in Fig. 1. Isovalue of $\delta\overline{g}^{\text{inter}}$ is set to be 0.01 a.u.

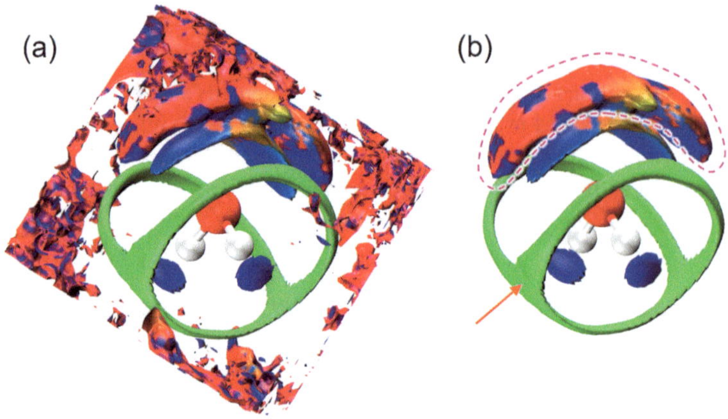

Fig. 5 For a pure water system during 1 ns MD simulation at room temperature, the average weak interactions between a selected water and other waters revealed by aNCI method before (**a**) and after (**b**) grid data manipulation to remove noise isosurfaces. Isovalue of aRDG is 0.3; the same color scale as in Fig. 1 is used but with lower and upper limits of −0.025 and 0.025 a.u., respectively

Fig. 4. The isosurfaces in the mIGM map are obviously inconsistent with the symmetry of the water molecule; the two O–H···O H bonds on the left side of this map appear to be much weaker than those on the right side. This observation arises from the fact that weak interactions are changing at every moment during the MD simulation; certainly, mIGM analysis on an instantaneous structure can only exhibit instantaneous weak interactions. This comparison reflects the unique advantage of amIGM in understanding the average weak interactions.

To demonstrate the remarkable advantage of amIGM over aNCI, the latter being another method aiming at revealing average weak interactions in a dynamic environment, an aNCI map of the above system is plotted as in Fig. 5a. aNCI employs isosurface of time-averaged RDG (aRDG) to exhibit weak interaction regions. It can be seen that the graphical effect of the raw aNCI map is extremely poor; there are countless noise isosurfaces distributed at every edge of the rectangle box containing the calculated grids. Fortunately, the grid data processing function in Multiwfn is able to screen out the isosurfaces far away from specific atoms. The aNCI map after this grid data manipulation is given in Fig. 5b. However, the map is still far from ideal. First, the bulgy isosurface highlighted by the dashed line is completely meaningless. Second, the meaning of the green isosurface indicated by the orange arrow is difficult to interpret. In addition, the edges of the isosurfaces in the aNCI map are somewhat jaggy. Both the aNCI map in Fig. 5 and the amIGM map in Fig. 4 were drawn with grid spacing of 0.15 Bohr; the latter looks much smoother.

5.2 Average Interaction Between a Solvated Phenol and Aqueous Environment

This section illustrates the use of the amIGM to reveal the average interaction between a solvated phenol and an aqueous environment. The MD simulation was carried out for 1 ns and maintained at room temperature. A phenol represented by the GAFF force field [33] is fixed at the center of the simulation box, while 443 SPC/E waters can move freely. The amIGM maps generated based on the 1,000 MD frames with different isovalues of $\delta \overline{g}^{\text{inter}}$ are shown in Fig. 6. It is expected that the hydroxyl group of phenol can act as both a significant H-bond donor and a H-bond acceptor during the MD simulation, and the mIGM map with a relatively large isovalue of $\delta \overline{g}^{\text{inter}}$ (0.005 a.u.) unambiguously confirmed this. If the isovalue of $\delta \overline{g}^{\text{inter}}$ is decreased to 0.003 a.u., the interaction regions related to the weaker H-bond acceptors, namely the π regions of the phenol, are unveiled. Finally, if the isovalue of $\delta \overline{g}^{\text{inter}}$ is further reduced to 0.0018 a.u., the main vdW interaction regions between the phenol and the surrounding waters become visible in the map. This example shows that by examining the amIGM map with different isovalues, the regions involving average weak interactions of different strengths can be clearly distinguished.

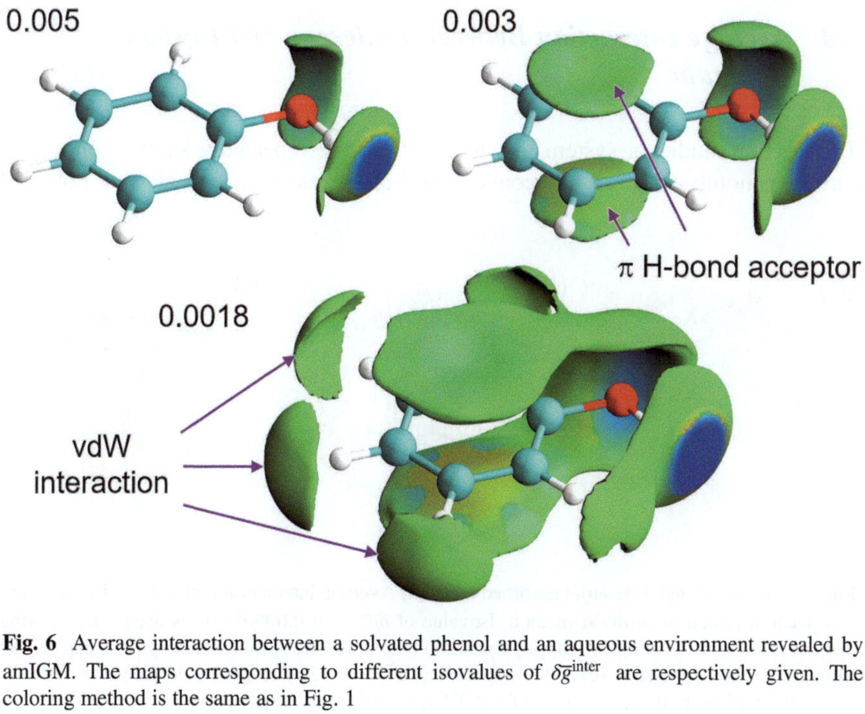

Fig. 6 Average interaction between a solvated phenol and an aqueous environment revealed by amIGM. The maps corresponding to different isovalues of $\delta \overline{g}^{\text{inter}}$ are respectively given. The coloring method is the same as in Fig. 1

5.3 Average Interaction Between Absorbed He Atom and Helicene

In Ref. [34], we performed MD simulation on the complex composed of a He atom and 7-helicene using the GFN0-xTB method [35] for 2,500 ps at 10 K. Here, we employ amIGM analysis to graphically exhibit the average interaction between the He atom and the 7-helicene, where 1,000 frames evenly extracted from the full MD trajectory were considered. In contrast to the previous examples, here the helicene in every frame was aligned to the initial frame to eliminate its overall motion. The amIGM map is shown in Fig. 7a. Because the He-helicene interaction is quite weak, a very small $\delta\overline{g}^{\text{inter}}$ (0.00005 a.u.) is adopted so that the isosurfaces characterizing the interactions are sufficiently visible. For ease of understanding this figure, the superposition map of the He atom in the 1,000 trajectory frames is provided in Fig. 7b, and the trajectory is colored according to time step. It can be clearly found that the amIGM method very successfully revealed the average interaction between the He atom and helicene; the areas where the $\delta\overline{g}^{\text{inter}}$ isosurface appears are exactly the main He-helicene interaction regions during the MD simulation. In contrast, the aNCI method completely fails to reveal the He-helicene interaction, as shown in Fig. 7c. The very poor graphical effect of the aNCI map in this case can be improved neither by fine-tuning the isovalue nor by applying isosurface screening.

5.4 Average Interaction Between Fullerene and Carbon Nanotube

This section studies a system in which a fullerene repeatedly shuttles through a carbon nanotube due to interconversion between kinetic energy and potential

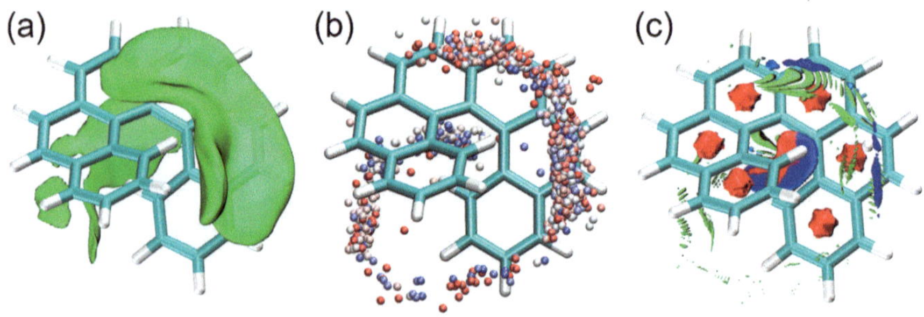

Fig. 7 7-Helicene and a He atom absorbed on it. (**a**) Average interaction between the He atom and 7-helicene revealed by amIGM method. Isovalue of $\delta\overline{g}^{\text{inter}}$ of 0.00005 a.u. is used. The coloring method is the same as in Fig. 1. (**b**) Superposition map of the He atom colored by time step of MD simulation (blue-white-red). (**c**) aNCI map with aRDG = 0.33. The same color bar as in Fig. 1 is used but with lower and upper limits of −0.025 and 0.025 a.u., respectively

energy; see Fig. 8a for graphical illustration of this system. In the initial structure, the fullerene is placed outside the nanotube at a certain distance from the tube mouth. The change in centroid distance between the fullerene and nanotube during the 100 ps MD simulation is shown in Fig. 8b. The CR1 atom type in the GROMOS54A7 force field [36] was adopted to represent the vdW interaction between the carbon atoms. Figure 8c shows the amIGM map calculated for 200 frames of the trajectory; two views are given. Since the fullerene shuttles back and forth repeatedly inside the tube, and the diameter of the fullerene is very close to the inner diameter of the nanotube, on average, atoms of the nanotube should exert a prominent π–π interaction with the fullerene. This fact is perfectly illustrated by the amIGM map, whose isosurface is evenly distributed on the inner wall of the nanotube, and the graphical effect is ideal. It is worth mentioning that although the nanotube has certain flexibility and thus its structure fluctuates noticeably during the MD simulation, this does not affect the reasonableness of the resulting amIGM map.

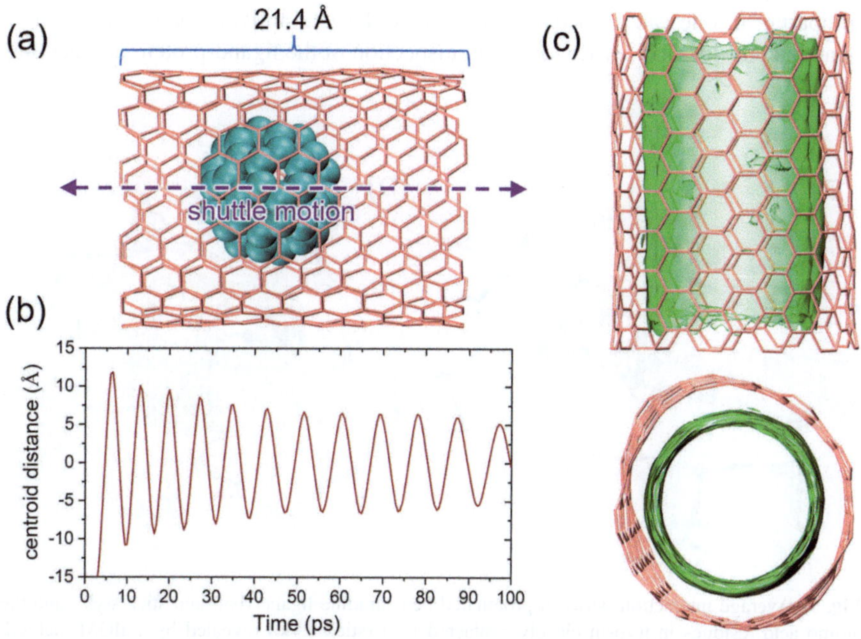

Fig. 8 Fullerene repeatedly shuttles through a carbon nanotube. (**a**) A snapshot of the system. (**b**) Centroid distance between the fullerene and nanotube during MD simulation. (**c**) amIGM map with isovalue of δg^{inter} of 0.0001 a.u.; two views are shown. The coloring method is the same as in Fig. 1

5.5 *Average Interaction Between Ligand and Surrounding Protein Residues*

The last example of the amIGM analysis is a protein-ligand complex (PDB ID: 3ATL). The frames used in the amIGM analysis were evenly taken from an MD simulation at 298.15 K for 1 ns within an explicit water box; the ligand and surrounding protein residues were extracted out for the amIGM analysis. Here, only the ligand was frozen during the MD simulation. The amIGM maps showing the average interaction between the ligand and the residues closely contacted to it are given in Fig. 9. Figure 9a employs a relatively large isovalue of $\delta\overline{g}^{\mathrm{inter}}$; in this case, the blue areas on the isosurfaces revealing the four strong H-bond interactions are clearly visible, and at the same time, some very weak H bonds and prominent vdW interactions are also exhibited. A smaller isovalue of $\delta\overline{g}^{\mathrm{inter}}$ was employed in plotting Fig. 9b, in which full vdW interactions between the ligand and protein are nicely revealed by the green isosurfaces. This example demonstrates the great value of amIGM analysis for studying average weak interactions in biological systems. The aNCI map of this system is as given in Fig. 17a of our previous review [21]; by comparing it, one can immediately recognize that the amIGM map is much more clear and considerably easier for visual inspection of the ligand-protein interactions.

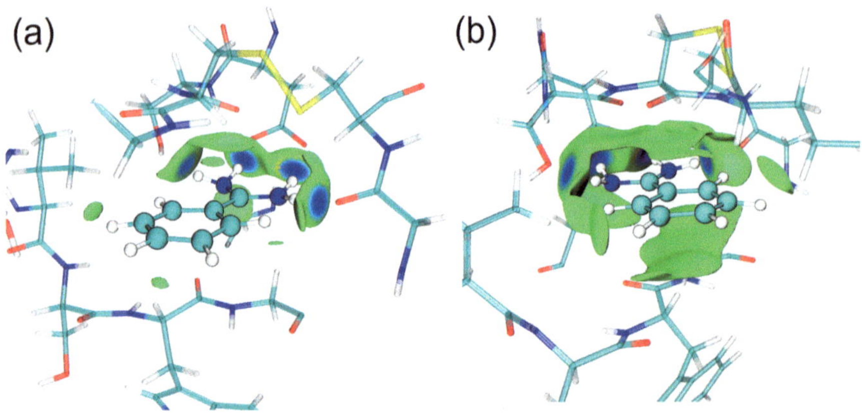

Fig. 9 Average interaction between protonated benzamidine ligand (ball-and-stick style) and the amino acid residues in trypsin closely contacted to it (stick style) revealed by amIGM method. Isovalue of $\delta\overline{g}^{\mathrm{inter}}$ is set to 0.006 a.u. (**a**) and 0.003 a.u. (**b**). The coloring method is the same as in Fig. 1. Note that in the maps, the coordinates of the residue atoms correspond to an arbitrary MD frame

6 Summary

In this chapter, we first briefly reviewed the popular IGM and IGMH methods, which are able to reveal weak interactions in a chemical system at a specific structure intuitively and graphically. Then we introduced a new method called mIGM. This method inherits various advantages of the IGM method and has evidently better graphical effects in describing the regions where weak interactions occur. Based on mIGM, we further proposed the amIGM method, which can fully exhibit the weak interactions in the dynamic environment simulated by MD techniques. The important practical value of the amIGM method is extensively demonstrated through a series of examples, and it is shown that amIGM has clear advantages over the aNCI method proposed earlier. Both mIGM and amIGM have been implemented recently in the comprehensive, popular, and freely available wavefunction analysis program Multiwfn. We believe that these two methods will be of significant value to chemists in the future in exploring weak interactions for a wide variety of chemical systems.

References

1. Lu T, Manzetti S (2014) Wavefunction and reactivity study of benzo[a]pyrene diol epoxide and its enantiomeric forms. Struct Chem 25:1521–1533
2. Lu T, Chen F (2013) Revealing the nature of intermolecular interaction and configurational preference of the nonpolar molecular dimers (H2)2, (N2)2, and (H2)(N2). J Mol Model 19: 5387–5395
3. Liu Z, Lu T (2024) Theoretical insight into complexation between cyclocarbons and C60 fullerene. Chem Eur J 30:e202402227
4. Liu Z, Wang X, Lu T et al (2023) Molecular assembly with a figure-of-eight nanohoop as a strategy for the collection and stabilization of cyclo[18]carbon. Phys Chem Chem Phys 25: 16707–16711
5. Leach AR (2001) Molecular modelling principles and applications. Pearson Education, Essex
6. Liu Z, Lu T, Chen Q (2021) Intermolecular interaction characteristics of the all-carboatomic ring, cyclo[18]carbon: focusing on molecular adsorption and stacking. Carbon 171:514–523
7. Byrd EFC, Rice BM (2006) Improved prediction of heats of formation of energetic materials using quantum mechanical calculations. J Phys Chem A 110:1005–1013
8. Murray JS, Brinck T, Lane P et al (1994) Statistically-based interaction indices derived from molecular surface electrostatic potentials: a general interaction properties function (GIPF). J Mol Struct (THEOCHEM) 307:55–64
9. Zheng W-F, Chen J, Qi X et al (2023) Modular and diverse synthesis of amino acids via asymmetric decarboxylative protonation of aminomalonic acids. Nat Chem 15:1672–1682
10. Johnson ER, Keinan S, Mori-Sánchez P et al (2010) Revealing noncovalent interactions. J Am Chem Soc 132:6498–6506
11. Wu P, Chaudret R, Hu X et al (2013) Noncovalent interaction analysis in fluctuating environments. J Chem Theory Comput 9:2226–2234
12. Lu T, Chen Q (2021) Interaction region indicator: a simple real space function clearly revealing both chemical bonds and weak interactions. Chem Methods 1:231–239
13. Lefebvre C, Khartabil H, Boisson J-C et al (2018) The independent gradient model: a new approach for probing strong and weak interactions in molecules from wave function calculations. ChemPhysChem 19:724–735

14. Lu T, Chen Q (2022) Erratum to "Independent gradient model based on Hirshfeld partition: a new method for visual study of interactions in chemical systems". ChemRxiv. https://doi.org/10.26434/chemrxiv-2022-g1m34
15. Lu T, Chen Q (2022) Independent gradient model based on Hirshfeld partition: a new method for visual study of interactions in chemical systems. J Comput Chem 43:539–555
16. Gao R, Liu Z, Liu Z et al (2023) Open-cage fullerene as a selective molecular trap for LiF/[BeF]+. Angew Chem Int Ed 62:e202300151
17. Gao W, Cai L, Kang F et al (2023) Bottom-up synthesis of metalated carbyne ribbons via elimination reactions. J Am Chem Soc 145:6203–6209
18. Zhang J, Zhang L, Xu Y et al (2018) Deciphering the binding behavior of flavonoids to the cyclin dependent kinase 6/cyclin D complex. PLoS One 13:e0196651
19. González-Veloso I, Carrazana-García JA, Josa D et al (2015) NCI analysis of the interaction cation$\cdots\pi$ in complexes with molecular bowls derived from fullerene. Comput Theor Chem 1053:123–129
20. Liu R, Qin X, Liu X et al (2022) Host-guest interactions between oleic acid and β-cyclodextrin: a combined experimental and theoretical study. Food Chem 387:132910
21. Lu T, Chen Q (2023) Visualization analysis of weak interactions in chemical systems. In: Comprehensive computational chemistry, vol 2. Elsevier, pp 240–264
22. Lu T, Chen F (2012) Multiwfn: a multifunctional wavefunction analyzer. J Comput Chem 33: 580–592
23. Lu T (2024) A comprehensive electron wavefunction analysis toolbox for chemists, Multiwfn. J Chem Phys 161:082503
24. Sedlak R, Janowski T, Pitoňák M et al (2013) Accuracy of quantum chemical methods for large noncovalent complexes. J Chem Theory Comput 9:3364–3374
25. Humphrey W, Dalke A, Schulten K (1996) VMD: visual molecular dynamics. J Mol Graph 14: 33–38
26. Sure R, Grimme S (2015) Comprehensive benchmark of association (free) energies of realistic host–guest complexes. J Chem Theory Comput 11:3785–3801
27. Dong Z, Ma Y (2020) Atomic-level handedness determination of chiral crystals using aberration-corrected scanning transmission electron microscopy. Nat Commun 11:1588
28. Mayer I (1983) Charge, bond order and valence in the AB initio SCF theory. Chem Phys Lett 97:270–274
29. Kühne TD, Iannuzzi M, Del Ben M et al (2020) CP2K: an electronic structure and molecular dynamics software package – quickstep: efficient and accurate electronic structure calculations. J Chem Phys 152:194103
30. Lu T, Chen Q (2020) mwfn: a strict, concise and extensible format for electronic wavefunction storage and exchange. ChemRxiv. https://doi.org/10.26434/chemrxiv-2021-lt04f-v6
31. Abraham MJ, Murtola T, Schulz R et al (2015) GROMACS: high performance molecular simulations through multi-level parallelism from laptops to supercomputers. SoftwareX 1:19–25
32. Berendsen HJC, Grigera JR, Straatsma TP (1987) The missing term in effective pair potentials. J Phys Chem 91:6269–6271
33. Wang J, Wolf RM, Caldwell JW et al (2004) Development and testing of a general amber force field. J Comput Chem 25:1157–1174
34. Lu T, Chen Q (2020) van der Waals potential: an important complement to molecular electrostatic potential in studying intermolecular interactions. J Mol Model 26:315
35. Pracht P, Caldeweyher E, Ehlert S et al (2019) A robust non-self-consistent tight-binding quantum chemistry method for large molecules. ChemRxiv. https://doi.org/10.26434/chemrxiv.8326202.v1
36. Schmid N, Eichenberger AP, Choutko A et al (2011) Definition and testing of the GROMOS force-field versions 54A7 and 54B7. Eur Biophys J 40:843–856

MIX
Papier aus verantwortungsvollen Quellen
Paper from responsible sources
FSC® C105338

If you have any concerns about our products,
you can contact us on
ProductSafety@springernature.com

In case Publisher is established outside the EU,
the EU authorized representative is:
Springer Nature Customer Service Center GmbH
Europaplatz 3, 69115 Heidelberg, Germany

Printed by Libri Plureos GmbH
in Hamburg, Germany